J.B.

ACS SYMPOSIUM SERIES **757**

Calixarenes for Separations

Gregg J. Lumetta, EDITOR
Pacific Northwest National Laboratory

Robin D. Rogers, EDITOR
The University of Alabama

Aravamudan S. Gopalan, EDITOR
New Mexico State University

American Chemical Society, Washington, DC

Library of Congress Cataloging-in-Publication Data

Calixarenes for separations / Gregg J. Lumetta, editor, Robin D. Rogers, editor, Aravamudan S. Goplan, editor.

p. cm.—(ACS symposium series, ISSN 0097-6156 ; 757)

Includes bibliographical references and index.

ISBN 0-8412-3660-7

1. Calixarenes—Congresses. 2. Separation (Technology)—Congresses.

I. Lumetta, Gregg J., 1960– . II. Rogers, Robin D., 1957– . III. Goplan, Aravamudan S., 1954– . IV. Series.

QD341..P5 C348   2000
547'.632—dc21   00-21861

The paper used in this publication meets the minimum requirements of American National Standard for Information Sciences—Permanence of Paper for Printed Library Materials, ANSI Z39.48–1984.

Copyright © 2000 American Chemical Society

Distributed by Oxford University Press

All Rights Reserved. Reprographic copying beyond that permitted by Sections 107 or 108 of the U.S. Copyright Act is allowed for internal use only, provided that a per-chapter fee of $20.00 plus $0.50 per page is paid to the Copyright Clearance Center, Inc., 222 Rosewood Drive, Danvers, MA 01923, USA. Republication or reproduction for sale of pages in this book is permitted only under license from ACS. Direct these and other permission requests to ACS Copyright Office, Publications Division, 1155 16th St., N.W., Washington, DC 20036.

The citation of trade names and/or names of manufacturers in this publication is not to be construed as an endorsement or as approval by ACS of the commercial products or services referenced herein; nor should the mere reference herein to any drawing, specification, chemical process, or other data be regarded as a license or as a conveyance of any right or permission to the holder, reader, or any other person or corporation, to manufacture, reproduce, use, or sell any patented invention or copyrighted work that may in any way be related thereto. Registered names, trademarks, etc., used in this publication, even without specific indication thereof, are not to be considered unprotected by law.

PRINTED IN THE UNITED STATES OF AMERICA

Foreword

THE ACS SYMPOSIUM SERIES was first published in 1974 to provide a mechanism for publishing symposia quickly in book form. The purpose of the series is to publish timely, comprehensive books developed from ACS sponsored symposia based on current scientific research. Occasionally, books are developed from symposia sponsored by other organizations when the topic is of keen interest to the chemistry audience.

Before agreeing to publish a book, the proposed table of contents is reviewed for appropriate and comprehensive coverage and for interest to the audience. Some papers may be excluded in order to better focus the book; others may be added to provide comprehensiveness. When appropriate, overview or introductory chapters are added. Drafts of chapters are peer-reviewed prior to final acceptance or rejection, and manuscripts are prepared in camera-ready format.

As a rule, only original research papers and original review papers are included in the volumes. Verbatim reproductions of previously published papers are not accepted.

ACS BOOKS DEPARTMENT

Contents

Preface xi

GENERAL

1. Calixarenes for Separations 2
   C. David Gutsche

CALIXARENE–CATION COMPLEXATION

2. Extraction of Cesium by Calix[4]arene-crown-6: From Synthesis
   to Process 12
   J.-F. Dozol, V. Lamare, N. Simon, R. Ungaro, and A. Casnati

3. Development of Process Chemistry for the Removal of Cesium
   from Acidic Nuclear Waste by Calix[4]arene-crown-6 Ethers 26
   Peter V. Bonnesen, Tamara J. Haverlock, Nancy L. Engle,
   Richard A. Sachleben, and Bruce A. Moyer

4. Radiation Stability of Calixarene-Based Solvent System 45
   R. A. Peterson, C. L. Crawford, F. F. Fondeur, and T. L. White

5. Behavior of Calix[4]arene-bis-crown-6 under Irradiation:
   Electrospray–Mass Spectrometry Investigation and Molecular
   Dynamics Simulations on Cs^+ and Na^+ Complexes
   of a Degradation Compound 56
   V. Lamare, J.-F. Dozol, F. Allain, H. Virelizier, C. Moulin,
   C. Jankowski, and J.-C. Tabet

6. Interfacial Features of Assisted Liquid–Liquid Extraction of Uranyl
   and Cesium Salts: A Molecular Dynamics Investigation 71
   M. Baaden, F. Berny, N. Muzet, L. Troxler, and G. Wipff

7. Benzyl Phenol Derivatives: Extraction Properties of Calixarene
   Fragments 86
   Lætitia H. Delmau, Jeffrey C. Bryan, Benjamin P. Hay, Nancy L. Engle,
   Richard A. Sachleben, and Bruce A. Moyer

8. Calixarene-Containing Extraction Mixture for the Combined
   Extraction of Cs, Sr, Pu, and Am from Alkaline Radioactive Wastes 107
   Igor V. Smirnov, Andrey Yu. Shadrin, Vasily A. Babain,
   Mikhail V. Logunov, Marina K. Chmutova, and Vitaly I. Kal'tchenko

9. Calix[4]arenes with a Novel Proton-Ionizable Group: Synthesis and Metal Ion Separations 112
   Richard A. Bartsch, Hong-Sik Hwang, Vladimir S. Talanov, Chunkyung Park, and Galina G. Talanova

10. Separations of Soft Heavy-Metal Cations by Lower Rim-Functionalized Calix[4]arenes 125
    Galina G. Talanova, Vladimir S. Talanov, Hong-Sik Hwang, Nazar S. A. Elkarim, and Richard A. Bartsch

11. CMPO-Substituted Calixarenes 135
    Volker Böhmer

12. Binding of Lanthanides(III) and Thorium(IV) by Phosphorylated Calixarenes 150
    F. Arnaud-Neu, S. Barboso, D. Byrne, L. J. Charbonnière, M. J. Schwing-Weill, and G. Ulrich

13. Lanthanide Calix[4]arene Complexes Investigated by NMR 165
    B. Lambert, V. Jacques, and J. F. Desreux

14. Bimetallic Lanthanide Supramolecular Edifices with Calixarenes 179
    Jean-Claude G. Bünzli, Frédéric Besançon, and Frédéric Ihringer

15. Molecular Recognition by Azacalix[3]arenes 195
    Philip D. Hampton, Si Wu, Panadda Chirakul, Zsolt Bencze, and Eileen N. Duesler

16. Metal Ion Complexation and Extraction Behavior of Some Acyclic Analogs of *tert*-Butyl-calix[4]arene Hydroxamate Extractants 208
    Timothy N. Lambert, Matthew D. Tallant, Gordon D. Jarvinen, and Aravamudan S. Gopalan

17. Calixarenes as Ligands in Environmentally-Benign Liquid–Liquid Extraction Media: Aqueous Biphasic Systems and Room Temperature Ionic Liquids 223
    Ann E. Visser, Richard P. Swatloski, Deborah H. Hartman, Jonathan G. Huddleston, and Robin D. Rogers

CALIXARENE–ANION COMPLEXATION

18. Calix[4]pyrrole-Functionalized Silica Gels: Novel Supports for the HPLC-Based Separation of Anions 238
    Jonathan L. Sessler, John W. Genge, Philip A. Gale, and Vladimír Král

19. Lower Rim Amide- and Amine-Substituted Calix[4]arenes
    as Phase Transfer Extractants for Oxyions between an Aqueous
    and an Organic Phase 255
    H. Fred Koch and D. Max Roundhill

CALIXARENE COMPLEXATION OF NEUTRAL MOLECULES

20. Deep Cavities and Capsules 270
    Dmitry M. Rudkevich and Julius Rebek, Jr.

21. Calixarene Metalloreceptors: Demonstration of Size and Shape
    Selectivity Inside a Calixarene Cavity 283
    Stephen J. Loeb and Beth R. Cameron

22. Synthesis and Structural Analysis of Thiacalixarene Derivatives 296
    Mir Wais Hosseini

23. Synthesis and Fullerene Complexation Studies of p-Allylcalix[5]arenes 313
    Charles G. Gibbs, Jian-she Wang, and C. David Gutsche

24. Hydrogen-Bonded Cavities Based upon Resorcin[4]arenes by Design 325
    Leonard R. MacGillivray and Jerry L. Atwood

Author Index 341

Subject Index 342

Preface

The unique molecular architecture of calixarenes makes them a suitable platform for constructing host molecules that can selectively bind a variety of guest substrates that range from cations and anions to fullerenes. Substrates can be bound within the hydrophobic bowl–cavity of the calixarenes through non-covalent interactions or outside the cavity by introducing suitable ligand systems on the upper or lower rim. Also, the self-assembly of some calixarene derivatives via non-covalent interactions leads to a new class of encapsulation agents, inclu-sion complexes, and molecular networks.

The fascinating conformational and chemical reactivity of the calixarene systems has led to their use in a variety of applications that range from de-velopment of efficient separation systems, specific sensors, and molecular switches to a new generation of catalysts. The scientific interest in the applications of this class of molecules to solve separation problems is demonstrated by the number of recent publications as well as the award of numerous patents that utilizes the chemistry of calixarenes. These patents include the use of calixarenes as extraction agents for separations, such as recovering uranyl ion from sea water and removing cesium from nuclear wastes. Calixarenes are also being used as selective agents for metal ion transport through supported liquid membranes. It is important to point out that efficient chemical separations are now integral not only to manu-facturing processes, but also for compliance with subsequent industrial waste-discharge requirements.

This book is the outcome of a recent symposium titled "Calixarene Molecules for Separations" that was held at the 217th American Chemical Society (ACS) National Meeting in Anaheim, California, March 23–25, 1999. Although a number of other books and reviews on calixarenes have appeared, they have not focused on the development of calixarene host–guest systems that have ap-plications to important challenges in the field of separations chemistry. Topics that are covered in this book include (1) synthetic methodology relevant to the development of calixarene-based receptors, (2) recent developments in the use of calixarenes for separations involving cations and anions, many of them important for radioactive waste remediation, and (3) binding of cations, anions, and neutral molecules by calixarenes and related molecules. The chapters pertaining to selec-tive metal ion extraction and coordination chemistry of some of the calixarene de-rivatives provide highly relevant discussions regarding the characterization of ligand–metal complexes using strategies such as molecular dynamics simulation, nuclear magnetic resonance spectroscopy, and electrospray mass spectroscopy. In addition, some of the chapters in this book address related systems such as heterocalix[3]arenes, thiacalixarenes, and calixpyrroles and their binding prop-erties. The assembly of calixarenes through non-covalent interactions to form deep cavities and capsules for encapsulation of target substrates is another area of discussion. This book will be useful for a wide range of scientists whose interests vary from the chemistry of calixarenes to the area of host–guest complexation, coordination chemistry, separation technologies, and environmental remediation.

Acknowledgments

We gratefully acknowledge the Donors of The Petroleum Research Fund, administered by the ACS, for support of the Calixarenes for Separations Symposium held in Anaheim, California. The financial support of the Separation Science and Technology Subdivision of the ACS Division of Industrial and Engineering Chem-istry, Inc. for the symposium is also acknowledged. We thank all the participants of this symposium and the authors and reviewers of the manuscripts published in this book for their cooperation and valuable contributions. Our sincere thanks to David Gutsche for his introductory review (Chapter 1) that gives the reader a broad picture of the exciting field of calixarenes. Finally, we thank the ACS for their encouragement and support in the publication of this book.

GREGG J. LUMETTA
Pacific Northwest National Laboratory
P.O. Box 999
Richland, WA 99352

ROBIN D. ROGERS
Department of Chemistry
The University of Alabama
Tuscaloosa, AL 35487

ARAVAMUDAN S. GOPALAN
Department of Chemistry and Biochemistry
New Mexico State University
Las Cruces, NM 88003

General

Chapter 1

Calixarenes for Separations

C. David Gutsche

Department of Chemistry, Texas Christian University, Fort Worth, TX 76129

The calixarenes, from their aborted birth in the 1870's and their successful but almost forgotten rebirth in the 1940's, through their hesitant resurrection in the 1970's, and into their robust state of good health in the 1990's, are presented in brief outline with an emphasis on the use of these molecules for chemical separations.

The separation of one kind of substance from another has occupied the attention of *Homo sapiens* for many eons. The digging of stones from the earth, the tearing of bark from a tree, the picking of berries from a bush are all examples of separations that, in one fashion or another, bestow a benefit to the one doing the separating. While these and other simple examples from antiquity continue to be an important part of modern life, a myriad of more sophisticated examples have been added in which chemistry often plays a key role. The particular separations that provide the focus for this book and for which this chapter provides a brief introduction involve ions and molecules as the entities being separated by a class of compounds called calixarenes as the tool for effecting the separations.

The calixarenes (*1*) may have first seen the light of day in the nineteenth century in the laboratory of Adolf von Baeyer when *p*-substituted phenols were treated with formaldehyde in the presence of acid or base to produce viscous, ill-defined products (*2*). However, the analytical tools in 1872 were far too feeble for the task of characterizing these materials, and it was not until half a century later that Alois Zinke in the early 1940's suggested that they possess a cyclic tetrameric structure (*3*). But Zinke, like Baeyer, also lacked sufficiently powerful analytical tools for really adequate characterization, and as a result he failed to realize that what he thought to be pure cyclic tetramers were actually mixtures. This was first recognized in the 1950's by Cornforth (*4*) and the oligomeric nature of the mixture then delineated in the 1970's by Gutsche (*5*) who gave the compounds their presently accepted name (*6*) of "calixarene" (Gr.*calix* meaning vase or chalice and *arene* indicating the presence of aryl residues in the macrocyclic array) and who developed procedures for obtaining good yields of the cyclic tetramer (*7*), cyclic hexamer (*8*), and cyclic octamer (*9*) from the base-induced reaction of *p-tert*-butylphenol and

cyclic tetramer

formaldehyde. By this time in the history of chemistry, the techniques of chromatography, spectral methods, and X-ray capabilities that have so profoundly transformed chemical analysis had arrived on the scene and provided reliable means for unequivocally establishing the structures of the calixarenes obtained by these one-step procedures. Of particular importance were the X-ray crystallographic determinations carried out by Andreetti, Ungaro, and Pochini (*10*). Concomitant with this sequence of events involving the one-step syntheses were the multi-step procedures originated by Hayes and Hunter (*11*) in the 1950's, exploited by Kämmerer (*12*) in the 1970's, and greatly improved and expanded by Böhmer (*13*) in the 1980's and 1990's.

A parallel sequence of events that also started in the 1940's involved the work of Niederl which revealed that the acid-catalyzed reaction of resorcinol with aldehydes (other than formaldehyde) yields cyclic tetramers similar in basic structure to those obtained from *p*-substituted phenols and formaldehyde (*14*). Niederl's procedures were optimized by Högberg (*15*) in the early 1980's leading to the ready accessibility of compounds that today are variously called "calix[4]resorcinarenes", "calix[4]resorcarenes (the preferred designation), "resorcin[4]arenes" or "resorcarenes" (*16*). These, along with the phenol-derived cyclic oligomers described above comprise the basic family of calixarenes to which numerous other cyclic structures, more or less-closely related to calixarenes, have been added in recent years. While

a calix[4]resorcarene

p-tert-butylcalix[4]arene   *p-tert*-butylcalix[6]arene   *p-tert*-butylcalix[8]arene

most of the examples of separations discussed in this book involve either the phenol-derived or resorcinol-derived calixarenes, there are also a few examples involving some of these other calixarene-like compounds. For the present discussion the term "separations" is broadly interpreted to include any one of a variety of complex-forming interactions between calixarenes as hosts with molecules or ions as guests.

The molecule-binding capability of a calixarene was first demonstrated in 1979 by Andreetti, Ungaro, and Pochini's X-ray analysis of *p-tert*-butylcalix[4]arene which showed that in the solid state a molecule of toluene is firmly embedded in its cavity (*10*). The ion-binding capability of a calixarene in solution was also first demonstrated by this same group (*17*) in 1982 when they detected the interaction of an oxyalkyl ether of *p-tert*-

a calixcrown

butylcalix[8]arene with guanidinium. In the following year they extended this work to the complexation of alkali earth cations, using a calix[4]arenecrown-5 compound as the host (*18*). The ion-binding properties of the parent, unaltered calixarenes were first demonstrated by Izatt's studies (*19*) with *p-tert*-butyl[4,6,8]calixarenes which showed that alkali metal cations in a strongly basic solution can be transported through a liquid membrane. Then, as more highly functionalized calixarenes started to become available the calixarenes began to attract increasingly widespread attention as entities for the separation of various ions and molecules. Much of the synthesis effort that started in the 1980's and continues to the present day has been directed to the devising of ways for adding functionality to the lower rim OH groups (*e.g.* esterification, etherification) and to the upper rim *p*-position of the aryl ring (*e.g.*,electrophilic substitution, Claisen rearrangement of O-allyl ethers, Mannich base route). These efforts have been highly successful, and today many hundreds of calixarenes carrying all manner of functional groups have been synthesized, providing the separation chemist with a rich and enticing array of hosts with which to work. Not only have these various synthesis procedures led to a wide variety of functionalized cavities, they have also introduced methods for establishing particular shapes of cavities by taking advantage of conformational control.

As already noted, among the earliest of the functionalized calixarenes to be synthesized and studied for ion-binding properties were Ungaro's calixcrowns (*18*). These continue to provide items of great interest to the separation chemist, as indicated by several of the contributions to this book. The calixcrowns are molecules containing a crown ether moiety, most often attached to the lower rim via etherification of the phenolic OH groups but sometimes attached to the upper rim. Not surprisingly the ion-selectivities of these hosts are found to be a function of the size of the crown moiety. For example, a calix[4]crown-5 shows a 12,000-fold selectivity (*20*) for K^+ over Na^+, while a calix[4]crown-6 shows a 33,000-fold selectivity (*21*) for Cs^+ over Na^+. Less expected, perhaps, is the sensitivity of the strength of complexation to the conformation of the calixarene. The aforementioned 12,000 value for K^+/Na^+, which is for the partial cone conformation, falls to about 100 for the cone conformation. Phenomena such as these have been extensively investigated in the 16 years since the first calixcrown was made, and many dozens of calixcrowns have now been constructed which possess a wide variety of combinations of the crown ether and calixarene moieties (including a few calix[4]resorcarenecrown compounds). In addition, a number of structurally related types of molecules have been constructed in which cyclic moieties other than crowns are affixed to the calixarene. Of special note are Reinhoudt's calixspherands, some of which form especially strong complexes with rubidium (*22, 23*).

a calixspherand

The strong ion-complexing ability of calixarenes, however, is not limited to those carrying crown ether or other cyclic moieties. Hundreds of calixarenes carrying simple pendant functionalities have also been studied, and many have been found to possess striking complexation properties. A particularly well investigated group focuses on calixarenes carrying OCH_2COR moieties (R = alkyl, OH, OR, NH_2, NR_2, etc) on the lower rim. McKervey, one of the earliest entrants into this area of calixarene chemistry, reported in 1985 on the ion-complexing characteristics of esters (R = OR) (24) and has continued his studies in the years following in which he has explored in great detail the complexation parameters of these kinds of compounds. Inevitably, he was soon joined by many other investigators who have entered the field. What this batallion of workers has shown (25), *inter alia*, is that the carboxamides (R = NH_2) are effective complexers for alkaline earth cations ($Ca^{+2} > Sr^{2+} > Ba^{2+} > Mg^{2+}$) and trivalent cations such as Pr^{3+}, Eu^{3+}, and Yb^{3+}, as also are the carboxylic acids (R = OH). On the other hand, calixarenes carrying OCH_2SNR_2 groups on the lower rim form strong complexes with Ag^+, Pb^{2+}, and Cd^{2+}, while those carrying OPR_2 groups form complexes with metal ions such as Pd^{2+}, Pt^{2+}, and Ru^{2+}. Readers will recognize interesting variations on these themes in several of the chapters in this book.

a calixarene uranophile

Among the more fascinating studies of ion complexation by calixarenes are those of Shinkai regarding UO_2^{++}, a species that exists in significant total amount but in vanishingly low concentration in sea water. An experiment that resulted in the successful extraction of a ponderable amount of UO_2^{++} from sea water employed calixarenes carrying sulfonic acid groups on the upper rim, a species first introduced by Shinkai (26).

Although the complexation of anions by calixarenes has been far less extensively studied than that of cations, increasing attention is being paid to this phenomenon. Pioneered by several research groups including those of Beer (27), Reinhoudt (28), and Puddephatt (29) and more recently those of Atwood (30), Roundhill (31), and others, interesting selectivities among inorganic as well as organic anions have been demonstrated. For example, a calixarene carrying cobalticinium moieties on the upper rim shows an almost 600-fold preference for acetate over chloride (32).

The first authenticated solid state complex of a calixarene containing a molecule embedded in its cavity was that of *p-tert*-butylcalix[4]arene and toluene, revealed, as noted above (10), by X-ray crystallography in 1979. The formation of solution state complexes of calixarenes with molecules was first studied in the 1980's by several research groups, including those of Shinkai (33), Gutsche (34), and Cram (35). However, the intensity of the effort devoted to the complexation of molecules in solution pales in comparison with that devoted to the studies of the complexation of ions in solution that have been briefly described above. With the exception of Cram's (35) and Sherman's (36) continuing investigations of the cavitands and carcerands, only sporadic examples of molecule complexation with calixarenes appear in the literature of the 1990's. Among these are the complexes of fullerenes with calixarenes, first discerned in solution by Williams and Verhoeven (37) and then in the solid state, serendipitously and almost simultaneously, by Shinkai and coworkers (*38*) and Atwood, Raston, and coworkers (*39*); subsequently a number of other investigators (*40, 41*) have entered the field.

Maturing during the same period that calixarene chemistry was emerging as a distinct area of study computational chemistry was becoming perfected, and inexorably the two fields have intersected. Today, the majority of calixarene chemists have facilities for carrying out the molecular mechanics calculations that frequently accompany the publications from their laboratories, those from Böhmer, Reinhoudt, and Shinkai being especially good examples. Of particular note are the contributions of the "full-time molecular computationalists" such as Jorgensen (*42*), Kollman (*43*), and Wipff (*44*).

To an ever increasing extent the study of calixarenes for chemical separations has become a mission- oriented undertaking, the goal of a research project often being defined with great specificity and practicality (*45*). And, to an increasing extent the results are showing up in the patent literature. One of the earliest examples is Izatt's 1984 patent (*46*) for the use of *p-tert*-butylcalix[8]arene in the removal of Cs^+ from nuclear wastes (*47*), and this topic continues to occupy an important place in calixarene technology as evidenced by the several contributions to this book. Of the more than 150 patents that have been issued dealing with a wide variety of calixarene applications, many relate to separations of one sort or another and often involve such specialized and sophisticated applications as ion selective electrodes (*48*), field effect transistors (*49*), chromogenic and fluorogenic sensors (*49*), and calixarene molecules for nonlinear optical devices (*50*).

a chromogenic calixarene

Calixarenes have come a long way from their tarry beginnings in the nineteenth century, their hesitant introduction in the 1940's, and their reinvestigation and resuscitation in the 1970's. Little did the early workers in the field anticipate the explosive growth that was to follow in the 1980's and 1990's, fueled with particular intensity by the applications of calixarenes in chemical separations. While other aspects of calixarene chemistry remain important and continue to command attention, there is little doubt that the complexation with ions and, perhaps to a somewhat lesser extent, with molecules will occupy center stage for some time to come. Thethree-day Symposium devoted to Calixarenes for Chemical Separations at the National Meeting of the American Chemical Society in Anaheim, California in April 1999 is clearly a testament to the validity of this conclusion.

References

1. For extensive reviews of the calixarenes *cf* (a) Gutsche, C. D. *Calixarenes Revisited* in "Monographs in Supramolecular Chemistry", Stoddart, J. F., Ed., Royal Society of Chemistry, London, **1998**; (b) Böhmer, V., "Calixarenes, Macrocycles with (Almost) Unlimited Possibilities", *Angew. Chem. Int. Ed. Engl.,* **1995**, *34,* 713 -745; (c) Vicens, J.; Böhmer, V. "Calixarenes. A Versatile Class of Macrocyclic Compounds", Kluwer Academic Publishers, Dordrecht, **1990**; (d) Gutsche, C. D. *Calixarenes* in

"Monographs in Supramolecular Chemistry", Stoddart, J. F., Ed., Royal Society of Chemistry, London, **1989**.
2. Baeyer, A. *Ber.* **1872**, *5*, 25, 280, 1094.
3. Zinke, A.; Ziegler, E. *Ber.* **1941**, *B74*, 1729; *ibid.* **1944**, *77*, 264.
4. Cornforth, J. W.; D'Arcy Hart, P.; Nicholls, G. A.; Rees, R. J. W.; Stock, J. A. *Br. J. Pharmacol.* **1955**, *10*, 73.
5. Gutsche, C. D.; Kung, T. C.; Hsu, M.-L. Abstracts of 11th Midwest Regional Meeting of the American Chemical Society, Carbondale, IL **1975**, No. 517.
6. Gutsche, C. D.; Muthukrishnan, R. *J. Org. Chem.* **1978**, *43*, 4905.
7. Gutsche, C. D.; Iqbal, M. *Org. Syn.* **1990**, *68*, 234.
8. Gutsche, C. D.; Dhawan, B.; Leonis, M.; Stewart, D. *Org. Syn.* **1990**, *68*, 238.
9. Munch, J. H.; Gutsche, C. D. *Org. Syn.* **1990**, *68*, 243.
10. Andreetti, G. D.; Ungaro, R.; Pochini, A. *J. Chem. Soc. Chem. Commun.* **1979**, 1005.
11. Hayes, B. T.; Hunter, R. F. *Chem. Ind.* **1956**, 193; *J. Appl. Chem.* **1958**, *8*, 743.
12. Kämmerer, H.; Happel, G.; Caesar, F. *Makromol. Chem.* **1972**, *162*, 179; Happel, G.; Matiasch, B.; Kämmerer, H. *ibid.* **1975**, *176*, 3317; Kämmerer, H.; Happel, G. *ibid.* **1978**, *179*, 1199; ;**1980**, *181*, 2049; **1981**, *112*, 759; Kämmerer, H.; Happel, G.; Böhmer, V.; Rathay, D. *Monatsh. Chem.* **1978**, *109*, 767.
13. Böhmer, V.; Chhim, P.; Kämmerer, H. *Makromol. Chem.* **1979**, *180*, 2503; Böhmer,V.; Marschollek, F.; Zetta, L. *J. Org. Chem.* **1987**, *52*, 3200; For more recent examples *cf*. Ref 1 and Böhmer, V. *Liebigs Ann./Recueil* **1997**, 2019.
14. Niederl, J. B.; Vogel, H. J. *J. Am. Chem. Soc.* **1940**, *62*, 2512.
15. Högberg, A. G. S. *J. Org. Chem.* **1980**, *45*, 4498; *J. Am. Chem. Soc.* **1980**, *102*, 6046.
16. For extensive reviews *cf*. (a) Cram, D. J.; Cram, J. M. "Container Molecules and Their Guests", "Monographs in Supramolecular Chemistry", Stoddart, J. F., Ed., Royal Society of Chemistry, London, **1994;** (b) Timmerman, P.; Verboom, W.; Reinhoudt, D. N. *Tetrahedron* **1996**, *52*, 2663.
17. Bocchi;, V.; Foina, D.; Pochini, A.; Ungaro, R.; Andreetti, G. D. *Tetrahedron* **1982** *38*, 373.
18. Alfieri, C.; Dradi, E.; Pochini, A.; Ungaro, R.; Andreetti, G. D. *J. Chem. Soc. Chem. Commun.* **1983**, 1075.
19. Izatt, R. M.; Lamb, J. D.; Hawkins, R. T.; Brown, P. R.; Izatt, S. R.; Christensen;, J. J. *J. Am.Chem. Soc.* **1983**, *105*, 1782; Izatt, S. R.; Hawkins, R. T.; Christensen, J. J.; Izatt, R. M. *ibid.* **1988** *110*, 6811.
20. Ghidini, E.; Ugozzoli F.; Ungaro, R.; Harkema, S.; El-Fadl, A. A.; Reinhoudt, D. N. *J. Am. Chem Soc* **1990**, *112*, 6979.
21. Ungaro, R.; Casnati, A.; Ugozzoli, F.; Pochini, A.; Dozol, J.-F.; Hill, C.; Rouquette, H. *Angew. Chem. Intl. Ed. Engl.* **1994**, *33*, 1506.
22. Reinhoudt, D. N.; Dijkstra, P. J.; in't Veld, P. J. A.; Bugge, K.-E.; Harkema, S.; Ungaro, R.; Ghidini, E. *J. Am. Chem. Soc.* **1987**, *109*, 4761;.
23. Dijkstra, P. J.; Brunink, J. A. J.; Bugge, K.-E.; Reinhoudt, D. N.; Harkema, S.; Ungaro, R.; Ugozzoli, G.; Ghidini, E. *ibid.* **1989**, *111*, 7657

24. McKervey;, M. A.; Seward, E. M.; Ferguson, G.; Ruhl, B.; Harris, S. J. *J. Chem. Soc. Chem Commun.* **1985**, 388.
25. For extensive reviews cf; (a) Wieser, C.; Dieleman, C. B.; Matt, D. 'Calixarene and Resorcinarene Ligands in Transition Metal Chemistry' *Coordination Chemical Reviews* **1997**, *165*, 93-161; (b) Ikeda, A.; Shinkai, S. *Chem. Rev.* **1997**, *97*, 1713-1734; (c) McKervey, M. A.; Schwing-Weil, M. J.; Arnaud- Neu, F. 'Cation Binding by Calixarenes' in *Comprehensive Supramolecular Chem*istry: Gokel, G. Ed Pergamon Press: Oxford: **1996**, Vol 1., pp 537-603; (d) Roundhill, D. M. 'Metal Complexes of Calixarenes', *Progr. Inorg. Chem.* **1995**, 533-592; (e) Ludwig, R. *Japan Atomic Energy Research Institute* **1995**, 1-55.
26. Shinkai, S.; Koreishi, H.; Ueda, K.; Arimura, T.; Manabe, O. *J. Am. Chem. Soc.* **1987**, *109*, 6371; Also cf patents issued to Shinkai *et al* (*Chem. Abstr.* **1987**, *108*, 64410q and 11638b)
27. Beer, P. D.; Dickson, C. A. P.; Fletcher, N.; Goulden, A. J.; Grieve, A.; Hodacova, J.; Wear, T. *J. Chem. Soc. Chem. Commun.* **1993**, 828
28. Boerrigter, H.; Grave, L.; Nissink, J. W. M.; Chrisstoffels, L. A. J.; van der Maas, J. H.; Verboom, W.; de Jong, F.; Reinhoudt, D. N. *J. Org. Chem.* **1998**, *63*, 4174;
29. Xu, W.; Vittal, J. J.; Puddephatt, R. J. *J. Am. Chem. Soc* **1995**, *117*, 8362;
30. Staffilani, M.; Hancock, K. S. B.; Steed, J. W.; Holman, K. T.; Atwood, J. L.; Juneja, R. K.; Burkhalter, R. S. *J. Am. Chem. Soc* **1997**, *119*, 6324;
31. Georgiev, E. M.; Wolf, N.; Roundhill, D. M. *Polyhedron* **1997**, *16*, 1581.
32. Beer, P. D.; Drew, M. G. B.; Hesek, D.; Nam, K. C. *J. Chem. Soc. Chem. Commun.* **1997**, 107.
33. Shinkai, S.; Mori, S.; Araki, K.; Manabe, O. *Bull. Chem. Soc. Jpn.* **1987**, *60*, 3679.
34. Gutsche, C. D.; Alam, I. *Tetrahedron* **1988**, *30* 4689.
35. Of particular note are the complexes of cavitands and carcerands extensively studied by Cram and coworkers and reviewed in ref 16a. Also *cf.* Schneider, H.-J.; Schneider, U. *J. Inclusion Phenom. Molec. Recognit. Chem.* **1994**, *19*, 67.
36. Chapman, R. G.; Olovsson, G.; Trotter, J.; Sherman, J. C. *J. Am. Chem. Soc* **1998**, *120*, 6252.
37. Williams, R. M.; Verhoeven, J. W. *Rec. Trav. Chim. Pays-Bas* **1992**, *111*, 531.
38. Suzuki, T. Nakashima, K. and Shinkai, S. *Chem. Lett.*, **1994**, 699.
39. Atwood, J. L.; Koutsantonis, G. A. and Raston, C. L. *Nature*, **1994**, *368*, 229.
40. Haino, T.; Yanase, M.; Fukazawa, Y. *Angew. Chem. Intl. Ed. Engl.* **1999**, 997; Yanase, M.; Haino, T.; Fukazawa, Y. *Tetrahedron Lett.* **1999**, *40*, 2781.
41. Wang, J.-S.; Gutsche, C. D. *J. Am. Chem. Soc.* **1998**, *120*, 122
42. McDonald, H. A.; Duffy;, E. M.; Jorgensen, W. L. *J. Am. Chem. Soc* 1998, 120, .
43. Miyamoto, S.; Kollman, P. A., *J. Am. Chem. Soc.* **1992**, *114*, 3668.
44. Fraternali, F.; Wipff, G. *J. Inclusion Phenom. Molec. Recognit.Chem.* **1997**, *28*, 63.
45. For reviews cf Perrin, R.; Harris, S. in ref 1c, p. 235; Perrin, R.; Lamartine, R.; Perrin, M. *Pure & Appl. Chem.* **1993**, *65*, 1549.
46. Izatt, R. M.; Christensen, J. J.; Hawkins, R. T. U. S. Patent 4,477,377, Oct 16, **1984**.
47. Schwing-Weill, M.-J.; Arnaud, Neu, F. *Gazz. Chim. Ital.* **1997**, *127*, 687.
48. Diamond, D.; McKervey, M. A. *Chem. Soc. Rev.* **1996**, *25*, 15; Diamond, D. *J. Inclusion Phenom. Molec. Recognit. Chem.* **1994**, *19*, 149.

49. Reinhoudt, D. N. *Recl. Trav. Chim. Pays-Bas* **1996**, *115*, 109; O'Conner, K. M.; Arrigan, D. W. M.; Svehla, G. *Electroanalysis* **1995**, *7*, 205.
50. Kenis, P. J. A.; Noordman, O. F. J.; Houbrechts, S.; van Hummel, G. J.; Harkema, S.; van Veggel, F. C. J. M.; Clays, K.; Engbersen, J. F. J.; Persoons, A.; van Hulst, N. F.; Reinhoudt, D. N. *J. Am. Chem. Soc* **1998**, *120*, 7875.

CALIXARENE–CATION COMPLEXATION

Chapter 2

Extraction of Cesium by Calix[4]arene-crown-6: From Synthesis to Process

J.-F. Dozol[1], V. Lamare[1], N. Simon[1], R. Ungaro[2], and A. Casnati[2]

[1]CEA Cadarache DCC/DESD, 13108 Saint Paul Lez Durance, France
[2]Università degli studi, Area Parco delle Scienze 17/A, I–43100 Parma, Italy

Cesium possesses two long lived isotopes ^{135}Cs and ^{137}Cs, the first one of which has a very long (2.3 × 10^6 y) half life and is one of the most mobile nuclides that would be present in a nuclear waste repository. In the framework of the French Actinex program, studies are under way to selectively remove Cs from high activity waste (HAW). ^{135}Cs would then be either destroyed by transmutation, or encapsulated in a very specific matrix. As for ^{137}Cs, it is interesting to remove it with strontium and actinides in order to decategorize high salinity medium activity waste (MAW) arising from dismantling of nuclear facilities. Cesium must be removed with a very high selectivity in the first case from a very acidic medium ([HNO$_3$] = 3-4 M), in the second one from acidic liquid waste at high sodium nitrate concentration ([NaNO$_3$] = 2-4 M). Calix[4]arene-crown-6 derivatives in the 1,3-alternate conformation exhibit outstanding efficiency and selectivity for cesium. These exceptional performances, much higher than those obtained with the best crown ethers, achieved on simulated waste were confirmed on a real high activity liquid waste in tests employing liquid-liquid extraction or supported liquid membranes.

Dismantling operations of reprocessing plants produce medium activity waste containing nuclides such as ^{90}Sr, ^{137}Cs, and actinides. Removal of these nuclides would enable waste to be decategorized and to be directed to a surface repository. These long lived nuclides, present in low amount in the waste (<10^{-3} M), could be sent to an interim storage or, after vitrification, to disposal in a geological formation.

© 2000 American Chemical Society

Within the framework of the French ACTINEX (ACTINide EXtraction) program, studies were launched for the recovery of minor actinides and long lived fission products from acidic dissolved spent fuel solutions, in order to destroy them by transmutation or to encapsulate them in specific matrixes. Efforts have to be directed towards ^{135}Cs, which is one of the most harmful elements, because of its long half life and its expected high mobility in a nuclear waste repository. There are a variety of liquid wastes which have to be treated :
- acidic or basic solutions arising from nuclear facility dismantling operations contain large amounts of inactive salts, particularly sodium nitrate, and ^{137}Cs at trace level (MAW).
- very acidic solutions arising from PUREX process (HAW).

In order to minimize the volume of waste, it is of paramount importance to find a compound able to remove cesium with high efficiency and selectivity, from acidic or basic high salinity media, and then to release this cation, if possible, in deionized water.

In this chapter we report on the synthesis, extraction properties, and transport measurements made on simulated and real radioactive waste of a very interesting class of ionophores, the calix[4]arene-crown-6, which shows an exceptional selectivity for cesium. Complexation studies in homogeneous solution (*1*) and molecular modelling (*2-5*) of the complexes between calix[4]arene-crown-6 and alkali metal ions were also performed. Molecular dynamics (MD) simulations were carried out by Wipff et al. on calix[4]arene-crown-6. They have first focused their attention on the influence of the calixarene conformation on alkali metal ion complex stabilities and hence on selectivity both in *vacuo* and in a water phase (*2*). Much work was also devoted to the role of the solvent, with a particular emphasis on the binding and extraction selectivities, by simulations performed in chloroform, at a water/chloroform interface, and in chloroform saturated with water molecules. (*2-5*)

Materials and Methods

Reagents : Dicarbollide, n-decylbenzo-21-crown-7, and dibenzo-21-crown-7 were synthesized by Katchem (Czech Republic), Chimie Plus (France), and Acros Organics (Belgium), respectively. All solvents and inorganic compounds were provided by Aldrich Chemical Company and Prolabo Company, respectively. Milli Q2 water was used to prepare solutions.

Extraction : All sodium and cesium distribution coefficients were measured by gamma spectrometry by using ^{22}Na (2.6 y) and ^{137}Cs (30 y) on two aliquots of 2 ml each. Measurement durations were adapted to obtain a reproducibility between ± 5 %. The selectivity of the different calixarenes was determined by measuring the distribution coefficients of cesium and sodium by contacting an aqueous solution with an organic phase containing the calixarene dissolved in 2-nitrophenyl hexyl ether (NPHE). The selectivity of cation M_2 over M_1 is defined as the ratio of the distribution coefficients of the two cations.

Transport : The thin flat sheet supported liquid membrane (SLM) device used was previously described by Stolwijk et al. (Figure 1) (*6*).

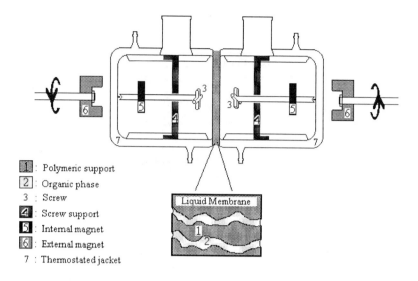

1 : Polymeric support
2 : Organic phase
3 : Screw
4 : Screw support
5 : Internal magnet
6 : External magnet
7 : Thermostated jacket

Figure 1. Flat sheet SLM device for transport experiments.

The transport of ^{137}Cs was followed by regular measurement of the decrease of the radionuclide in the feed solution, and its increase in the stripping solution, by γ spectrometry. This allowed graphical determination of the permeabilities P (cm.h^{-1}) of cesium through the SLM by plotting the logarithm of the ratio C/C° *versus* time, as described in the model of mass transfer proposed by Danesi (7) :

$$\mathrm{Ln}\left(\frac{C}{C^\circ}\right) = -\frac{\varepsilon.S}{V}P_f t$$

$$\mathrm{Ln}\left(1 - \frac{C'}{C^\circ}\right) = -\frac{\varepsilon.S}{V}P_s t$$

where
C : concentration of the cation in the feed solution at time t.
C' : concentration of the cation in the stripping solution at time t.
C° : initial concentration of the cation in the feed solution.
ε : volumic porosity of the SLM.
S : membrane surface area.
V : volume of feed and stripping solutions.
t : time.

The internal volumes of the cells varied from 45 to 50 cm^3 and the areas of the membranes from 15 to 16 cm^2 depending on the device used.

Synthesis of Calix[4]arene-Crown-6 Derivatives

Calix[4]arenes (**I**) are macrocycles made up of four phenolic units linked by methylene bridges (*8*). Their functionalization at the phenolic oxygen atoms or in their para position (*9*), together with the possibility to adopt different conformations [cone, partial cone, 1,2-alternate, 1,3-alternate (Figure 2)] leads to a large variety of ionophores with different sizes and shapes (*10*).

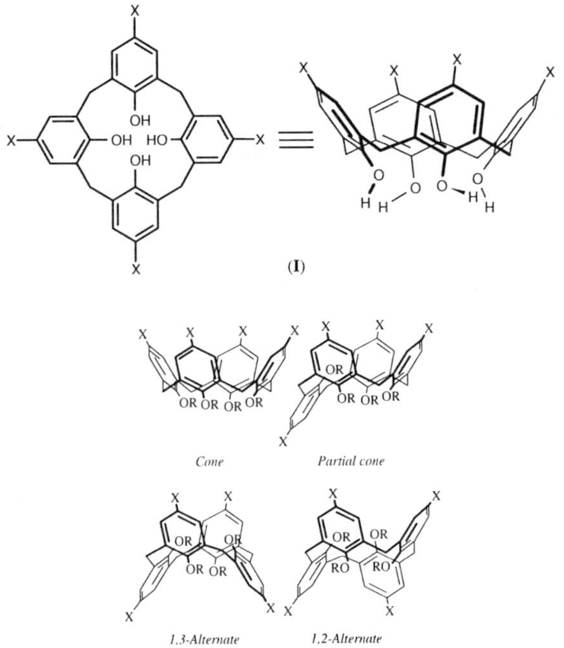

Figure 2. Calix[4]arenes and their conformations

The introduction of polyether bridges in 1,3 (distal) positions of calix[4]arenes, has recently afforded an interesting class of ionophores, the calix[4]arene crowns, which show interesting selectivity among alkali metal ions (*11-12*). 1,3-Dialkoxycalix[4]arene crowns can be obtained in a two steps synthesis starting from commercially available calix[4]arene (Scheme 1) (*12*). The first reaction exploits the selective 1,3 (distal) dialkylation of calix[4]arenes which can be obtained in nearly quantitative yields using a weak base like K_2CO_3 in acetone with an alkyl halide. Reaction of these 1,3-dialkoxycalix[4]arenes with the proper ditosylate and Cs_2CO_3 in dry CH_3CN affords in high yield (60-75%) the calix[4]arene crown derivatives. With the exception of the 1,3-dimethoxy calixcrown-6 (dimethoxycalix C6) and -crown-7 (dimethoxycalix C7) which, due to the reduced steric hindrance of the methyl group, are conformationally mobile, all the other derivatives are fixed in the 1,3-alternate structure (*12*).

Solvent Extraction Results

Solvent Extraction of Cesium and Cesium Selectivity Over Sodium

The distribution coefficients of cesium and sodium were determined by contacting an aqueous acidic solution (1 M HNO_3, 5×10^{-4} M MNO_3) with solutions of different calix[4]arene crowns (10^{-2} M) in NPHE (Table I). The extraction is mainly dependent on the conformational properties and on the size of the crown. Calix[4]arene-crown-6 derivatives fixed in the 1,3-alternate conformation display the highest cesium distribution coefficients (D_{Cs}), very low sodium distribution coefficients (D_{Na}), and hence the highest Cs/Na selectivity [α(Cs/Na)]. These values dramatically drop in the case of dimethoxy calix C6, since its conformation is mobile in solution and not preorganized for cesium binding. An even greater decrease of cesium extraction is observed in the case of dimethoxy calix C7, due to the non perfect complementarity between the size of the crown and that of the cation.

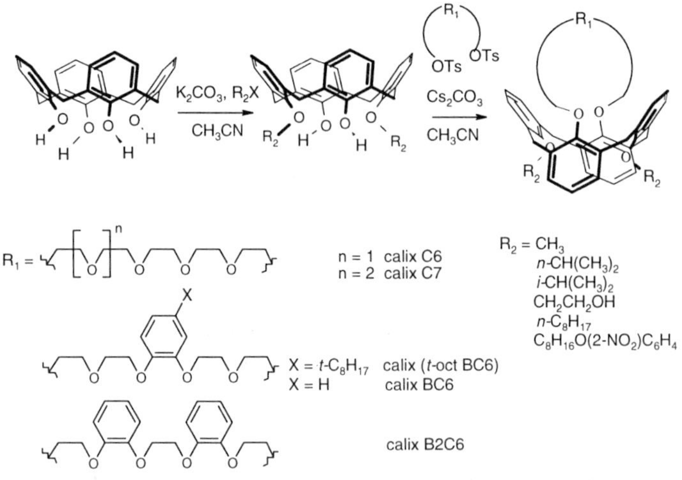

Scheme 1. Synthesis of 1,3-dialkoxycalix[4]arene-crown-6 and -crown-7

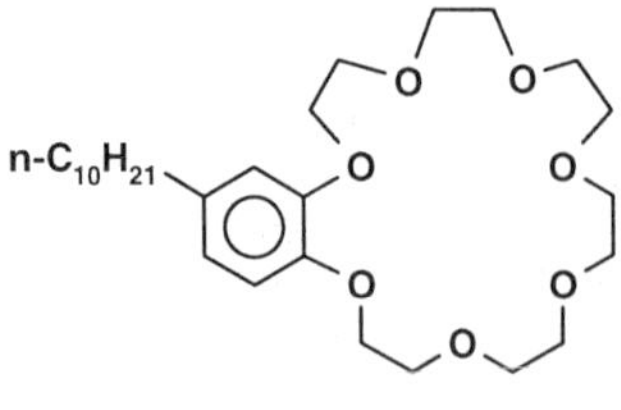

n-decylbenzo-21-crown-7

The exceptional affinity of dialkoxycalix[4]arene-crown-6 derivatives for cesium, much higher than those of the most efficient crown ethers, such as n-decyl benzo-21-crown-7, can be explained by the simultaneous operation of several effects such as the crown ether ring size, the polarity of the calix conformation and the possible occurrence of weak cation/π interactions with the aromatic nuclei close to the cesium ion. *(12-14)*

The selectivity of these calix-crown-6 derivatives is also confirmed by the competitive extraction of cesium from acidic solutions containing large amounts of sodium ($[NaNO_3] = 4$ M). Very high distribution coefficients are obtained (Table II) with calix-benzo-crown-6 (BC6), and especially with calix-dibenzo-crown-6 (B2C6), where the crown includes one or two benzene units, respectively (9).In fact, B2C6 presents an higher selectivity cesium over sodium due to a lesser affinity for sodium. This statement has been explained by the difficulty of complexation of Na^+-H_2O dehydration in a more hydrophobic complexation site (15). These results are confirmed by the cesium transport through supported liquid membranes. The permeabilities, (7) which convey the ability of the membrane to transfer cation, logically increase as the distribution coefficients increase.

Table I. Extraction of cesium and sodium – Selectivity Cs^+/Na^+

Crown ether	D_{Na}	D_{Cs}	$\alpha_{Cs/Na}$
Dimethoxy calix C6	3×10^{-3}	4×10^{-2}	13
Hydroxy ethoxy calix C6	4×10^{-3}	4.2	> 4200
Dipropoxy calix C6	2×10^{-3}	19.5	> 19500
Di isopropoxy calix C6	$< 10^{-3}$	28.5	> 28500
Di n octyloxy calix C6	$< 10^{-3}$	33	> 33000
Di iso propoxyoxy calix (t-oct BC6)	$< 10^{-3}$	34	> 34000
Di n octyloxy calix B2C6	$< 10^{-3}$	31	> 31000
Di methoxy calix C7	4×10^{-3}	7×10^{-3}	1.7
n-decylbenzo-21-crown 7	1.2×10^{-3}	0.3	250

NOTE: Aqueous solution : MNO_3 5×10^{-4} M - HNO_3 1 M
Organic solution : 1×10^{-2} M crown ether in NPHE

Table II. Competitive extraction and transport of cesium in presence of an excess of sodium

Crown ether	D_{Cs}	P_{Cs} $(cm.h^{-1})$
Di methoxy calix C6	0.034	$< 10^{-2}$
Di hydroxy ethoxy calix C6	5.2	0.4
Di propoxy calix C6	12	1.6
Di isopropoxy calix C6	18	1.3
Di octyloxy calix C6	25	1.9
Di iso propoxyoxy calix (t-oct BC6)	45	2.1
Di isopropoxy calix B2C6	56	4.3
Di methoxy calix C7	0.003	$< 10^{-2}$
n-decylbenzo-21-crown 7	0.12	9×10^{-2}

NOTE: Aqueous feed solution : $NaNO_3$ (4 M) HNO_3 (1 M) - Aqueous stripping solution : deionized water - Organic solution : 1×10^{-2} M carrier in NPHE

Extraction of Cesium from Nitric Acid Solutions

Cesium distribution coefficients, much higher than those of two dibenzo crown ethers 21C7, strongly rise with increasing nitric acid concentration until a value of 2-3 M, and then rapidly decrease (Figure 3). Interestingly the maximum of D_{Cs} is reached exactly for a nitric acid concentration close to that of typical fission products solutions.

The presence of benzene units on the crown bridge of the calixarene reduces the nitric acid extraction and strongly enhances that of cesium (Figure 3). For nitric acid concentration lower than 2 M, the Cs distribution coefficient is given by the following expression :

$$D_{Cs} = \frac{[Cs_{org}]}{[Cs_{aq}]} = K_{ex} [Calix].[NO_3^-]$$

where :
$[Cs_{org}]$: Cesium concentration in organic phase.
$[Cs_{aq}]$: Cesium concentration in aqueous phase.
K_{ex} : Extraction constant.
[Calix] : Calixarene concentration.
$[NO_3^-]$: Nitrate concentration.

Accordingly, at < 2 M HNO_3, increasing nitrate concentration favors Cs extraction, but at higher acidity, competing extraction of HNO_3 causes a decrease in D_{Cs}.

Distribution coefficients decrease with decreasing nitric acid concentration and therefore generally this interesting property is used to strip cesium in deionized water.

Extraction from High Salinity Media

For a wide range of pH (2-12), the increase of sodium concentration from 1 to 4 M, causes a significant decrease of cesium distribution coefficients (Table III). However the distribution coefficients, especially for calixarene dibenzo crown-6 (calix B2C6), are sufficiently high to allow cesium to be removed from these media.

Table III. Distribution coefficients of cesium from $NaNO_3$ solutions

Compound	$NaNO_3$			
	1 M	2 M	3 M	4 M
Di i-propoxy calix C6	18	12	8	7
Di n-octyloxy calix C6	12.5	9.7	8.2	6
Di n-octyloxy calix BC6	25	16	14	10
Di n-octyloxy calix B2C6	37	20	18.5	15

NOTE : Aqueous solution : $NaNO_3$ (1- 4 M) - $CsNO_3$ (1×10^{-6} M) - Organic solution : 1×10^{-2} M calix-crown-6 in NPHE

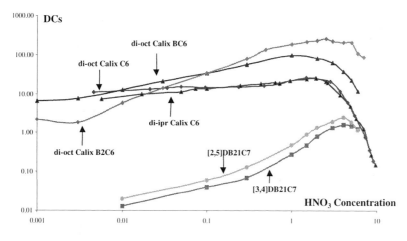

Figure 3. Extraction of cesium by calixarene-crown-6, [2,5] dibenzo crown-7 and [3,4] dibenzo crown-7 from acidic media (10^{-2} M extractant in NPHE).

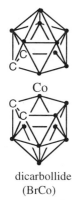

dicarbollide
(BrCo)

An interesting feature displayed by calixarene-crown-6 derivatives which is useful for the implementation of liquid-liquid extraction or of transport by supported liquid membranes, is that their extracting ability decreases by lowering the salinity of the medium. Therefore the cations can be easily stripped in deionized water. However, if necessary, the distribution coefficients can be strongly increased by adding to the organic phase the lipophilic cobalt bis dicarbollide which behaves as a cation exchanger. Mixing dicarbollide (BrCo) and calixarene-crown-6 leads to very high distribution coefficients and important synergistic factors, particularly with a stoichiometric mixture of the two components. However at high acidity the efficiency of dicarbollide decreases and completely disappears for a nitric acid concentration of 2 M (Figure 4).

Results From Test With Actual Radioactive Waste

Two types of measurements (distribution coefficients, transport through SLMs) were carried out on real high activity waste (HAW) arising from dissolution of a MOX Fuel (burn up 34 650 MWJ/tU) where uranium and plutonium have been previously extracted by TBP (Table IV). The experiments were performed in the CARMEN hot cell of CEA Fontenay aux Roses. Two calixarene-crown-6 were tested in the experiment carried out in a CARMEN cell (diisopropoxy-calix[4]arene-crown-6 and dinitrophenyl-octyloxy-calix[4]arene-crown-6).

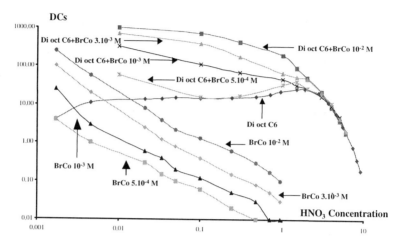

Figure 4. Extraction of cesium from acidic media to NPHE by the mixtures of dioctyloxy calix[4]arene-crown-6 ([OctC6] = 1 × 10^{-2} M) and dicarbollide (BrCo) at different concentrations.

Table IV. Composition of the high activity liquid waste

Elements	(mg L^{-1})	Nuclides	(GBqL^{-1})
Fe	23	^{106}Ru	89
Zr	450	^{125}Sb	17
Nb	< 5	^{134}Cs	240
Mo	427	^{137}Cs	710
Tc	< 5	^{144}Ce	145
Cs	580	^{154}Eu	52
Ce	430	^{241}Am	45
Nd	680	^{243}Am	1.1
U	< 10	^{242}Cm	220
Pu	< 5	^{244}Cm	1.85
Am	350		
Cm	73		

Distribution coefficient measurements were carried out, the HAW solution was contacted with 0.1 M calix[4]arene-crown-6 solution in NPHE then the Cs distribution coefficients were determined in duplicate by measuring the ^{137}Cs activity in each phase. High cesium distribution coefficients were obtained (Table V) although there was some discrepancy between the duplicate measurements. This discrepancy can be attributed to pollution of a sample in the hot cell. Moreover the high selectivity observed with the simulated waste was confirmed: for most of the elements and radionuclides (actinides or fission products: ^{154}Eu, ^{125}Sb, ^{144}Ce), Mo, Zr, Ce, Nd, the residual concentration or activity was less than 1% in the stripping solution, except for iron (2 %) and ruthenium (8 %).

Table V. Distribution coefficients of cesium from HNO$_3$ solutions

Compounds	D_{Cs}	
Di-i-propoxy calix C6	55 ± 5	78 ± 8

NOTE: Aqueous solution: HNO$_3$ 4 M + fission products + actinides - Organic solution : 1 × 10^{-1} M calix-crown-6 in NPHE

Diisopropoxy-calix[4]arene-crown-6 (*diiprcalixC6*) and dinitrophenyl-octyloxy-calix[4]arene-crown-6 (*diNPOEcalixC6*) were used as carriers in supported liquid membranes (0.1 M in NPHE). Respectively 77.5 % and 86.3 % of cesium were transported from liquid waste to the demineralized water in 9 hours. Higher percentages of transported cesium have been obtained by increasing the duration of the experiment. Only cesium was transported during these tests: molybdenum and zirconium were detected in the stripping phase at very low level, while with di NPOE calix C6, iron, neodymium, cerium were also detected. However less than one percent of these elements or radionuclides was transported.

According to the Danesi's model (7), cation fluxes were calculated from experiments carried out with a simulated waste at different nitric acid concentrations (2-6 M). A decrease of cation flux was observed with an increase of the acidity due to the competitive transport of HNO$_3$. For 4 M HNO$_3$ solution, the cesium flux with the real waste was slightly lower than that obtained with simulated waste (Table VI).

Table VI. Cesium fluxes (10^{-6} mol.cm^{-2}.h^{-1}) through calix-crown containing SLMs

	[HNO$_3$]	DiiprcalixC6	diNPOEcalixC6
Simulated waste	2 M	12.2	
	3 M	10.6	
	4 M	8.85	
	5 M	5.95	
	6 M	3.75	
H. A. W.	4 M	6.7	6

The success of this test with actual HAW and the excellent radiolytic stability of calixarene-crown-6 confirmed the high potentiality of calixarene crown derivatives and the feasibility of their application in high level liquid waste decontamination.

From Test to Process

In order to selectively remove cesium from PUREX raffinate, the general scheme of reprocessing depicted in Figure 5 can be designed.

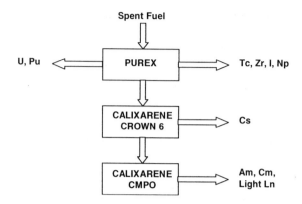

Figure 5. General scheme of improved reprocessing of spent fuel

To be consistent with all the steps of reprocessing, a process based on liquid-liquid extraction has to be defined. The system calixarene / NPHE, used in SLMs, had to be modified to use liquid-liquid contactors, due to its inappropriate hydraulic properties. For example, the density (d = 1.07) and the viscosity (η = 8.9 cP) of the organic solution (di-isopropoxy calix C6 1×10^{-2} M/ NPHE diluent), lead to such modifications. It was relatively easy to find a calix-crown containing organic phase with hydraulic properties (Table VII) close to those of the system used in PUREX by COGEMA La Hague, which requires the use of a mixed system TPH/NPHE (TPH: hydrogenated tetrapropylene).

Table VII. Hydraulic properties of some organic phase at 25°C.

Organic phase	Viscosity (cP)	density
di-isopropoxy calix C6 1×10^{-2} M/ NPHE	8.9	1.07
TPH (70 %)-NPHE (30 %)	2.07	0.8506
di-isopropoxy C6 TPH (70 %)-NPHE (30%)	2.45	0.8659
TBP (30 %) dodecane	1.675	0.825

It was sufficient to optimize the NPHE percent according to the following criteria :
- diluent – extractant compatibility;
- hydraulic properties : density, viscosity, interfacial tension;
- chemical criteria: sufficient cesium extraction (depending on the diluent), kinetics, third phase conditions,....;

- safety: flash point superior to 60°C.

Dioctyloxy-calix-C6 diluted in the mixture TPH (70%) NPHE (30%) was chosen since this system shows the best behavior.

Hydraulic data were compatible with those of the PUREX solvent, and assure the possibility of using liquid-liquid extraction contactors defined for the PUREX process (Table VIII). Conditions can be found to avoid the problem of third phase formation (Table IX), and extraction isotherms (Figure 6) show that cesium distribution coefficients are very satisfactory for the extraction and for the subsequent back-extraction.

Table VIII: Hydraulic properties of different organic phases

Organic phase	Surface tension
Dioctyloxy calixC6 (5×10^{-2}M) in TPH (70%)-NPHE (30%)	22.7
TBP (30%) in dodecane	24.8

Table IX : Third phase occurrence for solutions of dioctyloxy calix C6 in TPH (70%) /NPHE (30%) in contact with HNO_3 3M

Cesium concentration (M)	Temperature (°C)	3rd phase occurence
10^{-2}	25	No
10^{-2}	30	No
2.5×10^{-2}	25	Yes
2.5×10^{-2}	30	No
5×10^{-2}	30	Yes

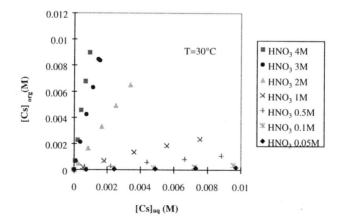

Figure 6. Extraction isotherms for dioctyloxycalix C6 (5×10^{-2} M) in TPH(70%)/NPHE(30%)

The flow-sheets were defined in close collaboration with CEA Valrho, Marcoule, France (Figure 7). It is intended to test such a flow-sheet with a simulated waste in the course of the year, and then with a real raffinate.

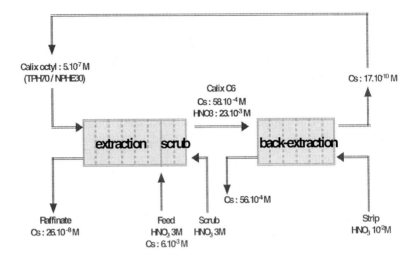

Figure 7. Selective cesium extraction flow-sheet

Conclusion

In a few years, thanks to a close collaboration of several European groups involved in design, synthesis, structural characterization, complexation, extraction, ion transport, an important breakthrough was achieved for cesium extraction. Dialkoxy calix[4]arene-crown-6 derivatives displays exceptional efficiency and selectivity for Cs over most other cations, especially sodium. They allow cesium to be removed from high salinity media having very acidic or basic pH. They can be used either for the recovery of cesium from the PUREX process raffinate or for the decategorization of alkaline, neutral, or acidic high salinity solutions. Moreover chemical and radiochemical stabilities well argue for future use of calixarenes and especially of di-alkoxy-calix[4]arene-crown-6 derivatives in radioactive waste treatment.

Acknowledgments

The authors thank the European Commission for its financial support in the framework of the research program on " Management and Storage of Radioactive Waste ".

Literature cited

1. Schwing-Weill, M.-J.; Arnaud-Neu, F. *Gazz. Chim. Ital.*, **1997**, *102*, 687.
2. Wipff, G.; Lauterbach, M. *Supramol. Chem.*, **1995**, *6*, 187.
3. Lauterbach, M.; Wipff, G. *Physical Supramolecular Chemistry*, Echegoyen, L., Kaifer, A. Ed. Kluwer, NATO ASI Series, Dordrecht, 1996, 1.
4. Lauterbach, M.; Engler, E.; Muzet, N.; Troxler, L.; Wipff, G. *J. Phys. Chem. B.* **1998**, *102*, 245.
5. Lauterbach, M.; Wipff, G.; Mark, A.; van Gunsteren, W. F. *Gazz. Chim. Ital.* **1997**, *127*, 699.
6. Stolwijk, T.B.; Sudhölter, E.J.R.; Reinhoudt, D.N. *J. Am. Chem. Soc.*, **1987**, *109*, 7042.
7. Danesi, P. *Sep. Sci. Technol.*, **1984-85**, *19*, 857.
8. a) Gutsche, C. D. *Calixarenes Revisited*, The Royal Society of Chemistry, Stoddart, J.F. Ed. Cambridge, 1998. b) Ikeda, A.; Shinkai, S. *Chem. Rev.*, **1997**, *97*, 1713. c) Pochini, A.; Ungaro, R. *Comprehensive Supramolecular Chemistry*, Vol. 2, Vögtle, F. Ed. Pergamon Press, Oxford, 1996, p. 103. d) Böhmer, V. *Angew. Chem., Int. Ed. Engl.* **1995**, *34*, 717. e) Casnati, A. *Gazz. Chim. Ital.*, **1997**, *127*, 637.
9. Arduini, A.; Casnati, A. "Calixarenes" in *Macrocyclic Synthesis: a Practical Approach*, Parker, D. Ed. Oxford University Press, Oxford, 1996, p. 145.
10. Andreetti, G. D.; Ugozzoli, F.; Pochini, A.; Ungaro, R. in *Inclusion Compounds*, Vol. 4, Atwood, J. L.; Davies, J. E. D.; Mac Nicol, D. D. Eds, Oxford University Press, Oxford, 1991, p.64. Ungaro, R.; Pochini, A, in *Topics in Inclusion Science. Calixarenes, a Versatile Class of Macrocyclic Compounds*, Böhmer, V.; Vicens, J. Eds., Kluwer Academic Publishers, 1991, p. 127. Schwing, M. J.; Mc Kervey, M. A. *ibid.* p. 149. McKervey, M. A.; Schwing-Weill, M. J.; Arnaud-Neu, F. in *Comprehensive Supramolecular Chemistry*, Vol. 1, Gokel, G. W. Ed., Pergamon, 1996, p. 537. Ungaro, R.; Arduini, A.; Casnati, A.; Pochini, A.; Ugozzoli, F., *Pure & Appl. Chem.*, **1996**, *68*, 1213.
11. Ghidini, E.; Ugozzoli, F.; Ungaro, R.; Harkema, S.; El-Fadl, A. A.; Reinhoudt, D. N. *J. Am. Chem. Soc.* **1990**, *112*, 6979. Casnati, A.; Pochini, A.; Ungaro, R.; Bocchi, C.; Ugozzoli, F.; Egberink, R. J. M.; Reinhoudt, D. N. *Chem. Eur. J.*, **1996**, *2*, 436.
12. Ungaro, R.; Casnati, A.; Ugozzoli, F.; Pochini, A.; Dozol, J. F.; Hill, C.; Rouquette, H. *Angew. Chem., Int. Ed. Engl.* **1994**, *33*, 1506. Casnati, A.; Pochini, A.; Ungaro, R.;, Ugozzoli, F.; Arnaud, F.; Fanni, S.; Schwing, M.-J.; Egberink, R. J. M.; de Jong, F.; Reinhoudt, D. N. *J. Am. Chem. Soc.* **1995**, *117*, 2767.
13. Ugozzoli, F.; Ori, O.; Casnati, A.; Pochini, A.; Ungaro, R.; Reinhoudt, D.N. *Supramol. Chem.*, **1995**, *5*, 179.
14. Lamare, V.; Dozol, J.-F.; Ugozzoli, F.; Casnati, A.; Ungaro, R. *Eur. J. Org. Chem.* **1998**, 1559.
15. Lamare, V.; Dozol, J.F.; Fuangswasdi, S.; Arnaud-Neu, F.; Thuéry, P.; Nierlich, M.; Asfari, Z.;Vicens, J. *J. Chem. Soc., Perkin Trans. 2*, **1999**, 271.

Chapter 3

Development of Process Chemistry for the Removal of Cesium from Acidic Nuclear Waste by Calix[4]arene-crown-6 Ethers

Peter V. Bonnesen, Tamara J. Haverlock, Nancy L. Engle, Richard A. Sachleben, and Bruce A. Moyer

Chemical and Analytical Sciences Division, Oak Ridge National Laboratory, Oak Ridge, TN 37831–6119

A solvent suitable for extracting cesium from acidic nitrate media, such as that stored at the U.S. Department of Energy's Idaho National Engineering and Environmental Laboratory (INEEL), has been developed. The solvent possesses good chemical stability, displays excellent cesium selectivity, and provides good extraction and stripping performance with satisfactory phase-coalescence behavior. The calix[4]arene-crown-6 ether used in this solvent (1,3-alt-bis-*n*-octyloxycalix[4]arene-benzo-crown-6) was selected from a series of mono- and bis-crown-6 derivatives of 1,3-alternate calix[4]arenes that were shown to possess good stability to a simulant of INEEL's Sodium Bearing Waste (SBW). Calixarene-benzo crown ethers possessing an alkyl substituent at the 4-position of the benzocrown, such as calix[4]arene-bis-*tert*-octylbenzo-crown-6, were shown to be much more susceptible to nitration by the SBW simulant or by 4 M nitric acid than calixarene-benzo crown ethers without the alkyl substituent. The cesium distribution behavior for a solvent comprised of 1,3-alt-bis-*n*-octyloxy-calix[4]arene-benzo-crown-6 at 0.01 M in an aliphatic diluent modified with a fluorine-containing alcohol was shown to be stable over the course of a 60 day continuous contact with the SBW simulant at 25 °C. Toward process development, cesium distribution ratios on extraction (SBW simulant), scrubbing (50 mM nitric acid), and stripping (1 mM nitric acid) operations demonstrated a functional solvent cycle using a solvent of this composition augmented by 0.001 M trioctylamine (TOA). Without this lipophilic amine, stripping is inefficient. The presence of TOA in the solvent had no adverse effect on the Cs distribution behavior during extraction and scrubbing operations, nor did TOA negatively impact the Cs/Na and Cs/K selectivity ratios.

© 2000 American Chemical Society

Introduction

Mono- and bis-crown-6 derivatives of 1,3-alternate calix[4]arenes possess high extractive strength for cesium (*1-15*) and, accordingly, are currently being investigated for possible use in separating radiocesium from nuclear wastes (*1,3,4,6,14-18*). These calixarene-crown ethers also possess excellent Cs/Na ($\geq 10^4$) and Cs/K ($\geq 10^2$) selectivities; however, they are currently quite expensive (> $100/g). Thus, for separation methods such as supported liquid membranes and solvent extraction employing calixarene-crown ethers to be economically viable, the consumption of these materials needs to be minimal. Ways to minimize the consumption of the calixarene-crown ether include optimizing the calixarene-crown ether's extracting power and reducing its loss (and therefore replacement costs) by a careful combination of good chemistry and engineering.

As discussed at length elsewhere (*17*), we have been examining solvent extraction in our separations-technology development, considering this separation method to be a proven high-throughput technique readily adaptable to remote operation for nuclear-waste cleanup. Process-suitable solvents for solvent extraction generally require high flash-point, low polarity, aliphatic hydrocarbon diluents (*19*). Unfortunately, the solubility and extractive power of many calixarene-crown ethers in such diluents is low. Toward a solution to this problem, we discovered that good solubility and Cs extraction power could be achieved when lipophilic calixarene-crown ethers such as calix[4]arene-bis-*tert*-octylbenzo-crown-6 (**1**, Figure 1) were combined with alkylphenoxy alcohol-based solvating components, or modifiers, in the kerosene diluent (*18*). In particular, a solvent composed of **1** at 0.01 M with the modifier 1-(1,1,2,2-tetrafluoroethoxy)-3-(4-*tert*-octylphenoxy)-2-propanol (Figure 2, ORNL "Cs-3") at 0.20 M in Isopar® L diluent was shown to be effective for removing cesium from alkaline nitrate waste stored at the U.S. Department of Energy's (DOE's) Westinghouse Savannah River Site (*17*), affording cesium distribution ratios above 12 for the first contact between pristine solvent and actual Savannah River Site High Level Waste (*20*). Thus, good extraction power at low (0.01 M) extractant concentration helps to minimize solvent-inventory costs. In addition, replacement costs of **1** should be small, as it is chemically stable to alkaline media and is of sufficient lipophilicity that losses to aqueous phases are negligible (*17*).

With respect to good engineering, the use of centrifugal contacting equipment allows rapid turnover of the solvent, minimal solvent inventory, and because of the short contact times, minimal exposure of the solvent to high radiation fields, thus decreasing losses of the (calixarene-crown ether) extractant by radiolytic degradation (*21-23*). Solvent and extractant losses due to entrainment are also minimized due to the forces developed in centrifugal contactors that separate the solvent and aqueous phases. Rapid turnover of the solvent means the most efficient use of the solvent possible, making it feasible to reduce the solvent inventory to just a few hundred gallons for a waste throughput equivalent to that of a full-scale plant capable of processing the 230,000 kL of alkaline radioactive waste stored at DOE's Hanford Site (*23*).

In this paper, we report the results of our initial efforts toward developing a process solvent suitable for extracting cesium from *acidic* high nitrate nuclear waste, such as that stored in waste tanks at the DOE's Idaho National Engineering and Environmental Laboratory (INEEL). With the change from alkaline to nitric acid

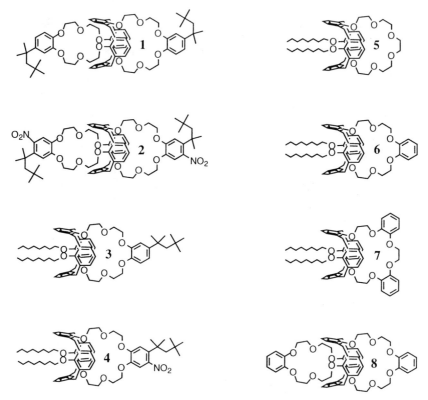

Figure 1. Calix[4]arene monocrown-6 (3-7) and biscrown-6 (1, 2, 8) ethers examined in this study.

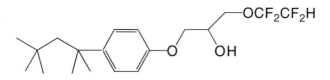

**1-(1,1,2,2-tetrafluoroethoxy)-3-
[4-(*t*-octyl)phenoxy]-2-propanol**

Figure 2. Solvent modifier ORNL "Cs-3" used in this study.

waste came new concerns with respect to the chemical stability of the solvent. On the alkaline-side it was discovered that, whereas the calixarene-crown ether extractant calix[4]arene-bis-*tert*-octylbenzo-crown-6 (**1**) possessed good long-term chemical stability to base (such as the 1.9 M free hydroxide present in the Savannah River simulant), the Cs-3 modifier did not. (A new class of modifiers that possess excellent alkaline stability have recently been developed.) In contrast, on the acid-side it was discovered that the Cs-3 modifier possessed good stability to the nitric acid (about 1.26 M) present in the INEEL waste simulant, but that **1** did not, instead becoming nitrated at an unacceptably fast rate. An explanation of the factors contributing to nitration susceptibility, along with an evaluation of calixarene-crown ethers and process solvents that promise to be suitable for an acid-side cesium extraction process, is presented below.

Experimental

Materials

All salts and solvents were reagent grade and were used as received, except trioctyl amine, which was distilled prior to use. Distilled, deionized water was obtained from a Barnstead Nanopure filtering system (resistivity 18 MΩ) and was used to prepare all aqueous solutions. Nitric and hydrochloric acids were Ultrex II grade (J.T. Baker). Isopar® L isoparaffinic diluent (lot# 0306 10967) was obtained from Exxon Chemical Company, Houston, Texas. The Sodium Bearing Waste (SBW) simulant was received from Dr. Donald J. Wood at DOE's Idaho National Engineering and Environmental Laboratory (INEEL). The ^{137}Cs radiotracer used for spiking the waste simulants was obtained as ^{137}CsCl in 1M HCl from Amersham (Arlington Heights, IL) and was used as received. The ^{22}Na radiotracer was obtained as ^{22}NaCl in water from Isotope Products Laboratories (Burbank, CA) and was used as received.

The calixarene crown ethers examined in this work are shown in Figure 1. Calix[4]arene-bis-*tert*-octylbenzo-crown-6 (**1**, (*18*)), 1,3-alt-bis-*n*-octyloxycalix[4]-arene 4'-*tert*-octylbenzo-crown-6 (**3**, (*24*)), 1,3-alt-bis-*n*-octyloxycalix[4]arene 5'-nitro-4'-*tert*-octylbenzo-crown-6 (**4**, (*24*)), bis-*n*-octyloxycalix[4]arene-crown-6 (**5**, (*5*)), 1,3-alt-bis-*n*-octyloxycalix[4]arene-benzo-crown-6 (**6**, (*24*)), and bis-*n*-octyloxycalix[4]arene-dibenzocrown-6 (**7**, (*25*)), were prepared as described previously. Calix[4]arene-bis-5'-nitro-4'-*tert*-octylbenzo-crown-6 (**2**), was prepared in the same manner used to prepare **4**. Calix[4]-bis-1,2-benzo-crown-6 (**8**) was purchased from ACROS Organics. The modifier 1-(1,1,2,2-tetrafluoroethoxy)-3-(4-*tert*-octylphenoxy)-2-propanol (Figure 2, ORNL "Cs-3") was prepared as previously described (*17*).

Batch-Equilibrium Experiments Using Cs-137 Tracer

As a general procedure, batch-equilibrium liquid-liquid contacting experiments were performed in polypropylene or Teflon® FEP tubes. An exception is the initial solvent stability experiments between the SBW simulant and the solvent containing calixarene-crown ether **1**, which were performed in deionized water-rinsed borosilicate glass vials with black-phenolic screw caps containing polyethylene inserts. In all contacting experiments, equal volumes of aqueous and organic phases were contacted for 30 min (or longer for stability experiments) at 25.0 ± 0.2 °C by end-over-end rotation at 35 ± 5 RPM using a Glass-Col® laboratory rotator placed inside a 25 °C constant temperature airbox. The vials were then centrifuged for three to five minutes at 2869 x g in a refrigerated centrifuge maintained at 25 ± 1 °C (Sanyo MSE Mistral 2000R) to ensure complete phase separation. Aliquots of each phase were removed for analysis, and the ^{137}Cs activity in each phase determined by standard gamma(γ)-counting techniques using a Packard® Cobra Quantum Model 5003 gamma counter equipped with a 3" NaI(Tl) crystal through-hole type detector. Distribution ratios (D_{Cs}) were determined as the ratio of the ^{137}Cs activity in the organic phase to the ^{137}Cs activity in the aqueous phase at equilibrium, and are reproducible to within ± 5%.

Nitric Acid Stability Experiments

The stability of selected calixarene crown ethers **1, 3, 5, 6**, and **8** to nitration by nitric acid was measured by contacting 20 mM deutero-chloroform solutions of each calixarene-crown with an equal volume of 4.0 M nitric acid in Nalgene Teflon® FEP 10-mL centrifuge tubes by end-over-end rotation as described above. Aliquots of the chloroform phase were periodically withdrawn, and the proton NMR (Bruker MSL 400 operating at 400.13 MHz for proton) spectrum recorded and compared with the corresponding spectra of both the pristine solution prior to contact, and the solution one hour after contact (some resonances shift upon extraction of nitric acid, but no nitration was observed in any case after 1 hour contact). A relaxation delay (D_0) of 20 seconds was employed to ensure the complete relaxation of all protons selected for integration. The resonances selected for intregration included the protons on the *tert*-octyl group for **1** and **3**, the aromatic protons on the benzo-crown portion of **1, 3, 6**, and **8**, and the aromatic protons on the "belt" of the calixarene for **5**, for both the parent calixarene-crown, and the nitrated product. The amount of nitrated calixarene-crown ether present as a function of time relative to the starting calixarene-crown ether was determined by comparison of the integration areas of the respective relevant peaks. Peak assignments for the nitrated calixarene-crowns were confirmed based on comparison to authentic samples: nitration of **1** gave only **2**, and nitration of **3** gave only **4** (as the nitrated products), under these nitration conditions.

Elemental Analyses by Inductively Coupled Argon Plasma (ICAP) Spectroscopy

Instrumentation

ICAP analyses were performed using a Thermal Jarell Ash (Franklin, MA) IRIS Inductively Coupled Argon Plasma Optical Emission Spectrometer (ICAP/OES), equipped with a charge-injection device (CID) capable of recording atomic emission lines in the wavelength range 177 to 780 nm.

Contacting and Analytical Procedures

The distribution behavior of the selected elements Al, B, Fe, Hg, K, and Na to selected solvents was investigated by contacting 3 mL of the solvent with an equal volume of the SBW simulant in 15-mL polypropylene centrifuge tubes for 30 minutes in the manner described above, in duplicate. To analyze for these metals using ICAP/OES, the loaded extraction solvents were first completely stripped into an aqueous phase. This was accomplished by combining 2.0 mL of the loaded organic phase with an equal volume of 1,3-diisopropylbenzene (Aldrich), and contacting the resulting organic solution with 8.0 mL of 2% HCl (Ultrex II grade) in 15-mL polypropylene centrifuge tubes as above. A second strip was then performed by taking 3.0 mL of the organic phase from the first strip (containing 1.5 mL of the original solvent) and contacting it with 6.0 mL of 2% HCl. The aqueous phases from the two strips, each representing a four-fold dilution of the metal ion concentration from the original solvent, were transferred directly to polypropylene tubes for ICAP analyses. The metal ion concentrations in the solvent were determined by multiplying the sum of the metal ion concentrations found in the first and second strip by four.

The instrumental detection limit (IDL) for the elements was set equal to three times the standard deviation of a 2% HCl blank based on 10 replicates. The detection limits for Al, B, Fe, Hg, K, and Na correspond to $D_{Al} < 8.4 \times 10^{-6}$, $D_B < 1.7 \times 10^{-4}$, $D_{Fe} < 2.4 \times 10^{-5}$, $D_{Hg} < 5.0 \times 10^{-4}$, $D_K < 2.7 \times 10^{-5}$, and $D_{Na} < 3.8 \times 10^{-6}$. The completeness of the metal ion stripping from the solvent using this method was validated by measuring the sodium distribution by ^{22}Na tracer in polypropylene vials. The sodium distribution values obtained using the tracer method were only 10-14% higher than the values obtained using ICAP, which, given an experimental uncertainty of ±5% in the distribution ratio obtained using either method, is reasonably good agreement.

The metal ion concentrations in the aqueous raffinate from the extraction and in the simulant itself were analyzed by diluting aliquots 100-fold with 2% HCl. The concentrations in the simulant and the raffinates for these metal ions were found to be essentially the same (within 5% experimental uncertainty) both to each other and to the concentrations listed in Table 1 for the data provided by INEEL (*26*). The metal ion concentrations found in the feed were used in the calculation of the metal ion distribution ratios, assuming 100% mass balance.

Results and Discussion

Initial Contacting Experiments with Sodium Bearing Waste Simulant

INEEL's Sodium Bearing Waste (SBW) comprises one class of nitric acid-containing high-level nuclear waste that is being targeted for Cs removal (*27*). The composition of the SBW simulant provided to us by Dr. Donald J. Wood at the INEEL is shown in Table I (*26*). The simulant contains a high concentration of nitrate (4.46 M), moderate concentrations of acid (1.26 M) and sodium (1.32 M), and lower concentrations of many other metals, including potassium (0.138 M), the main competing ion for complexation of cesium by the calixarene-crown ether (*15,17*). A

Table 1. Composition of Simulated Sodium-Bearing Waste (SBW)[a]

Component	M	Component	M
Acid (H$^+$)	1.26×10^0	K	1.38×10^{-1}
Al	5.56×10^{-1}	Mn (II)	1.42×10^{-2}
B	1.40×10^{-2}	MoO$_4^{2-}$	1.49×10^{-3}
Cd	2.05×10^{-6}	Na	1.32×10^0
Ca	9.83×10^{-2}	NO$_3^-$	4.46×10^0
Ce (III)	3.63×10^{-4}	Ni	1.63×10^{-3}
Cl	3.52×10^{-2}	Pb	9.27×10^{-4}
Cr (III)	5.63×10^{-3}	PO$_4^{3-}$	$<9.18 \times 10^{-3}$
Cs	7.52×10^{-5}	Sr	6.80×10^{-3}
F	9.66×10^{-2}	SO$_4^{2-}$	3.86×10^{-2}
Fe (III)	2.40×10^{-2}	Zr	8.76×10^{-3}
Hg (II)	1.93×10^{-3}		

[a]Courtesy of Dr. Donald J. Wood, INEEL. The composition shown here is essentially the same as that reported previously (26).

target cesium decontamination factor (DF) of 1250 from SBW has been recommended (28).

Prior experience at ORNL using solvents containing **1** (Figure 1) and the Cs-3 modifier (Figure 2) to extract Cs from alkaline simulants (17,18) served as a starting point for the acid-side extraction studies. The solubility of bis-crown calixarenes such as calix[4]-bis-1,2-benzo-crown-6 (**8**) and calix[4]-bis-crown-6 in alkane diluents, even with added modifiers, is too low for them to be of practical use in a process solvent (18); adding *tert*-octyl groups to **8** to give **1** affords the requisite solubility. Initial contacting experiments between a solvent composed of **1** (0.02 M), Cs-3 modifier (0.25 M) in Isopar® L diluent and the SBW simulant afforded D_{Cs} values of above eight. However, it was soon discovered that the cesium extraction behavior eroded with concomitant formation of a third phase following prolonged contact between the solvent and the SBW simulant. Figure 3 shows the % Cs in the solvent, the % Cs in the raffinate, and the Cs mass balance (the sum of the % Cs in the solvent and raffinate) as a function of hours the solvent was in continuous contact with the SBW simulant at 25 °C. Samples of the solvent taken at various times were also subjected to stripping contacts with water or 0.01 M nitric acid, and the stripping efficiency as % Cs stripped out of the solvent is also plotted as a function of solvent exposure time to the simulant. It can be seen that mass balance and % Cs in the solvent and aqueous raffinate were essentially stable for 168 hours (7 days). These contacting experiments were performed in borosilicate glass vials, and during this time it was observed that the solvent color changed from colorless to dark yellow. The stripping efficiency over this time interval did not remain constant, but instead dropped from around 60% to 20%. Over the 168-hour to 338-hour interval, an amber oily precipitate was observed on the walls of the glass vial. The % Cs in the solvent phase was observed to drop concomitantly with a drop in mass balance; this became even worse after 672 hours (28 days). The stripping efficiency, however, did not decrease much further after 168 hours.

A sample of the amber oily precipitate (from a duplicate vial containing no ^{137}Cs tracer) was rinsed with a small volume of deionized water, and dissolved in deutero-chloroform. Analysis of the proton NMR spectrum revealed that the amber oil contained **1** that had been nitrated at the 5'-position on each of the two *tert*-octylbenzo groups to give calix[4]arene-bis-5'-nitro-4'-*tert*-octylbenzo-crown-6 (**2**, Figure 1). The oily precipitate also contained small amounts of (non-nitrated) **1** and Cs-3 modifier that had co-precipitated with **2**. A sample of pure calix[4]arene-bis-5'-nitro-4'-*tert*-octylbenzo-crown-6 (**2**) was prepared by the reaction of **1** in methylene chloride with concentrated nitric acid for 1 hour at zero °C. NMR analysis of the authentic sample of **2** (an amber oil that is only sparingly soluble in the Cs-3-modified diluent) confirmed that **2** was the product of prolonged contact of **1** with the SBW simulant, and that nitration had taken place only at the 5'-position, *ortho* to the *tert*-octyl group. Thus, exposure of **1** to the nitric acid in the SBW simulant slowly generates **2**, which precipitates out of the solvent when its concentration exceeds its apparently low solubility limit. As a general guideline for minimal solvent durability for a process, the solvent (and all its components) should be stable to continuous contact with a feed simulant for at least 60 days (29). The fact that **1** is readily nitrated, with effective degradation of solvent performance following just seven days of continuous contact with the SBW simulant, indicates that **1** would be unacceptable for a process. The Cs-3 modifier appears to possess adequate stability to the SBW simulant, as solutions of it at 0.25 M in both Isopar® L and deutero-chloroform (the

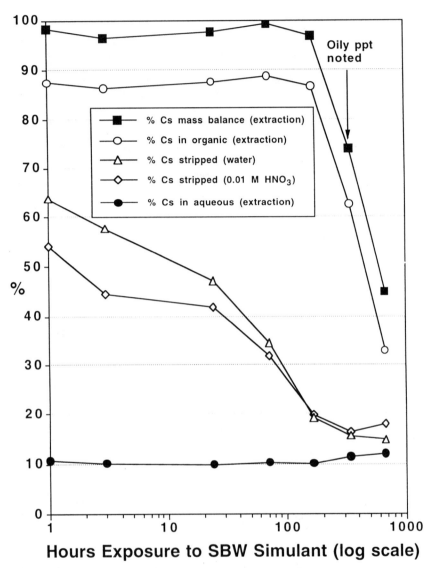

Figure 3. Cs extraction and stripping behavior for solvent Cs-3/225L (0.01 M 1, 0.25 M Cs-3 in Isopar® L diluent) as a function of solvent exposure time to SBW simulant. Following solvent exposure to simulant, solvents were stripped with deionized water or 0.01 M nitric acid (30 minute contacts, O/A = 1, 25 °C).

latter for proton NMR analysis) showed no signs of nitration following continuous contact with the SBW simulant over the course of 60 days.

Stability of Calixarene-Crown Ethers to Nitric Acid

In an effort to better understand why **1** so readily nitrated, and to evaluate the nitration stability of other calixarene-crown ethers, the direct reaction between a series of calixarene crown ethers at 20 mM concentration in deutero-chloroform and 4.0 M nitric acid at 25 °C was monitored by proton NMR as a function of contact time. Contacting was performed in 10-mL Teflon® FEP centrifuge tubes as described in the experimental section. Chloroform was selected since the nitrated products (such as **2**) were expected to remain soluble. (The absolute rate of nitration will likely be a bit different in a modified diluent than in chloroform, but in chloroform valid relative comparisons can still be made.) Calixarene-crown ethers **1**, **3**, **5**, **6**, and **8** (Figure 1) were examined. All of these calixarene-crown ethers except **8** are soluble in the modified diluent; **8** was examined as a non-alkylated control to **1**. The three bis-*n*-octyloxy mono crown calixes **5**, **6**, and **3** were examined to investigate the reactivity difference between a simple crown, a benzo crown, and an alkylated benzocrown with nitric acid within a similar series of calixarene-crown ethers.

The results are shown in Figure 4. The two calixarene-crown ethers possessing 4-*tert*-octylbenzo groups nitrated very quickly, with **1** giving **2** as the only nitrated product, and similarly **3** giving **4** (*24*). The other calixarene-crown ethers showed good stability to nitric acid; small amounts of nitration product were observed to grow in for **6** and **8** only after 30 days, with only about 3% nitrated product being observed for **6** after 72 days. The aromatic rings comprising the "belt" of the calixarene did not appear to nitrate under these conditions (as might have been suggested if **5** had shown signs of nitration).

An explanation as to why **1** and **3** are so readily nitrated can be found in the substituent effects controlling the degree of activation of an aromatic ring towards electrophilic aromatic substitution reactions (*30,31*). In comparing the 1,2-dialkoxy benzene fragment of a benzo crown (as in **6** and **8**) with a 1,2-dialkoxy, 4-alkyl benzene fragment (as in **1** and **3**), it can be seen that the *ortho,para*-directive effects of each activating alkoxy group do not reinforce one another. However, the weakly activating alkyl group in the 4-position directs nitration to the vacant 3- and 5-positions on the ring, which reinforces the directive effects of the alkoxy group at the 2-position. Thus, the rate of nitration at the 3- and 5-positions should be enhanced by the presence of an alkyl group at the 4-position. Since the 3-position is the more sterically congested, nitration would be favored at the 5-position, as is observed. This is further borne out by comparing the reaction between model compounds 4-*tert*-octylanisole and 4-*tert*-octylveratrole at 0.50 M in deutero-chloroform with 4 M nitric acid: the former is quite stable to nitration under these conditions, but the latter nitrates at a rate comparable to **6** and **8** (slightly faster, due to its higher concentration). In 4-*tert*-octylanisole, which is a model for the Cs-3 modifier with respect to substitution on the aromatic ring, there is again no reinforcement of directive effects. All these materials will of course nitrate when contacted with stronger (e.g., 15 M) nitric acid solutions, but the Cs-3 modifier and alkane-soluble calixarene-crown ethers **5** and **6** nitrate at a sufficiently slow rate when in contact with

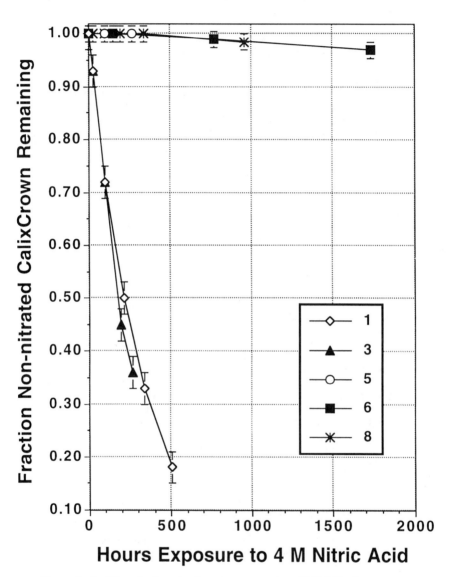

Figure 4. Stability of selected calixarene crown ethers (0.02 M in deuterochloroform) to 4 M nitric acid (equal volume contacts at 25 °C) as a function of exposure time.

4 M nitric acid (which is about the maximum nitric acid concentration of the INEEL wastes (*28*)), to be chemically stable enough to be potentially useful in a process. Conversely, the nitration rate for **1** with *dilute* nitric acid (e.g., 0.05 M) is sufficiently slow that solvents containing **1** can be subjected to dilute nitric acid scrubbing or stripping contacts as part of an alkaline-side extraction process (*17*). The rate of nitration for a given calixarene-crown ether in a particular diluent appears to be first order in both the calixarene-crown ether and nitric acid concentrations.

Cesium Extraction and Stripping Results from SBW Simulant

Having established that calixarene-crown ethers **5** and **6** appear to possess sufficient stability to 4 M nitric acid to be potentially useful in a process solvent for extracting Cs from the SBW simulant (1.26 M in acid), we examined their extraction behavior (at both 0.01 and 0.02 M) in Isopar® L diluent containing Cs-3 modifier at 0.50 M. One other calixarene-crown ether that should possess good nitric acid stability was examined: bis-*n*-octyloxycalix[4]arene dibenzocrown-6 (**7**) containing two benzo groups. The extraction behavior of the two calixarene-crown ethers **1** and **3** that "failed" the nitric acid stability test were also obtained for comparison. Calixarene-crown ether 1,3-alt-bis-*n*-octyloxycalix[4]arene 5'-nitro-4'-*tert*-octyl-benzo-crown-6 (**4**, the nitrated product of **3**), turned out to be sufficiently soluble in the modified diluent for solvents suitable for testing to be prepared. Though the presence of the nitro group in **5** is considered undesirable by potential end-users, and thus would preclude its consideration as a process-suitable extractant, **5** was nonetheless included in the test to learn what effect the presence of the nitro group would have on the Cs distribution behavior. The stripping behavior of the solvent following extraction with deionized water was also investigated. Prior to the 30-min extraction or stripping contact at 25 °C by end-over-end rotation, the vials were vortexed for 30 seconds, and the phase-coalescence behavior for each solvent recorded. All contacts were performed in duplicate.

As shown in Table II, the phase-coalescence behavior is highly dependent on the degree of lipophilicity on the crown portion of the calixarene–crown ether: **1** and **3**, with *tert*-octyl benzo groups, and **7**, with two benzo substituents, all displayed rapid phase-disengagement behavior on both extraction and stripping. The incorporation of the polar nitro group in **4** appears to cancel out the hydrophobicity of the *tert*-octyl group. A useful parameter for gauging the performance of a solvent in a process is the ratio of the extraction and stripping distribution ratios. It is beneficial to maximize this ratio: larger ratios lead to larger decontamination factors for a given number of contactor stages. Of the three candidate nitric acid-stable calixarene-crown ethers (**5**, **6**, and **7**), an attractive *combination* of good extraction strength, stripping efficiency, and phase coalescence behavior was obtained with **6** at 0.01 M; **5** and **7** were weaker extractants, and **5** exhibited longer phase-disengagement times on extraction. These three calixarene-crown ethers are expected to cost approximately the same to prepare, with **6** possibly being marginally cheaper. Though **7** displayed excellent phase coalescence behavior and stripping performance, its much lower extraction strength necessitates using it at a higher concentration. Indeed, **7** at 0.02 M performed similarly to **6** at 0.01 M; however **6**, at 0.01 M, would give about half the solvent inventory cost. The somewhat weaker extraction performance of **4** relative

Table II. Cs Extraction and Stripping Behavior for Various Calixarene-Crowns[a]

Calixarene-crown	Stability to 4 M Nitric Acid	Concentration, M	D_{Cs} Extraction	D_{Cs} Stripping	D_{Cs} Extr./D_{Cs} Strip.	Phase Coalescence Behavior on Extraction[b]
1	Poor	0.01	4.54	0.35	13.0	Excellent
		0.02	8.74	1.02	8.6	Excellent
3	Poor	0.01	3.86	0.23	16.8	Excellent
		0.02	8.54	0.68	12.6	Excellent
4	Good[c]	0.01	3.78	0.14	27.0	Good
		0.02	6.84	0.33	20.7	Fair
5	Good	0.01	2.58	0.12	22.0	Fair
		0.02	5.12	0.34	15.0	Fair
6	Good	0.01	4.42	0.22	20.1	Very Good
		0.02	8.59	0.65	13.2	Fair
7	Good[d]	0.01	2.00	0.08	25.0	Excellent
		0.02	3.98	0.22	18.1	Excellent

[a] Extraction is from SBW simulant, stripping with deionized water. Solvent contains calixarene-crown at noted concentration in Isopar® L, with Cs-3 modifier at 0.50 M in all cases. Extraction and stripping contacts were for 30 second vortex, followed by rotation for 30 minutes at 25 °C at O/A = 1.
[b] The phase-disengagement behavior on stripping was excellent in all cases.
[c] Predicted stability, based on the fact that this compound is already nitrated, and therefore deactivated to further nitration.
[d] Predicted stability, based on the good stability of the other non-alkylated benzo derivatives.

to 3 has been observed previously (*24*) and is consistent with a predicted decrease in the basicity of the ether oxygens on the benzene ring due to the electron withdrawing effect of the nitro group.

As an attractive overall calixarene-crown ether candidate for further process development work appeared to be **6**, a solvent comprised of it at 0.01 M with Cs-3 at 0.50 M in Isopar® L, designated as "Cs-3BOO/150L", was subjected to further study.

Chemical Stability, Process Development, and Cesium Selectivity Studies

The chemical stability of the solvent Cs-3BOO/150L to the SBW simulant was performed in the same manner as described above in the initial stability experiment that used **1**. Contact times were 30 minutes, 30 days, and 60 days, and the stripping efficiency of the solvent with deionized water was also measured. The cesium distribution ratios for the three times were 4.42 (30 min), 4.45 (30 days), and 4.45 (60 days), with no precipitation being observed, indicating that as expected this solvent possessed adequate stability. The stripping performance was also fairly stable, with stripping cesium distribution ratios of 0.22 (30 min), 0.41 (30 days), and 0.40 (60 days).

Toward process development, the cesium distribution behavior was investigated by subjecting the solvent to two successive (cross-current) extraction contacts with fresh SBW simulant, followed by two successive scrubbing contacts with fresh 0.050 M nitric acid, followed by three successive stripping contacts with fresh 0.001 M nitric acid. On the basis of previous process development work conducted for alkaline Cs extraction from High Level Tank Waste at DOE's Savannah River Site (*17*), we anticipated that the cesium stripping ratio would increase upon successive stripping contacts as the organic-phase concentration of cesium nitrate decreased beyond a certain level, and this is in fact what we observe (Figure 5). A proposed explanation for this increase in D_{Cs} is described in detail in Chapter 3 of reference 17. A brief explanantion as is follows: when the cesium nitrate concentration in the organic phase falls below about 1×10^{-5} M, the extraction mechanism is believed to change from being predominantly based on ion-pairs to being predominantly based on dissociated ions. In the latter mechanism, the predicted dependence on the organic phase nitrate concentration is such that a plot of $\log D_{Cs}$ vs. $\log[Cs]_{aq}$ should have a slope of $-1/2$; that is, $\log D_{Cs}$ will increase as $\log[Cs]_{aq}$ decreases. A solution to this stripping problem is to maintain the total organic-phase nitrate concentration in a concentration regime where the mechanism will be predominantly ion-paired. One way to do this is to dissolve a trialkyl amine, such as trioctylamine (TOA), into the solvent at a low concentration (e.g., ≤0.001 M). The amine will extract nitric acid from the strip solution into the solvent and essentially maintain a constant nitrate concentration such that D_{Cs} will not increase on successive stripping but instead will decrease to a low, level value.

Repeating the batch contacting with the solvent Cs-3BOO/150L containing TOA at 0.001 M (designated as "Cs-3BOO/150LT") gave no change in the cesium distribution ratios on extraction and scrubbing, however *a decrease toward a low, level value upon successive stripping contacts* was obtained (Figure 5). For development of an actual engineering flowsheet, the distribution ratios would be used

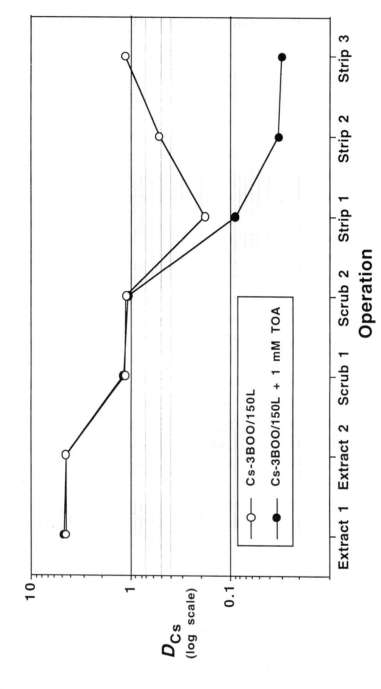

Figure 5. Cesium distribution behavior for solvent Cs-3BOO/150L (0.01 M 6, 0.50 M Cs-3 in Isopar® L, with and without trioctylamine at 0.001 M) following two successive extraction contacts with SBW simulant, two successive scrubbing contacts with 0.05 M nitric acid, and three successive stripping contacts with 0.001 M nitric acid.

to calculate and optimize countercurrent multi-stage contacting performance to meet explicit design goals regarding decontamination and concentration factors (23).

Finally, the Cs selectivity over selected elements Al, B, Fe(III), Hg(II), K, and Na in the SBW simulant was measured using both the Cs-3BOO/150L and Cs-3BOO/150LT solvents, to discern if the addition of TOA would negatively impact the selectivity performance of the solvent. The results, shown in Table III, indicate that with respect to metal ion selectivity, the solvents with and without TOA perform essentially the same. Interesting exceptions are Al and Fe, which appear to be extracted slightly better by the solvent that did not contain TOA. Mercury is extracted to some extent by both solvents, which may be beneficial if its removal from the waste is desired. Relative to Cs, the bulk waste elements Al, K, and Na are not significantly extracted; this is considered beneficial toward minimizing the final volume of the expensive waste form that must be produced and stored. The Cs/Na and Cs/K selectivities are excellent, and are in the range expected (10^4 and 10^2, respectively) for the calixarene-crown ether. The agreement between the sodium distribution ratios measured by ^{22}Na tracer and the values obtained using ICAP is good and helps to validate the ICAP distribution data.

Conclusions

A solvent (Cs-3BOO/150LT) suitable for extracting cesium from Sodium Bearing Waste has been developed which possesses good chemical stability, displays excellent cesium selectivity, and provides good extraction and stripping performance with satisfactory phase-coalescence behavior. Though **6** was selected as the calixarene-crown ether for this solvent, use of **5** or **7** may also provide adequate performance. It was shown that incorporation of an alkyl substituent at the 4-position of the aromatic ring of the benzocrown portion of a calixarene-crown ether has a detrimental effect on the susceptibility of the aromatic ring to nitration by nitric acid at concentrations that would be expected to be encountered in INEEL tank wastes (1-4 M). The unsubstituted benzocrown-derivatized calix[4]arenes, as well as the Cs-3 modifier, possess sufficient stability to nitric acid to be suitable for use in a process solvent. We believe that expensive calixarene-crown ethers can find practical use in a solvent extraction process, provided that chemically robust materials are selected and that solvents employing them are used in the most efficient manner possible. This includes maximizing the extraction power afforded by the calixarene-crown to minimize the extractant concentration required, while ensuring that the solvent can be effectively decontaminated (stripped of Cs), giving high decontamination factors in a small number of contacts. With regard to extraction power, addition of modifiers such as 1-(1,1,2,2-tetrafluoroethoxy)-3-(4-*tert*-octylphenoxy)-2-propanol (ORNL "Cs-3") to the solvent permits good Cs distribution ratios to be achieved at low calixarene-crown ether concentrations. Addition of trioctylamine (TOA) at 1 mM greatly improves the stripping efficiency of the solvent, allowing for essentially complete solvent decontamination. The addition of TOA at this level to the solvent appears to offer many benefits, without adverse affects on the extraction or scrubbing performance, nor is the excellent selectivity for Cs over the metals Al, B, Fe, Hg, K, or Na compromised.

Table III. Selected Metal Distribution and Cs Selectivity Data for Extraction from SBW Simulant Using Solvent Cs-3BOO/150L (0.01 M BOOCABC6, 0.50 M Cs-3 in Isopar® L) with and without Trioctylamine (TOA) at 0.001 M

Metal[a]	$D_M \pm 5\%$ (with TOA)	$\alpha_{Cs/M}$	$D_M \pm 5\%$ (without TOA)	$\alpha_{Cs/M}$
Cs (tracer)	4.76×10^0	1	4.55×10^0	1
Hg	3.88×10^{-2}	123	3.52×10^{-2}	129
K	1.52×10^{-2}	313	1.45×10^{-2}	314
B	2.53×10^{-3}	1880	1.98×10^{-3}	2300
Fe	1.95×10^{-4}	24400	3.12×10^{-4}	14600
Na (tracer)	1.79×10^{-4}	26600	1.74×10^{-4}	26100
Na	1.62×10^{-4}	29400	1.52×10^{-4}	29900
Al	8.22×10^{-6}	579000	1.70×10^{-5}	268000

[a] All metals determined by ICAP except where noted for Cs and Na (^{137}Cs and ^{22}Na tracers).

Acknowledgments

We wish to thank our collaborators Dr. Terry A. Todd and Dr. Donald J. Wood at the Idaho National Engineering and Environmental Laboratory for the SBW simulant, and for many helpful discussions. We also thank Drs. Jeffrey C. Bryan and Charles F. Coleman at Oak Ridge National Laboratory for a thorough review of the manuscript. This research was sponsored by the Efficient Separations and Processing Crosscutting Program, Office of Science and Technology, Office of Environmental Management, U.S. Department of Energy, under contract DE-AC05-96OR22464 with Oak Ridge National Laboratory, managed by Lockheed Martin Energy Research Corp.

Literature Cited

1. Ungaro, R.; Casnati, A.; Ugozzoli, F.; Pochini, A.; Dozol, J.-F.; Hill, C.; Rouquette, H. *Angew. Chem. Int. Ed. Engl.* **1994**, *33*, 1506.
2. Asfari, Z.;Wenger, S.;Vicens, J. *J. Inclusion Phenom.* **1994**, *19*, 137.
3. Hill, C.; Dozol, J.-F.; Lamare, V.; Rouquette, H.; Eymard, S.; Tournois, B.; Vicens, J.; Asfari, Z.; Bressot, C.; Ungaro, R.; Casnati A. *J. Inclusion Phenom. Mol. Recognit. Chem.* **1994**, *19*, 399.
4. Hill, C., Ph. D. Thesis, Universite Louis Pasteur de Strasbourg, 1994.
5. Casnati, A.; Pochini, A.; Ungaro, R.; Ugozzoli, F.; Arnaud, F.; Fanni, S.; Schwing, M.-J.; Egberink, R. J. M.; de Jong, F.; Reinhoudt, D. N. *J. Am. Chem. Soc.* **1995**, *117*, 2767.
6. Asfari, Z.; Bressot, C.; Vicens, J.; Hill, C.; Dozol, J.-F.; Rouquette, H.; Eymard, S.; Lamare, V.; Tournois, B. *Anal. Chem.* **1995**, *67*, 3133.
7. Arnaud-Neu, F.; Asfari, Z.; Souley, B.; Vicens J. *New J. Chem.* **1996**, *20*, 453.
8. Asfari, Z.; Nierlich, M.; Thuery, P.; Lamare, V.; Dozol, J.-F.; Leroy, M.; Vicens J. *Anales de Quimica Int. Ed.* **1996**, *92*, 260.
9. Lauterbach, M.; Wipff, G. In *Physical Supramolecular Chemistry*, Echegoyen, L., Kaifer, A. E., Eds.; Kluwer Academic Publishers: Netherlands, 1996; pp 65-102.
10. Thuery, P.; Nierlich, M.; Bressot, C.; Lamare, V.; Dozol, J.-F.; Asfari, Z.; Vicens, J. *J. Inclusion Phenom. Mol. Recognit. Chem.* **1996**, *23*, 305.
11. Thuery, P.; Nierlich, M.; Asfari, Z.; Vicens, J. *J. Inclusion Phenom. Mol. Recognit. Chem.* **1997**, *27*, 169.
12. Asfari, Z.; Naumann, C.; Vicens, J.; Nierlich, M.; Thuery, P.; Bressot, C.; Lamare, V.; Dozol, J.-F. *New J.Chem.* **1996**, *20*, 1183.
13. Thuery, P.; Nierlich, M.; Bryan, J. C.; Lamare, V.; Dozol, J. -F.; Asfari, Z.; Vicens, J. *J. Chem Soc., Dalton Trans.* **1997**, *22*, 4191.
14. Dozol, J.-F.; Bohmer, V.; McKervey, A.; Lopez Calahorra, F.; Reinhoudt, D.; Schwing, M.-J.; Ungaro, R.; Wipff, G. *New Macrocyclic Extractants for Radioactive Waste Treatment: Ionizable Crown Ethers and Functionalized Calixarenes;* Report EUR-17615; European Communities: Luxembourg, 1997.
15. Haverlock, T. J.; Bonnesen, P. V.; Sachleben, R. A.; Moyer, B.A. *Radiochim. Acta.* **1997**, *76*, 103.

16. Dozol, J.-F.; Simon, N.; Lamare, V.; Rouquette, H.; Eymard, S.; Toumois, B.; De Marc, D.; Macias, R. M. *Sep. Sci. Technol.* **1999**, *34*, 877.
17. Bonnesen, P. V.; Delmau, L. D.; Haverlock, T. J.; Moyer, B.A. *Alkaline-Side Extraction of Cesium from Savannah River Tank Waste Using a Calixarene-Crown Ether Extractant*; Report ORNL/TM-13704, Oak Ridge National Laboratory: Oak Ridge, TN, December, 1998.
18. Moyer, B. A.; Bonnesen, P. V.; Sachleben, R. A.; Presley, D. J. Solvent and Process for Extracting Cesium from Alkaline Waste Solutions. U.S. Pat. Appl. 60/057,974, September 3, 1998.
19. Ritcey, G. M.; Ashbrook, A. W. *Solvent Extraction, Principles and Applications to Process Metallurgy*; Elsevier: New York, 1984; Part I.
20. Peterson, R. A.; Fondeur, F.F. "High Level Waste Testing of Solvent Extraction Process", Report WSRS-TR-98-000368, Westinghouse Savannah River, Aiken, SC, October, 1998.
21. Leonard, R. A.; Wygmans, D. G.; McElwee, M. J.; Wasserman, M. O.; Vandegrift, G. F. *"The Use of a Centrifugal Contactor for Component Concentration by Solvent Extraction"*, Report ANL-92/26, Argonne National Laboratory, Argonne, IL, U.S.A., 1992.
22. Leonard, R. A.; Chamberlain, D. B.; Conner, C. *Sep. Sci. Technol.* **1997**, *32*, 193.
23. Leonard, R. A.; Conner, C.; Liberatore, M. W.; Bonnesen, P. V.; Presley, D. J.; Moyer, B. A.; Lumetta, G. J., *Sep. Sci. Technol.* **1999**, *34*, 1043.
24. Sachleben, R. A.; Bonnesen, P. V.; Descazeaud, T.; Haverlock, T. J.; Urvoas, A.; Moyer, B. A. *Solvent Extr. Ion Exch.* (1999, in press).
25. Lamare, V.; Dozol, J.-F.; Ungolopzi, F.; *Eur. J. Org. Chem.* **1998**, *8*, 1559.
26. Law, J. D.; Wood, D. J.; Herbst, R. S. *Sep. Sci. Technol.* **1997**, *32*, 223.
27. Todd, T. A.; Olson, A. L.; Palmer, W. B.; Valentine, J. H. In *Science and Technology for Disposal of Radioactive Tank Wastes*; Schulz, W. W., Lombardo, N. J., Eds.; Plenum, New York, NY, 1998; pp 219-230.
28. Todd, T. A.; Wood, D. J. Idaho National Engineering and Environmental Laboratory, Idaho Falls, ID. Private communication, 1997.
29. Leonard, R. A. Argonne National Laboratory, Argonne, IL. Private communication 1997.
30. Stock, L. M. *Aromatic Substitution Reactions;* Prentice-Hall, Inc.:Englewood Cliffs, NJ, 1968; pp. 40-58.
31. Fessenden, R. J.; Fessenden, J. S. *Organic Chemistry;* Willard Grant Press: Boston, MA, 1979; Chapter 10.

Chapter 4

Radiation Stability of Calixarene-Based Solvent System

R. A. Peterson, C. L. Crawford, F. F. Fondeur, and T. L. White

Westinghouse Savannah River Company, Aiken, SC 29808

The Savannah River Site recently evaluated four alternatives for processing the cesium rich High Level Waste stream at the Savannah River Site. One of the selected process uses a calixarene-based solvent extraction process to recover the cesium and produces a purified, lower volume acidic stream for vitrification. Previous work provided limited understanding of the stability of the solvent used in the process in radiation fields. During this work, samples of the solvent system, in contact with the caustic phase, were exposed to Co-60 gamma irradiation doses ranging from 0 to 27 Mrad. The performance of this solvent system for extraction of cesium and potassium was then determined. In addition, a High Performance Liquid Chromatography (HPLC) method was developed for the analysis of both the calixarene and a solvent modifier. The decrease in the concentration of the both of these materials was thereby determined.

Researchers at the Savannah River Site recently performed an evaluation of four selected processing alternatives for the treatment of the cesium rich High Level Waste (HLW) stream at the Savannah River Site. One of the selected process uses solvent extraction to recover the cesium and produces a purified, lower volume acidic stream for vitrification. This solvent system was developed by researchers at Oak Ridge National Laboratory for the processing of highly caustic (pH > 14) waste solutions common to the Department of Energy (DOE) complex.[1] This solvent systems consists of calix[4]arene-bis(t-octylbenzo-crown-6) (BOB Calix) and 1-(1,1,2,2,-tetrafluoroethoxy)-3[4-(t-octyl)phenoxy]-2-propanol (Cs-3 modifier) and a kerosene diluent. Prior to this work, Isopar had been used as the diluent during the development of the solvent system.

The typical HLW at the Savannah River Site contains high concentration of sodium (> 6 M) and relatively high concentrations of potassium (~ 20 mM) while containing a relatively low molar concentration of cesium (< 1 mM). However, the cesium exists in a mixture of isotopes with about ¼ of the cesium present as radioactive Cs-137.

Previous work provides little understanding of the stability of the solvent used in the process in radiation fields. Extensive experience in the nuclear industry with solvent systems has indicated that the degradation of solvent due to exposure to irradiation can significantly degrade the performance of the solvent.[2] In particular, experience with the PUREX (Plutonium URanium EXtraction) process has indicated that the degradation of both the extractant and the diluent will occur in the presence of high radiation fields.[3] Understanding of these degradation processes has lead to the successful development of equipment that minimizes this exposure and the development of methodologies for cleaning the degradation products from the solvent. To determine the viability of the new proposed solvent extraction system, similar understanding of the impact of radiation on this solvent system is required. This work investigates that aspect of the chemistry.

Experimental

The solvent used for this experiment involves a blend of 0.01 M BOB Calix, 0.5 M Cs-3 modifier, and Isopar L® as the balance. The BOB Calix was prepared by IBC Advanced Chemical Incorporated, (Provo, Utah). The Cs-3 modifier was provided by Oak Ridge National Laboratory. Exxon (Houston, Texas) provided the Isopar L®. The solvent was contacted with an equal volume of simulated HLW feed solution. Table I provides the composition of the simulated feed solution. All chemicals used in the preparation of the simulated feed solution, the scrub solution, and the strip solution were obtained from Aldrich Chemical (Milwaukee, Wisconsin). Figure 1 provides a flow diagram for these tests. The tests used a Co60 gamma source providing approximately 1 Mrad/h.

To facilitate the determination of the composition of the solvent system before and after exposure to gamma irradiation, a method was developed for determination of these compounds by HPLC. The instrument used was a Hewlett-Packard 1090 HPLC operated by HP ChemStation version 6.0 software. The HPLC grade solvents used are isopropanol (Acros) (Pittsburgh, Pennsylvania) and ultrapure water obtained for a Waters Milli-Q system.

Standard solutions were prepared by dissolving 100 mg of each compound in 10 mL of isopropanol/hexane (9:1) solvent and then diluting these stock solutions with isopropanol to the required concentrations. For the linearity study, concentrations of 10, 20, 50, 100, and 200 mg/L of a mixture of the compounds were prepared and analyzed (n=5). In addition, standards of 500, 1000, and 2000 mg/L were examined for the modifier. Isopar L samples were diluted with isopropanol until the analyte concentration was within the range of the linear calibration curve and then analyzed.

Of the different solvent systems tried for separation and resolution, an ODS column, isopropanol and water yielded the best results (Table II). Attempts to create an isocratic method that would separate all three compounds resulted in separation of calix and modifier with 4-(tertoctyl)phenol (85% IPA/15% water). Note that phenol

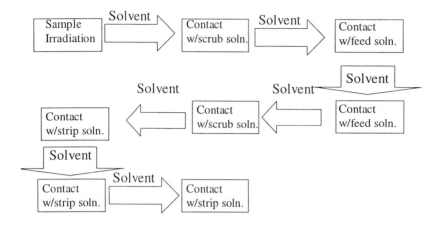

Figure 1. Schematic Diagram of Experimental Protocol

Table I. Simulant Composition

Species	Concentration (M)
Na^+	7.0
K^+	0.02
AlO_2^-	0.43
Cs^+	0.0007
OH^-	1.9
NO_3^-	2.7
NO_2^-	1.0
SO_4^{2-}	0.22
CO_3^{2-}	0.20
Cl^-	0.10
F^-	0.05
CrO_4^{2-}	0.015

compound was an anticipated degradation product of the modifier. The gradient method resulted in separation of all three compounds. A wavelength of 226 nm was selected for viewing modifier and 4-(tertoctyl)phenol while 205 provided the best sensitivity for calix. Figure 2 contains diagrams of two traces from the HPLC for standards containing a mixture of the three analytes.

Based on an estimated cesium[137] concentration in the 3.6×10^{-5} M, the anticipated yearly dose for the solvent is approximately 8 Mrad. Two sets of experimental conditions were utilized. The first set of tests investigated the impact of increasing exposure the gamma irradiation on the performance of the solvent system. In this testing, individual samples were exposed to 0, 1, 4.5, 9 and 27 Mrad in a static

Table II. Gradient reverse-phase HPLC method for Isopar L

Method	Conditions
Solvent system	Isopropanol-water
t_o to t_1 = 12 min	70%/30%
t_2 = 14 min	90%/10%
t_3 = 25 min	90%/10%
t_4 = 27 min	70%/30%
t_5 = 32 min	70%/30%
Column	Dychrom Chemcosorb 5 ODS-UH 3.2x250 mm, 5 μm pore size
Oven temperature	45°C
Flow-rate	0.25 mL
Stop time	32 min
UV	226 nm (modifier), 205 nm (calix)
injection volume	10 μL
Retention time for 4-t-octylphenol	9.4 min
Retention time for modifier	11.6 min
Retention time for calix	25.9 min
Linear calibration curve	10 mg/L to 200 mg/L, correlation = 0.999

environment. Note that due to experimental difficulties associated with irradiating these samples, it was not possible to agitate the samples during irradiation. After exposure the organic and aqueous phases were separated and the aqueous portion was analyzed for cesium by the ICP-MS (Inductively Coupled Plasma Mass Spectroscopy) and for potassium by AA (atomic adsorption). The organic phase then progressed as diagram in Figure 1, with analysis of each successive aqueous phase by ICP-MS and AA following contact. Note that the organic was initially scrubbed in attempt to remove any degradation products. The scrubbing solution contained 0.05 M HNO_3. The stripping solution shown in Figure 1 contained 0.0005 M HNO_3 and 0.0007 M $CsNO_3$. All samples were mixed by hand for a minimum of five minutes and then allowed to separate. These tests were performed at ambient laboratory temperatures (ca. 22 °C).

Distribution coefficients were determined for the extraction, scrub and strip stages by determining the cesium concentration in the organic phase based on the cesium material balance for the samples. The distribution coefficient is therefore defined as:

$$D = \frac{[Cs]_{org}}{[Cs]_{aq}}$$

Plate 1. Impact of irradiation on solvent appearance.

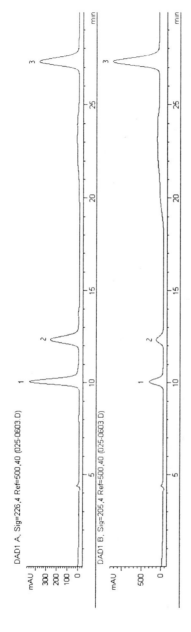

Figure 2. HPLC Trace for standard at two wavelengths.

Aliqouts of a portion of the sample exposed to 27 Mrad was also submitted for HPLC analysis.

The second set of tests investigated the impact of the diluent on the stability of the solvent. These tests used two distinct diluents. The first diluent (which was employed in the first set of tests) was Isopar L®, and branched kerosene. The second diluent used was n-parrafin, an unbranched kerosene. All of these tests were performed at 27 Mrad of exposed dose. Similar to the first set of tests, the extraction, scrubbing and stripping distribution coefficients were measured. In addition, the samples were submitted for HPLC analysis.

Discussion

Impact of Increasing Radiation Expososure on Extraction and Stripping Performance

Visual discoloration of the solvent system occurred upon exposure to gamma irradiation. Plate 1 contains a photograph of the samples following exposure. Inspection of this figure indicates a significant discoloration in the solvent upon exposure of up to 27 Mrad.

Similarly, degradation of the solvent performance was also observed. Figure 3 contains a plot of the extraction distribution coefficients. Note that in Figure 1, the solvent is in contact with the feed simulant during irradiation and that the second contact for the solvent system is with the scrub solution. This contact was anticipated to wash some of the deleterious compounds from the solvent system. Thus, Figure 3 contains distribution coefficients for both before and after this washing step. Note that no significant improvement in solvent performance following washing was observed. The sample submitted for HPLC analysis indicated approximately 33% degradation of the calixarene in the solvent. This loss of calixarene would likely be manifested in a lower distribution coefficient. Assuming that the loss of calixarene was linear over the range of exposure and that the distribution coefficient is linear with the amount of calixarene present, it is possible to estimate the distribution coefficient over the range of exposures (using the distribution coefficient at zero dose as a baseline). Figure 3 also contains this prediction. Inspection of the Figure indicates that a relatively good fit exists between the measured and predicted distribution coefficients. This result suggests that the assumption of linear decomposition is a reasonable approximation of these dose levels.

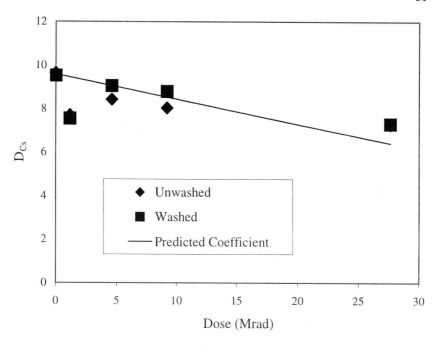

Figure 3. Impact of Dose on cesium distribution coefficient

For this system, potassium is simultaneously extracted with cesium. While potassium was present in much higher concentrations, the highly selective nature of the solvent system produces significantly lower distribution coefficients for potassium. Figure 4 contains the extraction data for potassium as a function of dose exposure. Note that the distribution coefficient for the unwashed solvent system at 27 Mrad is slightly higher than those at 5 and 10 Mrad. This difference in distribution coefficient is likely within the uncertainty of the analytical method. Inspection of this figure indicates that the potassium distribution coefficient dose not decline for the washed solvent with the exposure to gamma irradiation to the degree predicted based on the measured loss of calixarene (and thus does not appear to be a strong function of the concentration of the calixarene). Note that the concentration of potassium in the solvent phase following extraction is approximately 1/3 that of cesium. This difference in concentration may account for the lesser impact of loss of calixarene on potassium distribution coefficient. Note, however, that in contrast to the cesium, the potassium distribution coefficient appears to be significantly impacted by the washing process. In particular, washing the solvent appears to significantly improve the potassium distribution coefficient. This result suggests that a product of the irradiation decomposition inhibits the extraction of potassium.

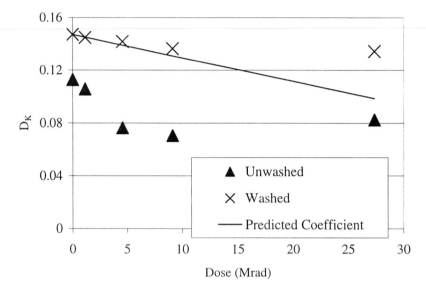

Figure 4. Impact of Dose on potassium distribution coefficient

Impact of Diluent on Solvent Radiolytic Stability

In addition to performing tests with the Isopar L® diluent, tests were performed with a normal parrafin diluent. Previous research has indicated that unbranched kerosenes are less susceptible to degradation in the presence of radiation fields in comparison to branched kerosenes. Samples were exposed to 27 Mrad and then were analyzed for the impact of the dose on both solvent composition and solvent extraction performance. Table III summarizes the differences in compositions and performance observed for these two diluents.

Table III. Impact of diluent of solvent stability

Impact	Isopar L	N-parrafin
Calixarene G-value	0.14 molecules/100 eV	0.10 molecules/100 eV
Cs-3 G-value	0.88 molecules/100 eV	0.48 molecules/100 eV
Decrease in distribution coefficient	37%	15 %

Table III presents the impact of solvent on calixarene and modifier stability in the form of G-values. These G-values are calculated based on the loss of calixarene and modifier at the end of exposure (in molecules lost) and the total energy exposure of the volume of solvent (in eV). Note that the decrease in the distribution coefficient for the isopar system is nearly twice that observed for the nparrafin system (in contrast to a 40% increase in the G-value for the calixarene). While the G-value for the modifier is significantly high for the Isopar L system, the modifier does not provide a one-for-one extraction of the cesium. Thus, it is not anticipated that the distribution coefficient will be a linear function of the modifier concentration. This result therefore suggests a synergistic impact of the decrease in calixarene and modifier concentration on solvent performance. Also note that different degradation products may form from decomposition of the diluent itself. None of these other degradation products were identified by HPLC analysis of the solvent system.

Impact of Stripping on Solvent System

For the solvent systems with either the Isopar L and the n-parrafin diluent, stripping of the solvent system upon exposure to high doses provided difficult. Note that for the second set of tests with the differing diluents, the solvent was not washed immediately after exposure to gamma irradiation. Subsequently, upon the initial contact of the solvent system with acidic solutions (following additional extraction with fresh caustic), the formation of a third phase was observed between the solvent and aqueous phases. This third phase likely formed from degradation products introduced during the irradiation of the solvent. These products likely were washed out during the initial scrubbing performed during the first phase of testing but were left in the system (and perhaps modified by the additional contacts with caustic) during the second phase of testing. This potential for the formation of a third phase will pose significant potential safety concerns for the operation of potential solvent extraction processes. Further studies of the impact of solvent washing on the potential to form this third phase are recommended.

Radiolysis of Calixarene System

Comparison can be made of our measured radiolytic decomposition yields for the two solutes of interests, calixarene and Cs-3, to other radiolytic decomposition studies reporting decomposition yields for loss of benzene and aliphatic hydrocarbons upon irradiation. Table IV shows data from the present work and includes several data from the literature. All radiolytic yields, or G-values, are expressed as the number of molecules formed per 100 eV of energy absorbed. Our range of decomposition yields for organic phase-dissolved calixarene and Cs-3 in the range of 0.1 to 0.9, are lower than decomposition yields reported for irradiation of analogous liquid-state organic compounds including benzene and select aromatic hydrocarbons. Note, however, that if the actual deposition of energy in the solvent system is considered, these G-values increase two-fold. Thus, the G-values for the modifier approach those of the other systems listed.

All data shown in Table IV are from gamma radiolysis studies. In the present studies, we used external Co-60 γ–radiation to simulate the effects of the Cs-137 β– radiation and (daughter) Ba-137 γ–radiation in the calixarene solvent system. Since these several radiations are all 'electron equivalent' from the viewpoint of track effects and ionizing radiation interaction with liquids, the results should be applicable to actual calixarene solvent systems containing radiocesium.

We did not attempt to measure radiolytic yields of higher molecular weight polymers formed in our tests. As noted above, we did observe a significant increase in yellow discoloring of the organic phases with increasing dose. Some radiolytic yields of polymeric materials from previous studies are shown in Table IV which indicate that significant yields of these materials can be formed upon organics irradiation. The last two polymeric yield values reported are for the cyclic ethers, dioxan and tetrahydrofuran. The cyclic ether grouping in calixarene could be expected to be more radiation resistant due incorporation of aromatic rings into the ring.

Table IV Radiolytic Yields from Gamma Radiolysis Studies

Hydrocarbon	G(-X)	G('polymer')	Reference
Calixarene(diluent 1)	0.14	Not measured	This work
Calixarene(diluent 2)	0.10	Not measured	This work
Cs-3 (diluent 1)	0.9	Not measured	This work
Cs-3 (diluent 2)	0.5	Not measured	This work
Toluene	1.8	--	[4]
Ethylbenzene	1.57	--	[5]
Xylene(o,m,p)	1.1	0.9 - 1.3	[6]
Isopropylene	1.8	1.7	[7]
Benzene	0.94	0.5	[8,9,10]
Dioxan	--	3.65	[11]
Tetrahydrofuran	--	5.0	[12]

Conclusions

The studies presented in this work indicate that the calixarene based solvent extraction system proposed for extraction of cesium from high level waste at the Savannah River Site should be sufficiently robust to withstand the anticipated gamma irradiation doses. It is anticipated that this stability is due to the deposition of most of the irradiation energy in the diluent and that the unsaturated diluent will be more susceptible to attack by the gamma irradiation. An HPLC technique was developed to provide the ability to determine the concentration of BOB Calix and Cs-3 modifier in the solvent system. Extraction experiments and HPLC measurements confirmed that little loss of either the BOB Calix molecule or the Cs-3 modifier occurred upon exposure of up to 3 years anticipated dose. However, additional work must be performed to determine the impact of the degradation products (likely to include polymeric degradation products from the diluent) on the ability to strip cesium from the solvent stream following extraction. Preliminary results indicate that the degradation products promote the formation of a third phase, thereby hindering the ability to strip the solvent of cesium.

References

[1] P.V. Bonnesen, L.H. Delmau, T.J. Haverlock, B.A. Moyer, "Alkaline-Side Extraction of Cesium from Savannah River Tank Waste Using a Calixarene-Crown Ether Extractant", Report ORNL/TM-13704, Oak Ridge Tennessee, 1998.

[2] W. Davis Jr., "Radiolytic Behavior", Science and Technology of Tributyl Phosphate Volume I, W. Schlutz Editors,

[3] J.P. Holland, J.R. Merklin and J. Razvi, Nucl. Instrum. Methods, 153, 598, 1978.

[4] J. Hoigne, W. G. Burns, W. R. Marsh and T Gaumann, Helv. Chim. Acta., Vol. 47, 247, 1964.

[5] H. Hofer and H. Heusinger, Z. Phys. Chem. (Frankfurt), Vol. 69, 47, 1970.

[6] D. Verdin, J. Phys. Chem., Vol. 67, 1263, 1963.

[7] R. R. Hentz, J. Phys. Chem, Vol. 66, 1622, 1962.

[8] M. Burton, J. Phys. Chem, Vol. 52, 564, 1948.

[9] W. G. Burns and R. Barker, Progress in Reaction Kinetics, Pergamon, Oxford, Vol. 3, 303, 1965.

[10] Aspects of Hydrocarbon Radiolysis, T. Gaumann and J. Hoigne, Editors, Academic, New York, 33, 1968.

[11] Y. Llabador and J. P. Adloff, J. Chim. Phys., Vol 61, 681, 1964.

[12] Y. Llabador and J. P. Adloff, J. Chim. Phys., Vol 61, 1467, 1964.

Chapter 5

Behavior of Calix[4]arene-bis-crown-6 under Irradiation

Electrospray–Mass Spectrometry Investigation and Molecular Dynamics Simulations on Cs^+ and Na^+ Complexes of a Degradation Compound

V. Lamare[1], J.-F. Dozol[1], F. Allain[2], H. Virelizier[2], C. Moulin[2], C. Jankowski[2], and J.-C. Tabet[3]

[1]CEA Cadarache, DESD/SEP, 13108 St Paul Lez Durance Cedex, France
[2]CEA Saclay, DPE/SPCP/LASO, 91191 Gif-sur-Yvette, France
[3]Université Paris VI, Laboratoire de Chimie Structurale Organique et Biologique, UMR 7613, 75252 Paris, France

1,3-alt-calix[4]arene-crown-6 derivatives are being studied to selectively remove cesium from radioactive liquid waste. In the framework of development studies for an extraction process, the possible degradation of these compounds under irradiation was investigated. The analysis by electrospray ionization mass spectrometry (ES/MS) of the irradiated organic phases, showed that these calixcrowns are stable although they form substitution products under radiolysis. Among them, are found nitrate and nitro derivatives when the irradiation was performed in the presence of nitric acid. Molecular Dynamics (MD) simulations were performed on cesium and sodium complexes of an unsymmetrical 1,3-alt-calix[4]arene-bis-crown-6, bearing one nitro group on the rim ($BC6-NO_2$). These simulations show that the addition of the nitro group does not significantly modify the BC6 behavior towards alkali cations, indicating that the exceptional selectivity towards Cs^+ is sustained even after nitration.

Calixarenes are versatile molecules whose extracting properties towards metallic cations can be tailored through an appropriate choice of cavity size, conformation, and functional groups grafting (1). In the field of spent nuclear fuel reprocessing, 1,3-alt-calix[4]arene-crown-6 derivatives were shown to be potential extractants for selectively removing cesium cation from radioactive liquid waste, such as the high activity acidic fission products solutions resulting from nuclear fuel reprocessing. Hence, the compounds were extensively studied through experimental and molecular dynamics investigations (*2-4*).

In the framework of developing a solvent extraction process, the radiolytic stability of these compounds and the characterization of their degradation products should now be investigated. Previous studies on the effect of radiolysis on crown ethers such as dicyclohexano-18-crown-6, have shown that these compounds are rather stable and that the degradation products are rather due to the fragmentation of the polyether chain (*5*).

Here, we report a study on 1,3-alt-calix[4]arene-bis-crown-6 (BC6, Figure 1) diluted in o-nitrophenyl octyl ether (NPOE), a solvent used to obtain stable supported liquid membranes and which could be present as a co-solvent in a future process. The behavior of BC6 under irradiation, both in terms of stability and degradation products was studied by Electrospray/Mass Spectrometry (ES/MS). ES/MS is a method of choice for such investigations since it is a soft ionization/desorption technique (*6,7*) that provides information on complexes in solution (*8-11*) and allows to study their stability in the gas phase (*7-12*) and to compare their respective selectivity towards cations.

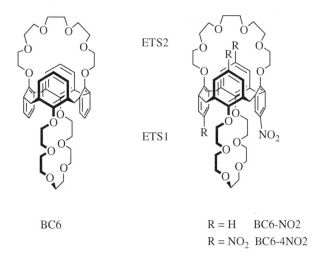

Figure 1. Reference calixarene and nitro compound modeled

After careful identification of BC6 substitution products, a molecular modeling study by means of molecular dynamics (MD) was undertaken in order to establish if these compounds could interfere on the process in particular by modifying the Cs^+/Na^+ selectivity in solution. This investigation was first motivated by the difficulty to obtain mono- or di-nitro calix[4]arene-bis-crown-6 derivatives, and to get an experimental answer to the behavior of such compounds before future process tests on real waste.

Materials and Methods

Materials

Calixarenes, NPOE, and NPHE (o-nitrophenyl hexyl ether) used in this work were supplied within the CEA network. Calixarenes were synthesized by Vicens (*13*) and were not further purified before use. Inorganic salts were purchased from Merck Chemical (Darmstadt, Germany) (NaCl) and Johnson-Matthey Chemical (Royston, England) (CsCl). The 65% nitric acid used for irradiation and extraction was obtained from Merck Chemical. The ES solvent acetonitrile (HPLC grade, 99.8 %) was purchased from Carlo Erba (Val de Reuil, France). All other solvents used in this work were from Merck Chemical.

Irradiation Process

BC6 was irradiated with a ^{60}Co source for 1500 hours, providing a dose equivalent to ten years operation in a reprocessing plant. The nominal absorbed dose was 3 MGy. The irradiation was performed on different static samples : 10^{-2} M BC6 in NPOE, and 10^{-2} M BC6 in NPOE in contact with 3 N HNO_3. Complementary irradiations with $H^{15}NO_3$ (isotopic enrichment in ^{15}N: 99%) were also performed.

Solid Phase Separation of Radiolysis Products

Crude organic phases were analyzed by ES/MS (e.g. Figure 4), but in order to prevent a loss of sensitivity due to the viscosity of the NPOE diluent, complementary analyses were performed after separation of the NPOE by solid phase extraction on a Si-column 500 mg, Lichrolut from Merck (Table II).

The column was washed with 8 ml of acetonitrile, then 10 µl of sample was deposed on the column. Elution with 3 ml of acetonitrile allowed extraction of the NPOE and its radiolysis products. Then elution with 3 ml of acetonitrile/water (85/15) and 2 ml of acetonitrile/water (70/30), allowed separation of the calixarene and its radiolysis products. Qualitative analyses of the different fractions obtained were then studied by ES/MS.

Electrospray/Mass Spectrometry Conditions

The ES mass spectra were performed on a triple quadrupole Quattro II (Micromass, France) analyser equipped with an electrospray source, with a mass to charge ratio (m/z) ranging from 20 th to 4000 th. The ES source was used without modification and typical operating conditions were as follows: the sample was introduced through a syringe pump (Harvard Apparatus 11, Cambridge, MA, USA) at a rate of 10 µl/min. ES was performed in the positive ion mode and the electrospray needle voltage was fixed at +3.35 kV. The source was heated at 80 °C, the lens and quadrupole voltages were optimized to obtain a maximun ion current for electrospray. The cone voltage was set at low voltage to avoid any fragmentation and the skimmer voltage at 1.9 V.

Before each analysis in ES/MS, a cation M⁺ was added to the solution of calixarene, a complex calixarene-cation was formed which enabled the detection of the calixarene by ES/MS (Figure 2). In the case of the crown fragmentation, formation of adducts between cation and fragments allowed their detection. In tandem mass spectrometry experiments (MS/MS), argon is used as a collision gas under single collision conditions.

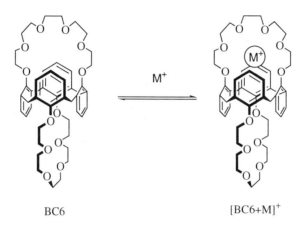

BC6   [BC6+M]⁺

Concentration ratio:
- BC6/M⁺ > 1, [BC6+M]⁺: main species
- BC6/M⁺ = 1, [BC6+M]⁺: main species, [BC6+2M]²⁺: minor species
- BC6/M⁺ < 1, [BC6+M]⁺: minor species, [BC6+2M]²⁺: main species.

Figure 2: Complexation of BC6 with a cation M⁺ (Cs^+, K^+, Na^+ or NH_4^+)

Molecular Dynamics Studies

All calculations were carried out on an Indigo2 R8000 SGI computer with the AMBER 4.1 software *(14)*, using the all-atom force field defined in the PARM91.DAT file, and the following representation of the potential energy *(15)*:

$$E_{pot} = \sum_{bonds} K_r(r - r_{eq})^2 + \sum_{angles} K_\theta(\theta - \theta_{eq})^2 + \sum_{dihedrals} \frac{V_n}{2}(1 + \cos(n\varphi - \eta)) +$$

$$\sum_{i<j} \left[\varepsilon_{ij} \left(\left(\frac{R*}{R_{ij}}\right)^{12} - \left(\frac{R*}{R_{ij}}\right)^{6} \right) \right] + \sum_{i<j} \left[\frac{q_i q_j}{\varepsilon R_{ij}} \right] + \sum_{H-bonds} \left[\varepsilon_{ij} \left(\left(\frac{R*}{R_{ij}}\right)^{12} - \left(\frac{R*}{R_{ij}}\right)^{10} \right) \right]$$

Details on parameters and protocols can be found in reference 4. For commodity, the calixarenes were divided into several residues: three "normal" phenoxy, one nitrophenoxy, and two crowns; the cation and the nitrate were also treated as residues. The atomic charge set is reported in Figure 3. Parameters for

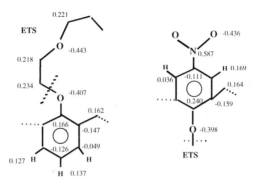

Figure 3. Atomic charge set used (MNDO charges scaled by 1.26)

the nitro group were taken from picrate anion (*16*). The following convention will be used in the text: the simulation NaETS1 means that the Na$^+$ cation is initially in the crown ETS1. Unless specified, average values were calculated on the entire MD run after discarding the first 10 ps of equilibration.

Results

Qualitative Identification of the BC6 Radiolysis Products

The analysis of the BC6 radiolysis products was performed on two solutions: 10^{-2} M BC6 in NPOE and 10^{-2} M BC6 in NPOE in contact with 3 N nitric acid. Figure 4 presents the different ES mass spectra obtained for BC6 before and after radiolysis on the crude organic phase to which was added dioctyloxy-calix[4]arene-crown-6 as an internal standard to compare peak intensities. In Table I are reported the different radiolysis products observed, after removing NPOE on a silica cartridge to increase the sensitivity of the analysis by ES/MS. Potassium cation is an impurity always present in the solvent. Considering that the response of the mass spectrometer is identical for all these radiolysis products, it should be noted that during irradiation of BC6 in NPOE, very few degradation products are formed, with peaks at m/z 825 and 841 attributed to crown fragmentation products. On the contrary, during irradiation of BC6 in presence of nitric acid, the formation of a larger number of products occurs, which corresponds to a nitration of the BC6 (Table I and Figure 4). The formation of BC6-NO$_2$ could take place according to the following SE or radical pathways *e.g.*:

$$BC6\text{-}H + {}^{\bullet}NO_2 \longrightarrow BC6\text{-}NO_2 + H^{\bullet}$$

Radiolysis of BC6 in presence H^{15}NO$_3$ (Figure 4c) leads to a shift of 1 mass unit of the peaks for each nitrogen 15 in the molecule. Therefore, it is possible to observe the shift by 1 or 2 mass unit for the following complexes [BC6+NO$_2$+Cs]$^+$

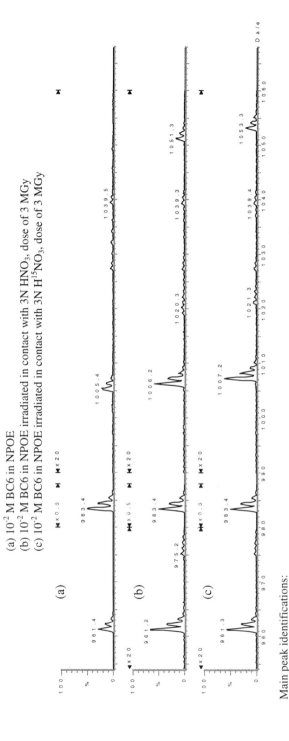

(a) 10^{-2} M BC6 in NPOE
(b) 10^{-2} M BC6 in NPOE irradiated in contact with 3N HNO_3, dose of 3 MGy
(c) 10^{-2} M BC6 in NPOE irradiated in contact with 3N $H^{15}NO_3$, dose of 3 MGy

Main peak identifications:
  m/z 961: $[BC6+Cs]^+$
  m/z 975: $[BC6+Cs+CH_3-H]^+$ or $[BC6+Cs+O-2H]^+$
  m/z 983: [internal standard+Cs]$^+$
  m/z 1005: [internal standard+Cs-H+Na]$^+$
  m/z 1006: $[BC6+Cs+NO_2]^+$
  m/z 1007: $[BC6+Cs+^{15}NO_2]^+$
  m/z 1020: $[BC6+Cs+NO_2+CH_3-H]^+$ or $[BC6+Cs+O-2H+NO_2]^+$
  m/z 1021: $[BC6+Cs+^{15}NO_2+CH_3-H]^+$ or $[BC6+Cs+O-2H+^{15}NO_2]^+$
  m/z 1051: $[BC6+Cs+2NO_2]^+$
  m/z 1053: $[BC6+Cs+2^{15}NO_2]^+$

*Internal standard: 1,3-di(n-octyl) calix[4]arene-crown-6

*Figure 4. ES mass spectra of BC6 in NPOE with internal standard**

Table I : Identification of the main irradiation products of BC6 by ES/MS

m/z	Irradiation of $10^{-2}M$ BC6 in NPOE	Relative abundance [a]	Irradiation of $10^{-2}M$ BC6 in NPOE in contact with HNO_3
825	$[BC6+K-(C_2H_4O)+2H]^+$?	*	-
841	$[BC6+K-(C_2H_4)+2H]^+$?	*	-
865	$[BC6+Na+CH_3-H]^+$ or $[BC6+Na+O-2H]^+$	**	$[BC6+Na+CH_3-H]^+$ or $[BC6+Na+O-2H]^+$
881	-	**	$[BC6+K+CH_3-H]^+$ or $[BC6+K+O-2H]^+$
883	$[BC6+K+O]^+$ or $[BC6+Na+2O]^+$	*	$[BC6+K+O]^+$ or $[BC6+Na+2O]^+$
896	-	***	$[BC6+Na+NO_2]^+$
910	-	**	$[BC6+CH_3+Na+NO_2]^+$
912	-	*	$[BC6+K+NO_2]^+$ or $[BC6+Na+NO_3]^+$
926	-	*	$[BC6+CH_3+Na+NO_3]^+$ or $[BC6+CH_3+K+NO_2]^+$
928	-	*	$[BC6+K+NO_3]^+$
941	-	***	$[BC6+Na+2NO_2]^+$
957	-	*	$[BC6+Na+NO_2+NO_3]^+$ or $[BC6+K+2NO_2]^+$
973	-	*	$[BC6+Na+2NO_3]^+$ or $[BC6+K+NO_2+NO_3]^+$

[a] relative abundance of radiolysis products compared to $[BC6+Na]^+$ (for the two columns of products).
' *** ' corresponds to an intensity of the complex > 10%
' ** ' corresponds to an intensity of the complex between 1 and 10%
' * ' corresponds to an intensity of the complex of 1% or lower.

and $[BC6+2NO_2+Cs]^+$ which allows to confirm the origin of the different nitro groups. To definitively identify all these different products, complementary experiments are now in progress with nitric acid labeled with oxygen 18 as well as with MS/MS measurements. The most probable structure for the product $BC6-NO_2$ is the one presented in Figure 1. In this case, steric hindrance of the nitro group in front of a complexing crown can eventually disturb cesium complexation and participate to a loss of cesium extraction efficiency by the calixarene. These different results show that BC6 is a very stable compound under radiolysis since the main products observed are due to addition of NO_2 or $O-NO_2$ on the calixarene. Hence, similar studies on crown ether report on the destruction of the crown *(5)*.

Influence of the Nitro Groups on the Stability of the Cation Complex in the Gas Phase and on the Extraction of a Cs-Na mixture.

Since the main degradation product is a mono-nitro derivative, supposed to be $BC6-NO_2$, its effect on the complexation efficiency of cesium cation was studied. Taking into account the difference of stability between the different complexes in gas and condensed phase (solvation) and since detection by ES/MS takes place in the gas phase, signals obtained are not totally representative of the nature and concentration of the complexes present in solution *(17)*. The species evolution from the condensed phase to the gas phase within the process of electrospray can eventually lead to the disappearance or reinforcement of some complexes according to their stability. It is therefore very important to compare the gas phase stability of the different complexes.

Gas phase stability of $[BC6+Cs]^+$, $[BC6-NO_2+Cs]^+$, and $[BC6-2NO_2+Cs]^+$:

Investigations by MS/MS on $[BC6+Cs]^+$, $[BC6-NO_2+Cs]^+$ and $[BC6-2NO_2+Cs]^+$ complexes were performed to study the relative stability in the gas phase of these three complexes. The cationized molecules were fragmented in the collision cell using argon as a collision gas. Figure 5 presents the relative intensity of each ion as a function of the collision energy and it can be seen that the stability of the cesium complexes decreases with an increasing of number of nitro groups (addition of two nitro groups leads to a loss of stability close to 7 eV which corresponds to an energy of mass center of 0.41 eV).

So, the relative peak intensities in the mass spectra shown in Figure 4 are not representative of the relative concentration of the different complexes, and one can expect that the actual concentration of nitro compounds is at least the same than BC6.

Comparison of BC6 and nitro derivatives for the complexation of cesium and selectivity of extraction

In obtaining reference compounds, the calixarene $BC6-4NO_2$ had been first synthesized. Even if it is known that it is better for quantitative measurement to use a calibration method *(18)*, a direct comparison (for simplification purposes) of the two calixarenes BC6 and $BC6-4NO_2$ was performed. The study of the complexation of cesium in acetonitrile by ES/MS shows that for the BC6, a doubly

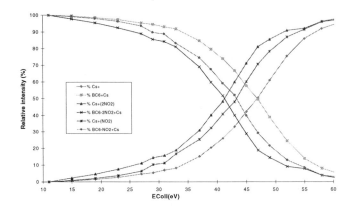

Figure 5. Representation of the stability in the gas phase of complexes [BC6+Cs]$^+$, [BC6-NO$_2$+Cs]$^+$ and [BC6-2NO$_2$+Cs]$^+$

charged species [BC6+2Cs]$^{2+}$ at m/z 547 (100%) and a minor monocharged species [BC6+Cs]$^+$ at m/z 961 (5% of the [BC6+2Cs]$^{2+}$) are formed (8). The complexation properties of BC6-4NO$_2$ are dramatically weaker and only a monocharged species with a weak intensity (5% of the [BC6+2Cs]$^{2+}$) corresponding to [BC6-4NO$_2$+Cs]$^+$ is observed at m/z 1141.

Due to the late availability of BC6-NO2, only preliminary solvent extraction experiments were performed on this compound (Table II), following protocol described in reference 13, which allowed the comparison of properties of mono and tetra-nitro derivatives and the irradiated organic phase.

Table II: Distribution coefficients D$_{M+}$ of Cs$^+$ and Na$^+$ with nitro compounds

	BC6	BC6-NO2	BC6-4NO2
D$_{Cs}$	19.5	8.5	6×10^{-3}
D$_{Na}$	1.3×10^{-2}	$< 10^{-3}$	$< 10^{-3}$

NOTE: 5×10^{-4} M alkali nitrate in 1 M HNO$_3$ for the aqueous phase, 10^{-2} M calixarene in NPHE

These results show that the presence of nitro groups on the calixarene strongly affects the cation complexation and extraction. In order to estimate if this loss of efficiency was only due to a lack of accessibility of the site of complexation, as in the case of p-tert-butyl derivatives, or if there was an other influence of the nitro group on the calixarene extraction efficiency and selectivity, molecular dynamics simulations were performed on the mononitro derivative (see below).

Extraction of cesium and sodium by the irradiated organic phases

Table III: Influence of irradiated NPOE on the Cs^+ and Na^+ extraction

	BC6 in irradiated NPOE	Irradiated organic phase
D_{Cs}	89	31
D_{Na}	7×10^{-3}	1.7×10^{-2}

The most striking effect of irradiation on the organic phase is the increase of cesium and sodium distribution coefficients (Table III). This phenomenon is not due to the calixarene, but to the NPOE diluent, which is chemically modified under a high dose rate irradiation.

When studied with ES/MS the irradiated NPOE showed some degradation. In particular, the octyl ether chain substitution with some specific groups such as hydroxyl, ketone, nitro or nitrate in different positions was observed. However the fragmentation of NPOE and the reduction of the side chain size does not take place. On the other hand, the hydroxylation of the octyl chain on six carbons was observed using the silylation and GC-MS method. It is necessary to point out that after radiolysis the viscosity of the solvent mixture increased considerably, mainly due to the formation of dimeric and trimeric species.

NPOE diluent is thus not suitable for an industrial process application, and a mixture NPHE/TPH 30/70 is currently under study for process development *(19)*.

Molecular Dynamics Simulations

MD simulations in the gas phase, taking into account the influence of the nitrate counter-ion, and in an explicit water phase were performed on BC6-NO2 Cs^+ and Na^+ complexes. The relative stability of these alkali complexes was studied with respect to the crowns, ETS1 or ETS2 (see Figure 1) in order to answer to the following questions: is there a preferred crown for the cation, does the cation affinity for one crown depend on the complex environment, and finally does the nitro group have any influence on the Cs^+/Na^+ selectivity? No direct comparison was done between BC6 and BC6-NO2 models, as there are too many differences between important parameters like atomic charges, highly perturbated on the paranitrophenoxy residue, to be able to compare both molecules *(4)*.

In the gas phase, no major structural or energetic modification was observed during 500 ps of MD run at 300 K. The structures fluctuate around average values and relax, leading to a final minimized conformation of lower energy than the minimized structure at the beginning of the MD calculation (Figure 6 and 7-Left). The only exception corresponds to simulation NaETS2, where the cation reached the crown ETS1 through the calixarene cavity at 90 ps of MD and stayed there until the end, showing an intrinsic preference for the crown opposite to the nitro group.

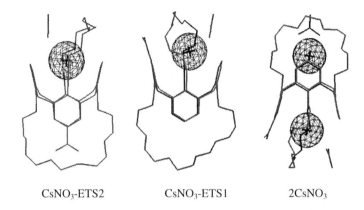

Figure 6. Cesium structures minimized after 500 ps of MD in the gas phase

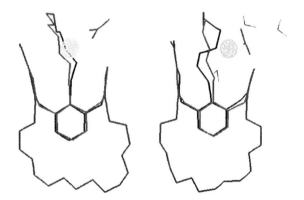

Figure 7. Left- Complex BC6-NO_2 with $NaNO_3$ in crown ETS1 minimized in the gas phase. Right- Snapshot of the complex at 40 ps of MD in water with the nearest water molecules around Na^+

The crown ETS1 shows the best complementarity for the cations simulated without nitrates (Table IV) but, in spite of the stabilizing influence of the nitro group, the calixarene displays a bad fit between Na^+ and the oxygen atoms of the crown (the individual dM^+_O fluctuation is at 0.7 Å for Na^+ and 0.3 Å for Cs^+). In any case, the average distance $<dM^+\_O_{NO2}>$ is much larger than the optimum $dM^+\_O_w$, corresponding to the first hydration sphere in pure aqueous phase (3.2 Å for Cs^+ and 2.5 Å for Na^+), indicating that the nitro group is not at the best location to efficiently participate in the complexation.

Table IV: Structural averages during MD simulations in gas phase at 300 K for BC6-NO$_2$

Å	METS1	METS2	MNO$_3$-ETS1	MNO$_3$-ETS2	2CsNO$_3$
$<dCs^+\_O_c>$[a]	3.29(2)	3.28(14)	3.42(16)	3.49(20)	3.39(21) ETS1 3.43(10) ETS2
$<dCs^+\_O_{NO2}>$	3.63(3)	-	4.89(2)	-	4.61(4)
$<dNa^+\_O_c>$	3.30(33)	3.71(1.08)[b]	3.46(87)	3.58(71)	-
$<dNa^+\_O_{NO2}>$	2.94(11)	-	4.39(15)	-	-

[a] :$<dCs^+\_O_c>$: average of 6 distances dCs^+_O ; $<dCs^+\_O_{NO2}>$: average of 2 distances $dCs^+\_O_{NO2}$ (fluctuation: $3.29(2) \equiv 3.29 \pm 0.02$).
[b]: average on 90 ps before moving to the crown ETS1.

Analysis of energetic data (Table V) shows the influence of the counter-ion on the relative stability of ETS1 and ETS2 complexes. Without nitrate, the alkali

Table V: Energy averages during MD simulations in gas phase at 300 K on BC6-NO$_2$

kcal.mol^{-1}	Etotal	$E_{int}M^+/calix$	$E_{int}M^+/NO_3$	$E_{int}M^+/NO_2$	$E_{int}NO_2/NO_3$
CsETS1	97(8)	-68(4)	-	-24(3)	-
CsETS2	101(8)	-59(3)	-	-9(0)	-
NaETS1	78(8)	-87(5)	-	-36(5)	-
NaETS2[a]	82(6)	-79(5)	-	-10(0)	-
CsNO$_3$ETS1	23(6)	-51(4)	-93(2)	-15(2)	12(1)
CsNO$_3$ETS2	17(6)	-46(4)	-93(2)	-9(0)	7(0)
NaNO$_3$ETS1	-4(6)	-61(7)	-117(2)	-21(7)	14(2)
NaNO$_3$ETS2	-10(6)	-56(4)	-116(4)	-9(0)	8(0)
2CsNO$_3$	-102(6)	-50(4)	-94(2)	-16(2)	12(1)
		-46(3)	-93(2)	-9(0)	7(0)

[a] calculated on 90 ps. E_{int}: total interaction energy between specified entities

complexes are more stable in ETS1, showing the optimum interactions between the cation and the calixarene. In the presence of the nitrate, the main interactions are between the ion pair, destabilizing the interactions between the calixarene and the cation. This effect is increased when the nitrate is near the nitro group, with an additional NO_2/NO_3 repulsion, leading to a preference of the cation for the crown ETS2.

From these simulations in the gas phase is concluded that, the preferred crown would depend on the proximity of the nitrate counter-ion. In a non-dissociating medium, the cation should be located in the crown far from the nitro group. Nevertheless, the binuclear $CsNO_3$ complex is stable and should be observed if the experimental conditions are favorable, *e.g.* a sufficient salt concentration is used.

Cesium and sodium complexes without counter-ion, as well as $NaNO_3$ complexes, were minimized after the MD simulation in the gas phase and immersed in a water box to be submitted to 100 ps of MD at 300 K in order to check the solvation and complex stability in presence of water. Both cesium complexes fluctuate around average values with a rapid hydration of the cation. This hydration is more important when the cation is in the crown ETS2. In this case, the radial distribution function (rdf) of water molecules around the cation during the dynamics indicates that 1.37 water molecules are in the first hydration sphere of Cs^+, in exchange with bulk water. On the contrary, for CsETS1 simulation, there is 1.00 water molecule complexed to the cation separated from the water phase. The hydrated cation is more stable in ETS1, and presents an optimum average distance $<dCs^+\_O_C> = 3.22(8)$ Å. In these simulations, the free crown ETS2 is hydrated although the free crown ETS1 stays water free. From these observations, an hydrated Cs^+, which is likely the species extracted, should be preferentially located in the crown ETS1.

The behavior of Na^+ is different. No cation hydration was observed from simulations of NaETS2 and $NaNO_3$ETS2, but an important contact with bulk water occurs when Na^+ is located in crown ETS1, leading to Na^+ decomplexation at 30 ps of MD run in simulation NaETS1. It has been previously observed that simulations in a water phase of sodium complexes with their nitrate counter-ion lead to more realistic structures when comparing to available X-ray data (20). Former simulations with BC6 and its benzocrown derivatives have shown that $NaNO_3$ complexes were stabilized by one molecule of water entering into the crown, the less hydrophobic crown corresponding to the calixarene showing the best affinity for Na^+, namely BC6 (4,20). In the simulation $NaNO_3$ETS1, the cation gets hydrated after 15 ps by a water molecule located in the crown (Figure 7-Right). The nitrate decomplexes at 40 ps and finally, Na^+ is stabilized by two water molecules. This is a behavior close to what was observed for BC6, that is a stabilization of Na^+ complex by one or two water molecules. So, in a water saturated phase, the location of Na^+ should also be in the crown ETS1, with a cation better hydrated than in crown ETS2. Like in simulations in the gas phase, the nitro group modifies the position of Na^+ in the crown but does not increase the affinity of that cation toward the calixcrown, as the O_{NO2} atoms are not well located to efficiently interact with Na^+ (average distance in $NaNO_3ETS1$ simulation in water: $<dNa^+\_O_{NO2}> = 4.96(30)$ Å).

Conclusion

This ES/MS study has shown that calixcrowns are radiolytically stable molecules. Very few degradation products are observed in the organic phase and mainly substitution products are found when the irradiation is performed in presence of nitric acid, leading to nitro derivatives. The nitro groups are likely located at the para position on the phenoxy. In the case of BC6, mono and dinitro substitution products were found. A representative molecule, 1,3-alt-calix[4]arene-bis-crown-6, bearing one nitro group on the rim, was studied by molecular dynamics in order to have a better understanding of its behavior towards cesium and sodium cations. It appears that the nitro group is not ideally located to participate efficiently to the binding of the cation in the crown. The preferred crown depends on the position of the nitrate counter-ion, and in the case of a dissociated ion pair, cesium cation should be located in the crown opposite to the nitro group. On the other hand, the behavior of sodium cation is not really different from its analogue in calixarene BC6. So, the addition of one nitro group opposite to a crown is not expected to significantly modify the selectivity of the organic phase. BC6-NO2 could thus be a good selective extractant for cesium, although experimental results show its weaker affinity for cesium in gas phase complexation or in liquid-liquid extraction when compared to BC6. The steric hindrance must be on the origin of this phenomenon due to more difficult access to the complexation site, as illustrated by results on the tetranitro derivative which has virtually no extraction properties towards cesium or sodium. The irradiated organic phase showed an increase in cation extraction, due to the chemical modification of NPOE under the high dose rate used in the experiment. This study suggests that calixcrowns should be sufficiently stable to extract cesium from radioactive liquid waste. An other diluent system, involving a mixture of NPHE/TPH, is currently under investigation for further process development.

Literature cited

1. Böhmer, V. *Angew. Chem. Int. Ed. Engl.*, **1995**, *34*, 713.
2. Hill, C.; Dozol, J.-F.; Lamare, V.; Rouquette, H.; Eymard, S.; Tournois, B.; Vicens, J.; Asfari, Z. Bressot, C.; Ungaro, R.; Casnati, A. *Calixarenes' 50th Anniversary*; Vicens, J.; Asfari, Z.; Harrowfield, J. M. Eds., Kluwer Academic Publishers 1995, pp. 399-408. Reprinted from *J. Incl. Phenom. Mol. Recognit. Chem.* **1994**, *19*, 399-408.
3. Dozol, J.-F.; Lamare, V.; Bressot, C.; Ungaro, R.; Casnati, A.; Vicens, J.; Asfari, Z. *Wo. Pat.*, WO9839321.

4. Lamare, V.; Dozol, J.-F.; Fuangswasdi, S.; Arnaud-Neu, F.; Thuery, P.; Nierlich, M.; Asfari, Z.; Vicens, J. *J. Chem. Soc. Perkin Trans. 2*, **1999**, 271 (and references quoted therein).
5. Draye, M.; Chomel, R.; Doutreluingne, P.; Guy, A.; Foos, J.; Lemaire, M. *J. Radioanal. Nucl. Chem., Letters* **1993**, *175* (1), 55-62.
6. Cole, R. B. *Electrospray Ionization Mass Spectrometry: Fundamentals Instrumentation and Applications*, Wiley Publisher: New York, 1997.
7. Cunnif, J. B.; Vouros, P. *J. Am. Soc. Mass Spectrom.* **1995**, *6*, 1175-1182.
8. Allain, F.; Virelizier, H.; Moulin, C.; Jankowski, C.; Dozol, J. F.; Tabet, J. C. submitted to *Anal. Chem. Acta*.
9. Blair, S. M.; Kempen, E. C.; Brodbelt, J. S. *J. Am. Soc. Mass Spectrom.*, **1998**, *9*, 1049-1059.
10. Peiris, D. M.; Yang, Y., Ramanathan, R.; Williams, K. R.; Watson, C. H.; Eyler, J. R. *Int. J. Mass Spectrom. Ion Processes*, **1996**, *157*, 365-378.
11. Leize, E.; Jaffrezic, A.; Van Dorsselaer, A. *J. Mass Spectrom.*, **1996**, *31*, 537-544.
12. Kebarle, P. *J. Mass Spectrom.*, **1997**, *32*, 922-929.
13. Asfari, Z.; Bressot, C.; Vicens, J.; Hill, C.; Dozol, J.F.; Rouquette, H.; Eymard, S.; Lamare, V.; Tournois, B. *Anal Chem.*, **1995**, *67*, 3139-3150.
14. Pearlman, D. A.; Case, D. A.; Caldwell, J. A.; Ross, W. S.; Cheatham III, T. E.; Ferguson, D. M.; Seibel, G. L.; Singh, U. C.; Weiner, P.; Kollman, P. A. AMBER 4.1, University of California, San Francisco, **1995**.
15. Weiner, S. J;. Kollman, P. A.; Nguyen, D. T.; Case, D. A. *J. Comput. Chem.* **1986**, *7*, 230-252.
16. Troxler, L. Ph.D. thesis, Université Louis Pasteur, Strasbourg, FRANCE, 1996.
17. Vincenti, M. *J. Mass Spectrom.*, **1995**, *30*, 925-939.
18. Young, D. S.; Hung, H. Y.; Liu, L. K. *J. Mass Spectrom.*, **1997**, *32*, 432-437.
19. Dozol, J.-F.; Lamare, V.; Simon, N. S.; Ungaro, R.; Casnati, A. In *Calixarene Molecules for Separations*, Lumetta, G.; Rogers, R. D.; Gopalan, A. S. Eds., ACS Symposium Series, under press.
20. Thuery, P.; Nierlich, M.; Lamare, V.; Dozol, J.-F.; Asfari, Z.; Vicens, J. *Supramol. Chem.*, **1997**, *8*, 319-332.

Chapter 6

Interfacial Features of Assisted Liquid–Liquid Extraction of Uranyl and Cesium Salts: A Molecular Dynamics Investigation

M. Baaden, F. Berny, N. Muzet, L. Troxler, and G. Wipff[1]

Institut de Chimie, Université Louis Pasteur, UMR CNRS 7551,
4, rue B. Pascal, 67 000 Strasbourg, France

We report molecular dynamics studies on the interfacial behavior of species involved in the assisted liquid-liquid extraction of cesium and uranyl cations. The distribution of uncomplexed Cs^+Pic^- and $UO_2(NO_3)_2$ salts is first described at the water / chloroform interface. This is followed by simulations of monolayers of calix[4]arene-crown6 Cs^+Pic^- complexes at the interface, where they are found to remain adsorbed, contrary to expectations from extraction experiments. The question of synergistic effects is addressed by simulating these calixarenes at a TBP saturated interface (TBP = tributylphosphate). Finally, in relation with the extraction of uranyl by TBP, we report a computer demixing experiment of a "perfectly mixed" ternary water / chloroform / TBP mixture containing 5 $UO_2(NO_3)_2$ molecules. The phase separation is found to be rapid, leading to the formation of a TBP layer between the aqueous and organic phases and to spontaneous complexation of the uranyl salts by TBP. The complexes formed are not extracted to chloroform, but remain close to the water / organic phase boundary. The simulations reveal the *importance of interfacial phenomena in the ion extraction and recognition processes*.

Despite the practical and fundamental importance of assisted liquid-liquid ion extraction, little is known on the detailed mechanism and microscopic events that take place. The extraction experiment generally proceeds by mixing an aqueous salt solution with an organic solution of the extractant molecule ("ionophore").(*1-4*) Then, phase separation proceeds, due to gravitational forces or centrifugation, and differences in surface tension between the two liquids. It has been stressed that the interfacial region between the aqueous and organic solvents plays a crucial role concerning the cation capture.(*1, 3, 4*) Computer simulations are becoming a valuable tool for providing microscopic insights into the properties of such interfaces in the presence of salts (*5-7*) or extractant molecules.(*8-14*) An important computational finding was the high surface activity of

[1]Corresponding author (wipff@chimie.u-strasbg.fr).

extractant molecules in their free or complexed states, which led us to suggest that the cation complexation and recognition takes place at the interface.(*10, 11*)

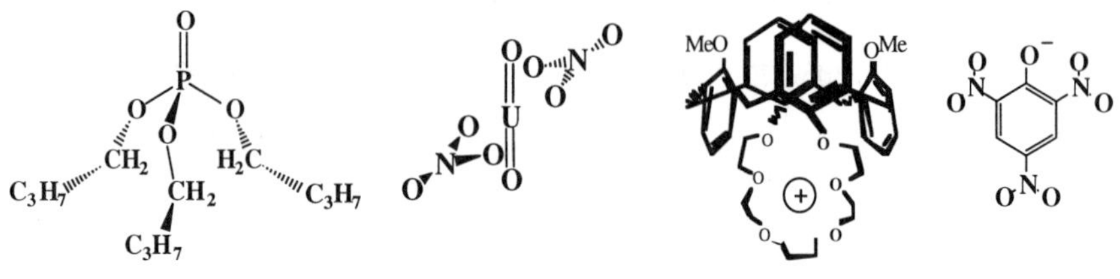

TBP   **UO$_2$(NO$_3$)$_2$**   ***CalixC6* Cs$^+$**   **Picrate**

Chart 1. Simulated species.

In this paper we report molecular dynamics (MD) studies on the interfacial behavior of systems involved in the extraction of cesium and of uranyl cations from water to an organic phase (chloroform). The first simulations deal with the uncomplexed Cs$^+$Pic$^-$ (Pic$^-$ = Picrate anion; Chart 1) and UO$_2$(NO$_3$)$_2$ salts at a water / chloroform interface. Our goal is mainly to assess to what extent these cations approach the interface, a prerequisite for their complexation by the adsorbed ionophores. The second series of computer experiments relates to Cs$^+$ extraction by a recently developed calixarene **L** (= 1,3-alternate calix[4]arene-crown6; *CalixC6* in Chart 1). In previous simulations on a single LCs$^+$Pic$^-$ complex, the latter was found to remain adsorbed at the interface.(*10, 12-14*) We therefore decided to simulate a layer of LCs$^+$Pic$^-$ complexes at the interface, either alone, or in the presence of free hosts and salts, or with interfacial TBP molecules, in order to model possible synergistic and salting out effects on extraction. We wanted to test whether these conditions would lead to "extraction", i.e. to diffusion of some complexes from the interface to the organic phase.

Finally, as a first step towards the simulation of uranyl extraction by TBP in the PUREX process,(*15, 16*) we report a "computer demixing experiment" on a "perfectly mixed" water - chloroform mixture containing 30 TBP's and 5 UO$_2$(NO$_3$)$_2$ species as solute. In this system, TBP is not only a potential ligand for uranyl (which is extracted as UO$_2$(NO$_3$)$_2$(TBP)$_2$ complex (*15*)), but acts as a surfactant or co-solvent.(*17*) We thus wanted to assess whether TBP complexes spontaneously form and whether they are extracted to the organic phase. Another question concerns the status of free TBP's: do they adsorb at the interface or do they dissolve in the organic phase ? How complete is the phase separation ? Is there solvent mixing and is water "dragged" into the organic phase ? We notice that in our simulated system the TBP concentration is two or three times lower than that actually used in the PUREX process and that the aqueous phase is neutral, instead of being highly acidic. These simplifications were imposed by computational difficulties in directly simulating the "real" PUREX mixture.

METHODS

The simulations were performed with AMBER4.1(*18*) where the potential energy is described by a sum of bond, angle and dihedral deformation energies, and pairwise additive 1-6-12 (electrostatic + van der Waals) interactions between non-bonded atoms. Details are given in ref. 11. The parameters for the **L** ligand, for Pic$^-$, NO$_3^-$ and Cs$^+$ ions are the same as in previous studies.(*11, 12*) Data on TBP come from ref. 19. As solvent models, we used TIP3P for water (*20*) and OPLS for chloroform.(*21*) The simulated solvent systems and simulation conditions are described in Table I. All C-H, O-H, H$\cdots$H, C-Cl and Cl$\cdots$Cl "bonds" were constrained with SHAKE,(*18*) using a time step of 1 fs. In most cases, the temperature was monitored by independently coupling the water subsystem and the remaining subsystem to a thermal bath at the reference temperature. In the case of uncomplexed salts, the whole system was coupled to the bath. Details on the simulated systems are given in Table I.

Table I: Simulation conditions at the interface:
Box size $V_x*V_y*(V_{z-chlor}+V_{z-wat})$. Number of chloroform and water molecules. Periodicity (2D / 3D) of the simulated system and simulated time. Cation concentration with respect to the box size.

	Solute	Box Size $Å^3$	Solvent $N_{chl} + N_{wat}$	2D / 3D	Time (ns)	Mol/l
A	10[UO$_2^{2+}$(NO$_3^-$)$_2$]	50*47*(37+37)	637+2727	2D	1.3	0.10
B	10[UO$_2$(NO$_3$)$_2$]	50*47*(37+37)	637+2727	2D	1.3	0.10
C	10(Cs$^+$Pic$^-$)	45*44*(35+32)	497+2051	2D	1.0	0.13
D	10(LCs$^+$Pic$^-$)	47*52*(34+46)	572+3515	3D	0.36	0.08
E	10(LCs$^+$Pic$^-$)	47*52*(43+37)	734+2673	3D	1.12	0.08
F	8L, 9(LCs$^+$Pic$^-$), 5(Cs$^+$Pic$^-$)	47*46*(46+49)	636+3822	3D	0.85	0.11
G	30(TBP), (LCs$^+$Pic$^-$), 2L, 8(Cs$^+$NO$_3^-$)	40*39*(37+35)	341+1448	3D	1.0	0.13
H	30(TBP), 5[UO$_2$(NO$_3$)$_2$]	40*40*(35+32)	357+1459	3D	5.0	0.08

The water / chloroform interface has been built as indicated in ref. 10, starting with two adjacent boxes of pure water and chloroform, respectively (Chart 2). The 30(TBP), 9(LCs$^+$Pic$^-$), 10(Cs$^+$Pic$^-$) or 10[UO$_2$(NO$_3$)$_2$] solutes were initially placed at the interface, equally shared between the two solvent phases. In order to build the mixed 30(TBP) / 5[UO$_2$(NO$_3$)$_2$] solution, we first prepared a "perfectly mixed" ternary water / chloroform / TBP solution containing 30 TBP's. This was achieved by starting from a bilayer of TBP molecules, adsorbed at the water / chloroform interface and equilibrated for 1 ns at constant pressure, in order to reach densities close to 1.0 for the "bulk" water and 1.42 for the chloroform phases, respectively.

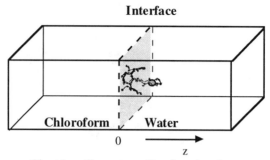

Chart 2. The water - chloroform interface.

After energy minimization, the mixing was run for 1 ns at 500 K using modified charges on TBP ($q_P = 0.4$; $q_O = -0.4$) and on chloroform (OPLS charges scaled by 4.0). Five $UO_2(NO_3)_2$ complexes were added randomly to this "perfectly mixed" ternary solution, removing the solvent molecules at less than 2 Å from the solute. The mixing was pursued for 100 ps at 700K with modified charges on TBP, chloroform, UO_2^{2+} ($q_U = -0.70$; $q_O = 0.35$) and NO_3^- ($q_N = +0.75$; $q_O = -0.25$). The demixing MD simulation started after energy minimization, resetting the temperature to 300 K and all charges to their reference values. Most systems were represented with 3D periodic boundary conditions, while for others periodicity was applied along the X and Y directions only, using a restraining potential at the Z boundaries to avoid evaporation of solvent molecules (see Table I, and ref. 8 for details).

The results were analyzed as described in ref. 10 and 11. The position of the interface (noted ITF in short) was recalculated every 0.2 ps, as the intersection between the water and chloroform density curves.

RESULTS

1- The uncomplexed salts at the water / chloroform interface.

Two simulations of 1.3 ns were performed with the $10[UO_2(NO_3)_2]$ salt as solute, where the ions were either free of constraints ($UO_2^{2+}(NO_3^-)_2$; model A), or constrained to form a neutral intimate $UO_2(NO_3)_2$ complex, where the two NO_3^- anions display a bidentate coordination in the equatorial plane of the uranyl (model B). Model B mimics the situation where the salt is mostly associated, due to an excess of NO_3^- anions. With both models, $UO_2(NO_3)_2$ is found to be "repelled by the interface" and to migrate to the bulk aqueous phase. Only 1 % of the U atoms are within 7 Å from the interface (models A and B; Figure 1). The salts free of constraints (A) rapidly dissociate, leading to $UO_2(OH_2)_5^+$ species, in which five water molecules coordinate UO_2^{2+} in its equatorial plane. In water, most of the constrained complexes (B) are equatorially coordinated by two additional water molecules, leading to a coordination number of six for uranyl. In addition, oligomers of $[UO_2(NO_3)_2]_n$ type (n = 2 to 4) are observed for time periods as long as 0.5 ns: one nitrate anion builds a relay between the U atoms of two uranyls

(Figure 2), which retain a six coordination. The average distribution of the salts peaks near the center of the water slab at about 18 Å (cases A and B) from the interface. One also notices that in system A, the NO_3^- anions come closer to the interface than the UO_2^{2+} cations (Figure 1).

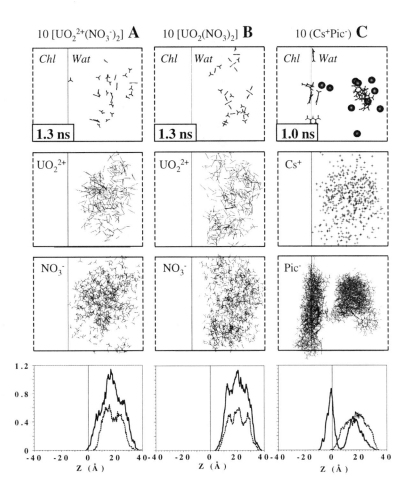

Figure 1. The $10[UO_2(NO_3)_2]$ (models A and B) and $10(Cs^+Pic^-)$ (C) salts simulated at the interface. From top to bottom: snapshots at the end of the simulations; cumulated positions of the cations; cumulated positions of the anions; density profiles along the Z axis for the cations (dotted lines) and for the anions (full lines) (see Chart 2). Cumulated views and density profiles calculated during the last 0.7 ns.

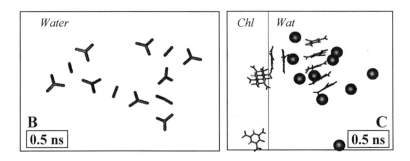

Figure 2. Snapshots of a [$UO_2(NO_3)_2$] tetramer in the water phase (model B; left) and of the Cs^+Pic^- salt at the interface (right), forming stacks of five Pic^- in water and of two Pic^- at the interface. Solvent not shown for clarity.

The Cs^+Pic^- ions behave quite differently during the dynamics. First, from the beginning of the simulations until the end (1 ns), the anions display a marked tendency to form labile stacks in water, ranging from two to five Pic^- units. Some of these negatively charged stacks are surrounded by Cs^+ cations (Figures 1 and 2), as found previously for the Pic^-K^+ salt.(22) Other Pic^- anions are found at the water / chloroform interface where they remain during the whole simulation, some of them forming stacked dimers. About 50 % of the anions are, on the average, within 7 Å from the phase boundary. They create a negative potential which attracts Cs^+ cations: 8% of them are in the same region. This scenario is analogous to the one observed for $Guanidinium^+Cl^-$ salts, where the hydrophobic $Guanidinium^+$ cations also form stacks in water and accumulate at the interface, while the Cl^- counterions are immersed in water.(9) Thus both Pic^- and $Guanidinium^+$ species are surface active although, unlike classical amphiphiles, they do not possess a "polar head" flanked by a hydrophobic chain. Once at the interface, they attract hydrophilic counterions which would otherwise be "repelled" by the interface.(8, 9, 11)

2- Adsorption of the Cs^+ calixarene complexes at the interface. Effect of concentration and of synergism with TBP molecules.

In this section we describe simulations (noted D to F) of LCs^+Pic^- complexes, as the calixarene **L** belongs to a class of remarkable extractants for Cs^+.(23-26) The systems D to F all contain ten LCs^+Pic^- species, nine of them being initially placed at the interface, close to a position corresponding to the free energy minimum for a single LCs^+ complex.(10) As these complexes are surface active and reduce the surface tension of water, we wanted to test whether their accumulation leads to extraction to the organic phase. This turned out not to be the case.

In the simulation starting with the tenth LCs^+Pic^- complex on the water side of the interface (model D; Figure 3), the latter migrated to the interface in less than 0.3 ns, and remained there until the end. When it was initially placed on the chloroform side (at a $Cs^+\cdots ITF$ distance of about 13 Å; model E; Figure 3), it also moved towards the interface (at about 5 Å) despite its coulombic attraction with the accompanying Pic^- counterion which remained in chloroform. At about 1 ns, one Cs^+ cation decomplexed and moved to water, while the corresponding host moved about 11 Å deeper into chloroform.

The system F is more complex than D - E and likely more representative of a real extraction equilibrium, as the nine interfacial LCs^+Pic^- complexes are initially flanked by a second layer of 8 free **L** hosts in chloroform and by 5 Cs^+Pic^- ion pairs in water (Figure 3). Again, all complexes remained adsorbed at the interface, while two to three Pic^- oscillated between "bulk water" and the interface, forming stacked arrangements. One **L** host migrated to the interface, while two others moved deeper into chloroform (Figure 3).

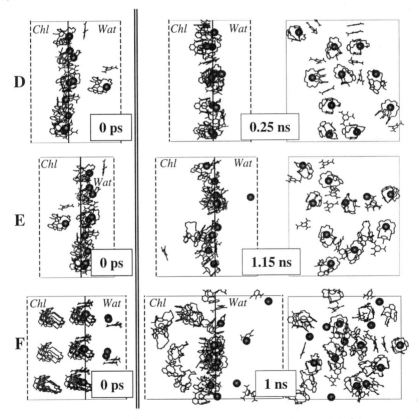

*Figure 3. The $9(LCs^+Pic^-)$ system at the interface, plus one LCs^+Pic^- complex initially either in water (system D; top) or in chloroform (system E; middle). System F (bottom): $9(LCs^+Pic^-)$ layer + 8 **L** hosts in chloroform + 5 Cs^+Pic^- pairs initially in water. Starting (left) and final arrangements (right; orthogonal views).*

In all cases, the interfacial LCs$^+$ complexes are positioned with the twofold symmetry axis either perpendicular, or parallel to the interface, with their cationic moiety attracted by water (Figure 3). The interfacial Pic$^-$ counterions achieve π stacking interactions with phenolic groups of L, or form stacked dimers. As shown in Figure 3, the interface is unsaturated, and the complexed Cs$^+$ cations form an irregular layer, as do the Pic$^-$ anions. These species are not static, but oscillate about 5 Å away from the interface during the dynamics, which confirms that they sit in a potential well.

Towards the simulation of "synergistic + salting out effects" ?

In the above simulations no complex was extracted to the organic phase, presumably due to the attraction by water. Thus, we built another model (G), where one LCs$^+$Pic$^-$ complex is already extracted in chloroform, but has no contact with water. This was achieved by intercalating two unsaturated layers of 15 TBP's each between the complex and the interface (Figure 4). The TBP's were placed in an inverted orientation, i.e. with their phosphoryl dipole pointing to chloroform, rather than to water. This allowed us to investigate whether they would reorient in the expected amphiphilic orientation, or mix with the organic phase where TBP is soluble. In addition, two L hosts were placed in chloroform and eight Cs$^+$NO$_3^-$ ion pairs were in water. After 1 ns of simulation (Figure 4), the TBP's formed a disordered bilayer at the interface. Most of them rotated their P=O dipole towards water, attracting uncomplexed Cs$^+$ cations near the interface (Figures 4 and 5). These cations formed one to three labile 1:1 and 1:2 complexes with TBP. On the average, about 20% of the Cs$^+$ cations sit within 7 Å from the interface, in marked contrast to the distribution observed in the absence of TBP.(*11*) The two uncomplexed L ligands and the Pic$^-$ anion remained in the organic phase, while the LCs$^+$ complex first migrated to the "TBP layer", where one to two TBP's coordinated to its Cs$^+$ cation and pulled it right at the interface (Figure 5). Thus, although the LCs(TBP)$_2^+$ complex is more lipophilic than the LCs$^+$ one, it is still highly surface active and does not migrate to the organic phase.

To summarize, *in none of the simulations, spontaneous extraction of the LCs$^+$Pic$^-$ complex is observed*, contrary to expectations from the extraction results.(*23, 26*) As observed in a demixing simulation,(*11*) this complex is more surface active than the free calixarene L, because the complexed Cs$^+$ moiety is more polar and attracted by water than the uncomplexed crown or calixarene moieties.

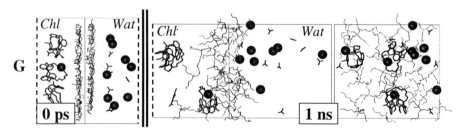

Figure 4. System G at 0 ps (left) and at 1 ns (right; orthogonal views)

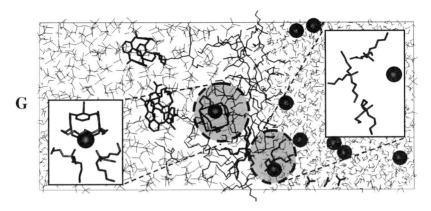

Figure 5. System G at 1 ns. Snapshot of the $LCs(TBP)_2^+$ complex (left) and of a $Cs(TBP)_2^+$ "complex" at the TBP - water interface (right).

3- Demixing of a ternary water / chloroform / TBP solution: Assisted extraction of the $UO_2(NO_3)_2$ salt ?

In this section, we describe the phase separation of a "perfectly mixed" water / chloroform / TBP solution containing five $UO_2(NO_3)_2$ salts (system H). The latter have been constrained as intimate ion pairs, in order to model a situation close to experimental conditions where NO_3^- anions are in excess and the uranyl salt is therefore mostly associated, and to possibly facilitate the extraction of uranyl as a neutral species to chloroform.

At the beginning of the demixing simulation (0 ps), the water and chloroform molecules are equally diluted and "perfectly mixed" in the solvent box, together with the 30 TBP and 5 $UO_2(NO_3)_2$ species (Figure 6). Rapidly, water separates from chloroform. At 0.2 ns, two "bubbles" of water form in chloroform, while most of the TBP's sit close to the boundary region between the two liquids. All $UO_2(NO_3)_2$ salts are surrounded by water molecules and sit either inside, or near the periphery of the "bubble". After 0.5 ns, the two liquids are almost separated, and seem on the pathway leading to the formation of two "horizontal" interfaces, where all TBP's are adsorbed. However, the liquids reorganize in the next 0.5 ns, and the interfaces become "vertical", which leads to a reduction of their area. At 1 ns, the water and organic phases are almost completely separated, but the interface is very rough and mobile, as the TBP's facilitate the water - "oil" mixing. Between 1 and 5 ns, no major solvent reorganization takes place. Schematically, the final system is composed of "vertical" slabs of water / TBP / chloroform / TBP / water, replicated by the imposed 3D periodicity. The interface becomes somewhat more regular and one to two hydrated TBP's sit in chloroform. A closer look at the system reveals however the *spontaneous complexation of all $UO_2(NO_3)_2$ salts by one to two TBP's*.

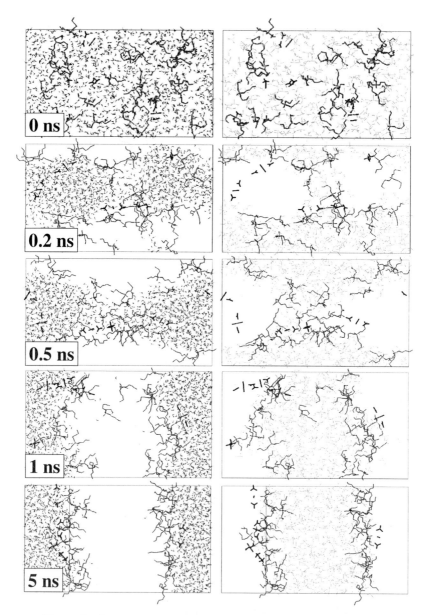

Figure 6. System H. Demixing of the "perfectly mixed" water / chloroform binary mixture, containing 30 TBP's and 5 UO$_2$(NO$_3$)$_2$ salts. For clarity, the water (left) and chloroform (right) solvents are displayed separately.

The first complex, of 1:1 type, formed early (at 0.2 ns) at the periphery between chloroform and water. A 1:2 complex appeared at about 1.0 ns (Figure 6), again at the liquid boundaries. This is, to our knowledge, the first observation of spontaneous complexation of cations in a "computer experiment." The $UO_2(NO_3)_2(TBP)_2$ complex has the structure (pseudo D_{2h} symmetry) and stoichiometry of the species extracted experimentally.(15) Although the U atom is fully shielded from water, the salt does not migrate from the interfacial to the bulk organic region, presumably because it is still attracted by water (by -50±10 kcal/mol, mostly due to the anion contribution). The four other salt species sit finally on the water side of the interface, forming $UO_2(NO_3)_2(TBP)(H_2O)$ species (Figure 7). The corresponding water / $UO_2(NO_3)_2$ attraction is -100±10 kcal/mol. As a result, the distribution of the $UO_2(NO_3)_2$ salts in the presence of TBP is very different from the one observed for the "pure salt" described above.

At the water-TBP interface, most uncomplexed TBP's display the expected amphiphilic orientation and point their phosphoryl oxygen to water, forming one to two hydrogen bonds per TBP. However, the TBP "layer" is unsaturated and highly disordered. Some "inversed orientations" are also observed. The corresponding TBP's form $TBP \cdot H_2O$ "supermolecules" in chloroform, near the interface (Figures 6 and 7). No free water molecules are found in the bulk organic phase.

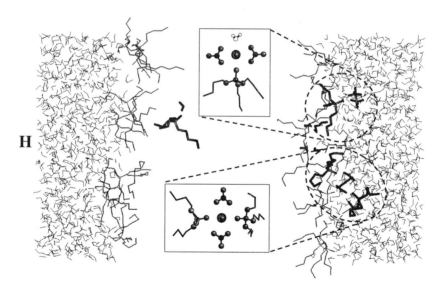

Figure 7. System H at 4.5 ns, showing a $TBP \cdot H_2O$ "supermolecule" in chloroform (left) and 1:1 and 1:2 complexes formed at the TBP - water interface (right). The chloroform molecules are omitted for clarity.

DISCUSSION AND CONCLUSIONS

We report MD simulations on the interfacial behavior of species involved in the unassisted / assisted liquid-liquid extraction of cations, with a particular focus on Cs^+ and UO_2^{2+} cations. These new results point out a number of important features concerning the mechanism of ion recognition and extraction.

The water / organic liquids are never completely mixed, but separated by an interface. The demixing simulation reported here, like the previous ones,(9, 11) shows that the "perfectly mixed" binary water-organic mixture separates rapidly, on a nanosecond timescale at the microscopic level. This means that, in a real extraction experiment, the liquids are never completely mixed, but form metastable microdroplets, which display a large interfacial area where extractant molecules and their complexes strongly adsorb.(27) This raises a number of issues, addressed in the following discussion.

Complexation and ion recognition take place at the liquid-liquid interface. From a mechanistic point of view, the presently and previously reported simulations (8, 9) strongly suggest that *cation capture by neutral ligands takes place at the liquid-liquid boundaries.* Indeed the uncomplexed neutral ionophores and their complexes are "repulsed" by the bulk water phase, because of the high cavitation energy (34, 35) and surface tension of water, compared to the organic phase.(36) Thus, complexation cannot take place in the water phase. At the interface, the extractants sit almost entirely in the organic phase (whose cavitation energy is lower) and still enjoy specific attractions by interfacial water molecules ("water fingers" (28)). As shown earlier, microhydration facilitates the adsorption of symmetrical ligands like 18-crown6,(29) TPTZ (8) or the [222] cryptand at the interface.(13, 30) Calixarenes are more polar, amphiphilic and surface active.(10, 12-14) Once complexed, their amphiphilicity and surface activity increase, which is consistent with current mechanistic speculations (1) and experimental observations at the water-air interface.(31-33) According to simulations (10) and experiments,(31-33) *ion recognition* by calixarenes also takes place at the water surface.

How can cations approach the interface? In the absence of assistance, the uncomplexed cations cannot approach the interface sufficiently close to be captured. Inspired by our results, we propose three mechanisms that increase the concentration of uncomplexed cations at the interface. The first one concerns counterion effects. Lipophilic anions, like Pic^-, ClO_4^-,(9) BPh_4^- (37) or dicarbollides are highly surface active, although they are more or less centrosymmetrical and do not display an amphiphilic topology like fatty acid derivatives. They therefore induce a negative potential which attracts cations at the interface, as observed in the Cs^+Pic^- system. Other examples are given in ref. 8 and 9. The second mechanism involves amphiphilic extractant molecules, illustrated by our simulations on TBP. In two simulations, the interfacial TBP's are found to attract cations at the interface. Uranyl salts are partially complexed, while Cs^+ cations are more loosely "bound". The third aspect concerns the "salting out" effect, which relates to the content of the source phase. On the computational side, it would be desirable to model a high concentration of salts and acid in the aqueous phase to probe if ions are "pushed" to the surface, and the complexes pushed to the organic phase. We did not report such simulations, because of the inherent computational difficulties and possible artifacts, related to the treatment of long range electrostatic forces.(38) We suggest that an important feature of the salting out effect concerns the status of ion pairs, especially for divalent or trivalent cations. Increasing the anion concentration also increases the concentration of formally neutral MX_n species, which are less hydrophilic than the M^{n+} cations, and therefore more concentrated in the interfacial region. This also suggests that they will be complexed and extracted by neutral ligands as MX_n species. Thus the binding mode of the ligands and the stoichiometry of their extracted complexes should be anion dependent.

Synergistic effects. In two simulations, addition of TBP molecules revealed important features, related to the high TBP concentration at the interface: (i) these molecules increase the cation concentration at the interface, and (ii) they co-complex the cation to be extracted. We suggest that these are important aspects of synergism at the liquid-liquid interface.

How are the complexes extracted from the interface to the organic phase? Like in simulations with monomeric complexes,(*8, 9*) none of the complexes diffuse from the interface to the organic phase. This apparent contradiction with macroscopic extraction data may be due to the fact that the interface is still unsaturated, and the interfacial pressure too high. The nature of assemblies at the interface remains to be elucidated, as it likely involves an equilibrium between complexes, extractants, and ions, which depends on concentration and salting out effects. The situation observed with LCs^+Pic^- differs from the one simulated with the 18-crown6 K^+Pic^- complexes, where some cations decomplexed and others were extracted to chloroform.(*29*)

In real demixing experiments, the macroscopic phase separation is driven by gravitational or centrifugational forces. This leads to a collapse of microdroplets and decrease of the interfacial area, likely accompanied by desorption of the complexes from the interface and their extraction to the organic phase. These events involve timescales and dimensions which are presently not amenable to MD computational approaches.

Concerning the uranyl extraction by TBP, we note that our simulated system differs from the experimental one by two important features. (i) First, the TBP concentration in the PUREX process is higher (about 30% in volume) than the one simulated. Increasing the concentration likely yields a water / TBP interface which favours the formation of 1:2 complexes and their shielding from water. Saturation of the interface by TBP's also prevents overlap between the organic and the water phase. (ii) Second, in the real experiment, the source phase is not neutral, but contains about 3-6 M in HNO_3.(*16*) This leads to a reduction of the water activity in the interfacial region, and presumably increases the TBP concentration at the interface due to enhanced hydrogen bonding interactions. Both factors facilitate the capture and extraction of uranyl. The nature of the organic phase (chloroform *vs* kerosene) is likely less important for uranyl or plutonyl extraction than features (i) and (ii). In real systems, the nature of the "organic phase" may also range from homogeneous solutions to micellar systems up to third phase separation. In all cases, liquid-liquid interfaces are involved.(*39*)

Computational aspects. A number of computational issues concerning the energy representation of the system and the treatment of long range interactions have been addressed in previous papers.(*10-12,14*) Although the solvent parameters fitted for pure liquids may inaccurately describe their mutual interactions,(*40*) we showed that several models of chloroform, some of them including explicit polarization terms, lead to similar results in demixing simulations.(*11*) The representation of the system boundaries (2D *vs* 3D) has also little effect on the interfacial properties.(*14*) Although there is room for methodological improvements, we believe that no major bias exists in our simulation results.

We hope that such "computer experiments" will stimulate further theoretical and experimental studies on interfacial phenomena. Beyond ion separation, the latter have bearing on other fields like interfacial electrochemistry, phase transfer catalysis, drug transport and availability at water phase boundaries near membranes, micelles, or polymeric surfaces.

Acknowledgements.

The authors are grateful to CNRS-IDRIS and to Université Louis Pasteur for allocation of computer resources, and to EEC (F14W-96CT0022 contract) for support. MB, FB and NM thank the French Ministery of Research for a grant.

REFERENCES

1. Moyer, B. A., in *Molecular Recognition: Receptors for Cationic Guests*; **1996**, J. L. Atwood; J. E. D. Davies; D. D. McNicol; F. Vögtle and J.-M. Lehn Ed.; Pergamon, New York, pp 325-365 and references cited therein.
2. Koenig, K. E.; Lein, G. M.; Stuckler, P.; Kaneda, T.; Cram, D. J., *J. Am. Chem. Soc.* **1979**, *101*, 3553.
3. Danesi, P. R., in *Principles and Practices of Solvent Extraction*; **1992**, J. Rydberg; C. Musikas and G. R. Chopin Ed.; M. Dekker, Inc., New York, pp 157-207.
4. Danesi, P. R.; Chirizia, R.; Coleman, C. F., in *Critical Reviews in Analytical Chemistry*; **1980**, B. Campbell Ed.; CRC Press, Boca Raton, Florida, p. 1.
5. Benjamin, I., *Acc. Chem. Res.* **1995**, *28*, 233-239 and references cited therein.
6. Benjamin, I., *Annu. Rev. Phys. Chem.* **1997**, *48*, 407-451 and references cited therein.
7. Torrie, G. M.; Valleau, J. P., *J. Chem. Phys.* **1980**, *73*, 5807-5816.
8. Berny, F.; Muzet, N.; Schurhammer, R.; Troxler, L.; Wipff, G., in *Current Challenges in Supramolecular Assemblies, NATO ARW Athens*; **1998**, G. Tsoucaris Ed.; Kluwer Acad. Pub., Dordrecht, pp 221-248.
9. Berny, F.; Muzet, N.; Troxler, L.; Wipff, G., in *Supramolecular Science: where it is and where it is going*; **1999**, R. Ungaro, E. Dalcanale Ed.; Kluwer Acad. Pub., Dordrecht, pp 95-125.
10. Lauterbach, M.; Engler, E.; Muzet, N.; Troxler, L.; Wipff, G., *J. Phys. Chem. B* **1998**, *102*, 225-256; Lauterbach, M.; Wipff, G., in *Physical Supramolecular Chemistry*; **1996**, L. Echegoyen, A. Kaifer Ed.; Kluwer Acad. Pub., Dordrecht pp 65-102.
11. Muzet, N.; Engler, E.; Wipff, G., *J. Phys. Chem. B* **1998**, *102*, 10772-10788.
12. Wipff, G.; Lauterbach, M., *Supramol. Chem.* **1995**, *6*, 187-207. Lauterbach, M.; Wipff, G., in *Physical Supramolecular Chemistry*; **1996**, L. Echegoyen, A. Kaifer Ed.; Kluwer Acad. Pub., Dordrecht pp 65-102.
13. Wipff, G.; Engler, E.; Guilbaud, P.; Lauterbach, M.; Troxler, L.; Varnek, A., *New J. Chem.* **1996**, *20*, 403-417. Guilbaud, P.; Wipff, G., *New J. Chem.* **1996**, *20*, 631-642.
14. Varnek, A.; Sirlin, C.; Wipff, G., in *Crystallography of Supramolecular Compounds*; **1995**, G. Tsoucaris Ed.; Kluwer Acad. Pub., Dordrecht pp 67-100. Varnek, A.; Wipff, G., *J. Comput. Chem.* **1996**, *17*, 1520-1531. A. Varnek, G. Wipff, *Solv. Extract. Ion Exch.*, in press.

15. Horwitz, E. P.; Kalina, D. G.; Diamond, H.; Vandegrift, G. F.; Schultz, W. W., *Solv. Extract. Ion Exch.*; **1985**, pp 75-109.
16. Cecille, L.; Casarci, M.; Pietrelli, L. *New Separation Chemistry Techniques for Radioctive Waste and other Specific Applications.*; **1991**, Commission of the European Communities Ed.; Elsevier Applied Science: London New York.
17. Marcus, Y. *Ion Solvation.*; **1985**, Ed.; Wiley: Chichester.
18. Pearlman, D. A.; Case, D. A.; Caldwell, J. C.; Ross, W. S.; Cheatham III, T. E.; Ferguson, D. M.; Seibel, G. L.; Singh, U. C.; Weiner, P.; Kollman, P. A., *AMBER4.1* **1995**
19. Beudaert, P.; Lamare, V.; Dozol, J.-F.; Troxler, L.; Wipff, G., *Solv. Extract. Ion Exch.* **1998**, *16*, 597-618.
20. Jorgensen, W. L.; Chandrasekhar, J.; Madura, J. D., *J. Chem. Phys.* **1983**, *79*, 926-936.
21. Jorgensen, W. L.; Briggs, J. M.; Contreras, M. L., *J. Phys. Chem.* **1990**, *94*, 1683-1686. The only exception concerns the G system for which the all atom model of ref. 40 was used.
22. Troxler, L.; Harrowfield, J. M.; Wipff, G., *J. Phys. Chem.* **1998**, *102*, 6821-6830.
23. Ungaro, R.; Casnati, A.; Ugozzoli, F.; Pochini, A.; Dozol, J.-F.; Hill, C.; Rouquette, H., *Angew. Chem. Int. Ed.* **1994**, *33*, 1506-1509.
24. Hill, C.; Dozol, J.-F.; Lamare, V.; Rouquette, H.; Eymard, S.; Tournois, B.; Vicens, J.; Asfari, Z.; Bressot, C.; Ungaro, R.; Casnati, A., *J. Inclus. Phenom. Mol. Recog. Chem.* **1995**, *19*, 399-408.
25. Casnati, A.; Pochini, A.; Ungaro, R.; Bocchi, C.; Ugozzoli, F.; Egberink, R. J. M.; Struijk, H.; Lugtenberg, R.; de Jong, F.; Reinhoudt, D. N., *Chem. Eur. J.* **1996**, *2*, 436-445.
26. Casnati, A.; Pochini, A.; Ungaro, R.; Ugozzoli, F.; Arnaud, F.; Fanni, S.; Schwing-Weil, M.-J.; Egberink, R. J. M.; Reinhoudt, D. N., *J. Am. Chem. Soc.* **1995**, *117*, 2767-2777.
27. Watarai, H., *Trends in Analytical Chemistry* **1993**, *12*, 313-318.
28. Benjamin, I., *Science* **1993**, *261*, 1558-1560.
29. Troxler, L.; Wipff, G., *Analytical Sciences* **1998**, *14*, 43-56.
30. Varnek, A.; Troxler, L.; Wipff, G., *Chem. Eur. J.* **1997**, *3*, 552-560.
31. Ishikawa, Y.; Kunitake, T.; Matsuda, T.; Otsuka, T.; Shinkai, S., *J. Chem. Soc. Chem. Commun.* **1989**, 736-737.
32. Dei, L.; Casnati, A.; Nostro, P. L.; Baglioni, P., *Langmuir* **1995**, *11*, 1268-1272.
33. Nostro, P. L.; Casnati, A.; Bossoletti, L.; Dei, L.; Baglioni, P., *Colloids and Surfaces A* **1996**, *116*, 203-209.
34. Pierotti, R. A., *Chem. Rev.* **1976**, *76*, 717-726.
35. Pohorille, A.; Pratt, L. R., *J. Am. Chem. Soc.* **1990**, *112*, 5066-5074.
36. Adamson, A. W. *Physical Chemistry of Surfaces. 5th ed.*; **1990**, Wiley: New York.
37. Schurhammer, R.; Wipff, G., *New J. Chem.* **1999**, 381-391.
38. Allen, M. P.; Tildesley, D. J. *Computer Simulation of Liquids.*; **1987**, W. F. van Gunsteren, P. K. Weiner Ed.; Clarendon press: Oxford. Smith, P. E.; van Gunsteren, W. F., in *Computer Simulations of Biomolecular Systems*; **1993**, W. F. van Gunsteren, P. K. Weiner, A. J. Wilkinson Ed.; ESCOM, Leiden pp 182-212.
39. Rao, P. R. V.; Kolarik, Z., *Solv. Extr. Ion Exch.* **1996**, *14*, 955-993. Osseo-Asare, K., *Colloid Interface Sci.* **1991**, *37*, 123-173. Edwards, H. G. M.; Hughes, M. A.; Smith, D. N., *Vibrationnal Spectrocopy* **1996**, *10*, 281-289.
40. Chang, T.-M.; Peterson, K. A.; Dang, L. X., *J. Chem. Phys.* **1995**, *103*, 7502-7513.

Chapter 7

Benzyl Phenol Derivatives: Extraction Properties of Calixarene Fragments

Lætitia H. Delmau[1], Jeffrey C. Bryan[1], Benjamin P. Hay[2], Nancy L. Engle[1], Richard A. Sachleben[1], and Bruce A. Moyer[1]

[1]Chemical and Analytical Sciences Division, Oak Ridge National Laboratory, Oak Ridge, TN 37831-6119
[2]EMSL, Pacific Northwest National Laboratory, Richland, WA 99352

The remarkable self-assembly of benzyl phenol in binding Cs^+ has been shown in this work to exhibit sensitivity to alkyl substitution effects in liquid-liquid extraction experiments. Substituted benzyl phenols, particularly 4-sec-butyl-2-(α-methylbenzyl)phenol (BAMBP), have been known for over 30 years for their high cesium extraction efficiency and selectivity. The selectivity demonstrated by BAMBP has only been exceeded by calix[4]arenes functionalized with crown ethers. Some structural analogy exists between these two classes of molecules, as substituted benzyl phenols can be considered as calix[4]arene fragments. Additionally, both classes bind Cs^+ through oxygen donor and π-arene interactions. Based on computational results and crystal structures, a series of benzyl phenol derivatives were synthesized to allow insightful comparisons of extraction efficiency and selectivity as influenced by substituents on key sites. Solvent extraction data are presented regarding cesium extraction efficiencies of various benzyl phenol derivatives from alkaline media along with their separation factors for Cs^+ vs. Na^+. Substitution of alkyl groups at para and meta positions of the phenol ring of benzyl phenol and substitution of a second benzyl group at the ortho position all lower cesium extraction strength. Methyl substitution at the methylene carbon at best slightly increases extraction strength. Selectivity for for Cs^+ vs. Na^+ varies from 25 to 3800 with no apparent trend. Since the results do not reflect the expected effects on conformational strain predicted from molecular mechanics, it is concluded that effects of substitution on the self-assembly properties of the free ligands and their complexes are more important than effects on conformational strain.

Introduction

Historically, among the most efficient molecules ever tested to selectively extract Cs^+ from sodium salt media are substituted benzyl phenols (specifically 4-*sec*-butyl-2-(α-methylbenzyl)phenol, BAMBP), which were first investigated about thirty years ago.[1,2] Both extraction efficiency and Cs^+/Na^+ selectivity for these extractants depend strongly on the pH of the solution, with the extraction mechanism occurring through a cation-exchange type process. Among the few molecules known to give better cesium extraction efficiency and selectivity over Na^+ than BAMBP are calix[4]arenes in the 1,3-alternate conformation functionalized with one or two crown ether groups. Calixarene molecules attracted active interest about 20 years ago,[3] but the selectivity for Cs^+ over Na^+ and other alkali metal ions was discovered less than ten years ago.[4,5] Continuing efforts since then have focussed on the origins and improvement of the observed selectivity. The structural similarity between calixarenes and benzyl phenol derivatives is striking, since benzyl phenol can be considered as half of a calix[4]arene. One might expect that reducing the number of phenol groups on calix[4]arene from four to two might enhance the comparison to 2-benzyl phenol. Indeed, recent calculations performed on di-dehydroxycalix[4]arene-crown ether extractants suggest that they would be even stronger Cs^+ extractants than the calix[4]arene crown molecules already proven to be excellent Cs^+ separation agents.[6] As can be seen in Figure 1, the di-dehydroxycalix[4]arene moiety essentially consists of two 2-benzylphenol molecules linked by two methylene groups. In addition, it has been shown that the arene interactions with Cs^+ play a major role in complexation of this cation by the calixarene crowns,[7] even if most emphasis is usually given to the role of the crown.[8] As part of an ongoing effort to adjust on the calixarene framework so that cesium selectivity is increased over not only Na^+ but also all alkali metal ions, we have studied various substituent effects on the extraction properties of benzyl phenol. Study of these compounds was suggested by the results obtained by molecular modeling, particularly using the strain energy associated with the torsion angles involving the methylene bridge, and also by crystal structures of the Cs^+ and K^+ complexes with the unsubstituted benzyl phenol.[9] It was reasonable to consider that the effects found with some functional groups would give useful insight into corresponding substitution on the calixarenes.

The present study involved the synthesis of a series of substituted benzyl phenols in order to compare the effects of one type of substitution at a time. Four major substitution sites were considered: the para and meta positions on the phenol, the methylene bridge (ortho to the phenol), and the ortho positions of the benzyl group. For comparison, we also studied some substituted phenols that did not have an ortho benzyl group, to understand the effect of this arene ring. In addition, dibenzyl phenol derivatives were also examined.

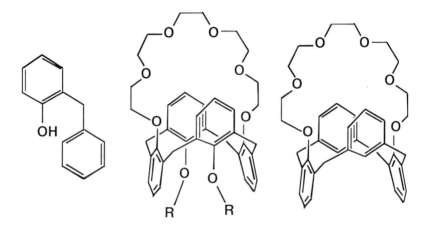

Figure 1. Structures of 2-benzyl phenol, calix[4]arene-crown-6, and di-dehydroxycalix[4]arene-crown-6.

Experimental Section

Materials

Benzyl phenol (**BPh**), 4-*sec*-butyl phenol (**Ph 1**), 4-*tert*-octyl phenol (**Ph 2**), 3,4,5-trimethyl phenol, and other solvents for the syntheses were reagent grade, provided by Aldrich, and used as received. Benzyl-α-methyl-*sec*-butyl phenol (BAMBP, **BPhM 1**) was double distilled and its purity was checked by NMR. The other phenols used in this study were synthesized according to the procedures described below. Line drawings of all compounds are shown in Table 1. Organic phases were prepared by dissolving a precisely weighed amount of ligand into distilled chloroform (99.9% GC grade, EM). Aqueous phases were prepared by weighing precise quantities of sodium or cesium hydroxide and adjusting the volume with deionized water. Initial concentrations of cesium or sodium hydroxide were measured by titration with hydrochloric acid. The radioisotopes ^{137}Cs or ^{22}Na were obtained in the chloride form from Isotope Products Laboratories (Burbank, California).

Methods

Experiments were performed by contacting 1.5 mL of organic and aqueous phases in stoppered polypropylene tubes. The aqueous phases were spiked with radioactive solutions before contact. Extraction was carried out by rotating at 50-60 rpm the tubes (Glass-Col® laboratory rotator) for 1 hour at 25±0.2 °C and centrifuging (Sanyo MSE Mistral 2000R) for 3 minutes at 4000 rpm at 25 °C. Aliquots (1 mL) of each phase were removed and counted using a NaI(Tl) crystal detector (Packard Cobra model S003). (Some experiments involved a smaller volume of both organic and aqueous phases, namely 0.9 mL. Only 0.65 mL of each phase was counted.) The analysis for potassium and rubidium extraction was performed by ion chromatography (DIONEX DX-500 with Ionpac CS12A column).

Upon extraction, a loss of 5% of **BPh** was determined in the aqueous phase by proton NMR. No loss was detected for the other benzyl phenol derivatives. Consequently, distribution of the benzyl phenols to the aqueous phase was considered to be a negligible factor in the observed differences in their extraction properties.

Syntheses

General approach

Synthesized compounds (Table I) were obtained via two main reaction pathways as illustrated in Figure 2. One reaction (equation 1) leads to compounds similar to BAMBP (**BPhM 1**), with a methyl group on the methylene bridge between the phenol and the phenyl groups. The other (equation 2) gives derivatives of benzyl phenol with an unsubstituted methylene bridge. In both cases, the syntheses produced the desired compound along with a by-product of interest containing benzyl groups on the two ortho positions of the phenol. A general description of the synthetic procedures is described below, followed by specific product characterizations.

Synthesis of benzyl α methyl phenol derivatives

The substituted phenol (1 equiv.) and 1-phenyl ethanol (1 equiv.) were mixed in dichloromethane. BF_3/acetate (1 equiv.) was added dropwise. The solution was stirred for three hours at room temperature. Both mono and bis derivatives were formed in this reaction (2-benzyl α-methyl and 2,6-dibenzyl α-methyl). Though the consumption of the substituted phenol is not complete, no more phenyl ethanol should be added, since its addition would drive the reaction toward the formation of the bis derivative. The solution was washed with saturated $NaHCO_3$, dried over anhydrous Na_2SO_4, and the solvent evaporated to reduce the solution volume to 5-10 mL. Column chromatography (silica, gradient elution with hexane/ethyl ether up to 15% ether) allowed separation of the two derivatives, with the dibenzyl derivative eluting first.

Synthesis of benzyl phenol derivatives

Benzyl chloride (1 equiv.) and the substituted phenol (1 equiv.) were mixed in dichloromethane. $SnCl_4$ (1 equiv.) was added dropwise. The solution was stirred for three hours at room temperature. As above, both mono and bis derivatives were formed in this reaction (2-benzyl and 2,6-dibenzyl), and excess benzyl chloride should not be added, since its addition would drive the reaction further toward the formation of the bis derivative. Isolation and purification of the products were achieved as described above. All benzyl phenols were prepared by this procedure except *tert*-octyl benzyl phenol. An alternate procedure, described below, was utilized for this material to obtain a better yield.

Table I: Structures of the Compounds Used in this Study

	Ph 1	Ph 2	
BPh	BPh 1	BPh 2	BPh 3
	BPhM 1	BPhM 2	BPhM 3
			DBPh 3
		DBPhM 2	DBPhM 3

Equation 1

Equation 2

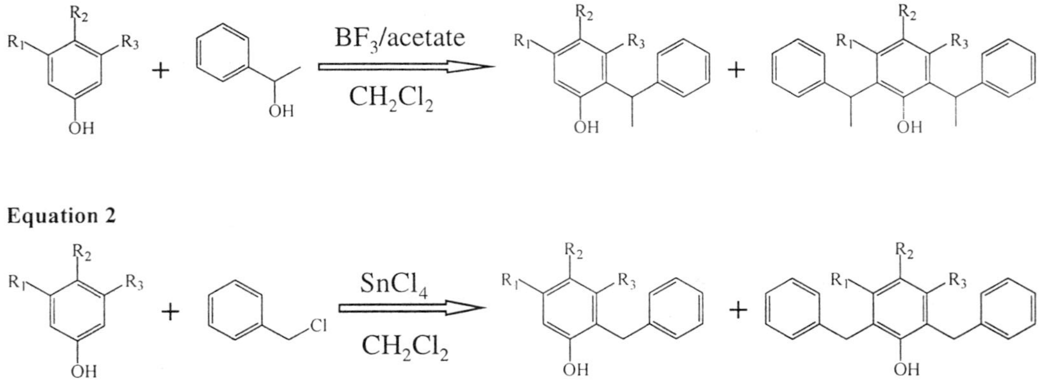

Figure 2. Synthesis schemes for the 2-benzyl phenol derivatives and 2-benzyl-α-methyl phenol derivatives.

Synthesis of tert-octyl Benzyl Phenol (BPh 2)

2-Hydroxy-diphenyl methane (1 equiv.) was mixed with 2,4,4-trimethyl-1-pentene 95% (1.10 equiv.) at room temperature. Trifluoroacetic acid was added until everything dissolved (about 6 equiv.). The solution was allowed to stand for two hours with occasional swirling. Trifluoroacetic acid was evaporated, and toluene was used to co-evaporate the remaining initial solvent. The crude product was obtained as a pale brownish yellow liquid. Chromatography using gradient elution with hexane/ethylacetate (max. ethyl acetate 20%) afforded the pure compound.

Product Yield and Characterization

2-Benzyl –sec butyl phenol (BPh 1). Yield = 10 %. ^{1}H (400.13 MHz; CDCl$_3$): δ 0.82 (3H, t, 8Hz, -CH$_2$-C$\underline{H}_3$), 1.21 (3H, d, 7Hz, -CH-C$\underline{H}_3$), 1.55 (2H, d-q J = 8Hz, CH$_3$-C$\underline{H}_2$-CH-), 2.52 (1H, t-q, J = 7Hz, CH$_2$-C$\underline{H}$-CH$_3$), 3.99 (2H, s, ArOH-C$\underline{H}_2$-Ar)), 4.53 (1H, s, ArOH), 6.96 (2H, m, ArH), 7.08-7.34 (6H, m, ArH).

2-Benzyl –tert octyl phenol (BPh 2). Yield = 91 %. ^{1}H (400.13 MHz; CDCl$_3$): δ 0.70 (9H, s, C(CH$_3$)$_3$), 1.32 (6H, s, C(CH$_3$)$_2$), 1.66 (2H, s, C-CH$_2$-C), 4.00 (2H, s, ArOH-C$\underline{H}_2$-Ar), 4.55 (1H, s, ArOH), 6.69 (1H, d, J = 8Hz, ArOH$\underline{H}^6$), 7.06-7.32 (8H, m, ArH).

2-Benzyl -3,4,5 trimethyl phenol (BPh 3). After chromatography, a dark yellow oily liquid that subsequently solidifies was obtained. The compound was recrystallized from pure hexane. Yield = 22 %. ^{1}H (400.13 MHz; CDCl$_3$): δ 2.13 (3H, s, ArOH(C$\underline{H}_3$), 2.17 (3H, s, ArOH(C$\underline{H}_3$), 2.25 (3H, s, ArOH(C$\underline{H}_3$)), 4.07 (2H, s, ArOH-(C$\underline{H}_2$)-Ar), 4.46 (1H, s, ArOH), 5.31 (1H, s, ArOHH^6), 6.54 (1H, s, ArH^4), 7.16 (4H, m, ArH).

2-Benzyl-α-methyl-tert-octyl phenol (BPhM 2). Clear oily liquid. Yield = 35 %. ^{1}H (400.13 MHz; CDCl$_3$): δ 0.71 (9H, s, C(CH$_3$)$_3$), 1.35 (6H, s, C(CH$_3$)$_2$), 1.63 (3H, d, J = 7Hz, Ar-CH(C$\underline{H}_3$)-Ar), 1.69 (2H, s, C-CH$_2$-C), 4.36 (1H, q, J = 7Hz, Ar-C$\underline{H}$(CH$_3$)-Ar), 4.66 (1H, s, ArOH), 6.64 (1H, d, J = 8Hz, ArOH$\underline{H}^6$), 7.08 (1H, d-d, J$_1$ = 8Hz, J$_2$ = 2Hz, ArOH$\underline{H}^5$), 7.17-7.30 (6H, m, ArH, ArOH$\underline{H}^3$).

2-Benzyl-α-methyl-3,4,5 trimethyl phenol (BPhM 3). Yield = 19 %. ^{1}H (400.13 MHz; CDCl$_3$): δ 1.80 (3H, d, J = 7Hz, Ar-CH(C$\underline{H}_3$)-Ar), 2.27 (3H, s, ArOH(C$\underline{H}_3$), 2.34 (3H, s, ArOH(C$\underline{H}_3$), 2.37 (3H, s, ArOH(C$\underline{H}_3$), 4.83 (1H, q, J = 7Hz, Ar-C$\underline{H}$(CH$_3$)-Ar), 5.32 (1H, s, ArOH), 6.57 (1H, s, ArOH$\underline{H}^6$), 7.33 (1H, m, ArH^4), 7.43 (4H, m, ArH).

2,6 dibenzyl-α-methyl-3,4,5 trimethyl phenol (DBPh 3). After chromatography, a dark yellow oily liquid that subsequently solidifies was obtained. The compound was recrystallized from pure hexane. Yield = 33 %. ^{1}H (400.13 MHz; CDCl$_3$):): δ 2.22 (3H, s, ArOH(C$\underline{H}_3$)^4), 2.24 (6H, s, ArOH(CH$_3$)^3, ArOH(C$\underline{H}_3$)^5), 4.11 (4H, s, ArOH-((C$\underline{H}_2$)-Ar)$_2$), 4.50 (1H, s, ArOH), 7.16 (4H, m, ArH), 7.19 (2H, d, J = 8Hz, ArH), 7.27 (4H, m, ArH).

2,6 dibenzyl-α-methyl-*tert* octyl phenol (DBPhM 2). Yield = 5 %. ^{1}H (400.13 MHz; CDCl$_3$): δ 0.69 (9H, 2 singlets, C(CH$_3$)$_3$), 1.35 (6H, 2 singlets, C(CH$_3$)$_2$), 1.59 (6H, m, Ar-CH(C$\underline{H}_3$)-Ar), 1.55-1.64 (2H, s, C-CH$_2$-C), 4.36 (3H, m, Ar-C$\underline{H}$(CH$_3$)-Ar + ArOH), 7.09 (1H, s, ArH), 7.19 (7H, m, ArH), 7.27 (4H, d-d, J = 7Hz, ArH).

2,6 dibenzyl-α-methyl-3,4,5 trimethyl phenol (DBPhM 3). Yield = 7 %. ^{1}H (400.13 MHz; CDCl$_3$): δ 1.77 (6H, 2 doublets, 7Hz, Ar-CH(C$\underline{H}_3$)-Ar), 2.33 (9H, 3 singlets, ArOH(C$\underline{H}_3$)$_3$), 4.87 (2H, 2 quartets, 7Hz, Ar-C$\underline{H}$(CH$_3$)-Ar), 7.23-7.43 (10H, m, ArH).

Results and Discussion

This section is divided into three main parts: crystallographic results obtained on Cs$^+$ and K$^+$ complexes with benzyl phenol, molecular modeling calculations on the complexation of the different alkali metals by benzyl phenol molecules, and finally solvent extraction experiments presenting a comparison of the differently substituted molecules.

Crystal Structures

X-ray crystal structures of 2-benzylphenol (**BPh**), and its complexes with either K$^+$ (complex [K(2-benzylphenol)$_3$][2-benzylphenolate]) or with Cs$^+$ (complex [Cs(2-benzylphenol)$_2$][2-benzylphenolate]) were reported separately.[9] The latter two structures are depicted in Figure 3. Both structures demonstrate that 2-benzylphenol binds the metal ion through both its oxygen atom and its benzo group. This phenomenon can be considered as a "wrap around" effect, since the benzyl group tends to have its ring oriented so that the arene interaction with the cation is optimal. In the next section, this effect will be related directly to the torsion angle between the two benzene rings. The phenolate oxygen atom does not coordinate directly to either K$^+$ or Cs$^+$; instead, it is stabilized by hydrogen bonds to the phenolic hydroxyl groups bound to the cation. This aggregation is consistent with extraction studies performed with 2-benzylphenol and its derivatives.[1] While these structures exhibit a different

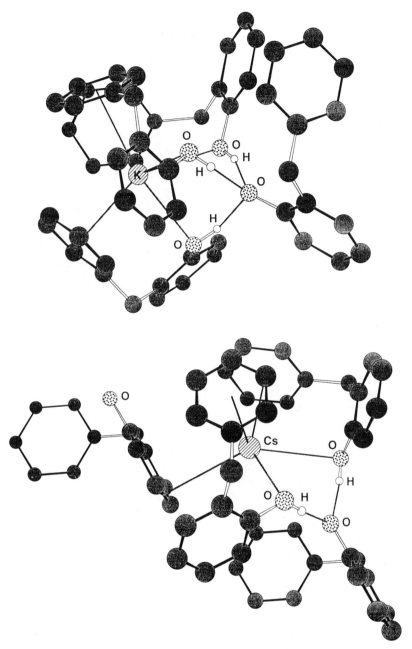

Figure 3. Crystal structures of complexes ([K(2-benzylphenol)$_3$][2-benzylphenolate]) and ([Cs(2-benzylphenol)$_2$][2-benzylphenolate]).

ligand environment, it was suggested that the potassium structure may be more representative of the supramolecular cesium-BAMBP complex in solution.[9]

Molecular Modeling

Molecular modeling studies were performed in tandem with the X-ray crystal structure investigations. The behavior of the **BPh** molecule was modeled due to the simplicity of the molecule along with the fact that X-ray crystal data were available. In addition, the modeling for this compound provided a good initial basis for understanding the behavior of more complicated derivatives. Preliminary molecular modeling studies focussed on the behavior of a *single* molecule of **BPh** interacting with each alkali metal ion, since intraligand steric effects can influence complex stability.[10] This was done to understand donor group orientation, which, in five- and six-member chelate rings of amines[11] and ethers[12,13] leads to inherent metal size-preferences that are structurally based. It has been shown that the strain that develops in a ligand on metal ion complexation provides a useful measure of the degree of donor site organization for a specific metal ion.[12,14] In a previous paper,[9] we performed such an analysis on isolated **BPh** chelates with the alkali cations. This analysis revealed that the ligand had to undergo significant structural reorganization to reach the binding conformation. This is illustrated in Figure 4.

The structural changes to **BPh** involve rotation about both $C(sp^2)$-$C(sp^3)$ bonds. However, the change in the 2-3-4-5 dihedral angle is much larger than the change in the 1-2-3-4 dihedral angle. The amount of structural reorganization varied with cation size where the least distortion occurred with Cs^+ and the largest occurred with Li^+. One strategy for increasing the binding affinity of **BPh** for Cs^+ is to attempt to preorganize the ligand through the addition of substituents. One way that this could be done is to add alkyl substituents in various locations on the molecule.

Using the MM3 model,[15] we examined the influence of adding methyl substituents to two locations of **BPh**. These were on the methylene linkage or on the phenol ring adjacent to the CH_2 group (the "3 position", counting the phenol as 1 and benzyl as 2). Rotational potential energy surfaces (PES's) for rotation about the 2-3-4-5 bond in these two derivatives are compared with that of **BPh** in Figure 5. The PES's are symmetric about 180° for **BPh** and for the arene-substituted case, but asymmetric about 180° for the methylene-substituted case. In the latter case, the 0 to 180° region is when the hydroxy group is on the opposite side of the methyl group and the 180 to 360° region is when the hydroxy group is on the same side as the methyl group. In **BPh** complexes, the 2-3-4-5 dihedral angle ranges from 100° for Cs^+ to 126° for Li^+.[9] Examination of Figure 5 reveals that both substitution patterns lower the ligand strain in this region. Thus, on the basis of this analysis, either substitution is predicted to result in stronger binding affinities for the alkali cations. For example, alkylation of the methylene group shifts the minimum to ~50°, and at 100°, the angle found in the **BPh**-Cs^+ complex, the ligand strain is lowered by 0.8

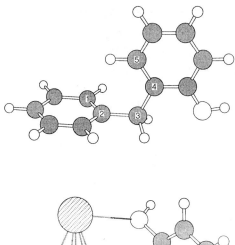
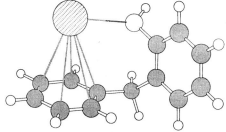

Figure 4. Calculated structures of the free ligand 2-benzyl phenol and the cesium-benzyl phenol complex (a full discussion of the structural details is provided in reference 7).

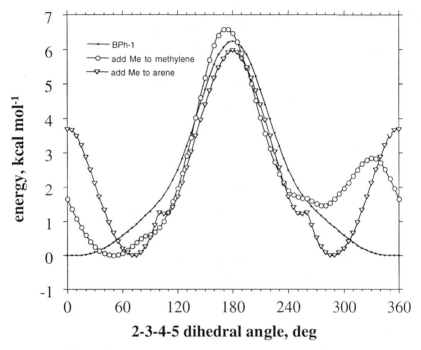

Figure 5. Rotational potential energy surfaces (PES's) for rotation about the 2-3-4-5 bond for 2-benzyl phenol, 2-benzyl 3-methyl phenol, and 2-benzyl-α-methyl phenol.

kcal mol^{-1}. Similarly, alkylation of the arene shifts the minimum to ~75°, and lowers the ligand strain by 0.4 kcal mol^{-1} at 100°. Based on these observations, the corresponding molecules (substitution on the methylene group and on the arene ring) were synthesized and tested by solvent extraction.

Solvent Extraction

Solvent extraction experiments were performed first using simply benzyl phenol (**BPh**) and BAMBP (**BPhM 1**) in order to determine the best conditions for comparison of extraction and selectivity properties. To maintain high ionic strength for the purpose of avoiding activity problems, a salt concentration of 0.1 M was chosen. To check the influence of the solution pH, Cs^+ was extracted from different mixtures of cesium chloride and cesium hydroxide, and the pH measured at equilibrium. Figure 6 shows, as expected, that at low pH, extraction of Cs^+ is barely detectable. On the contrary, at higher pH, much better results are obtained, and in spite of the steep increase, duplicate values are in excellent agreement. For the subsequent experiments, the total concentration of hydroxide was chosen to be 0.1 M. As shown in Figure 7, a ligand concentration of 0.75 M proved adequate to obtain experimentally convenient cesium distribution ratios. A concern for this system is to avoid saturation (i.e. > 25 % loading) of the benzyl phenol derivatives, since that tends to lead to higher levels of aggregation in the organic phase and loss of selectivity. However, under all conditions reported here, loading (defined as [Cs]$_{org}$/[ligand]$_{total}$ x 100%) was less than 8%. Also shown in Figure 7 (■ and ▲) is the increase in D_{Cs} values when other alkali metal cations are present in the aqueous phase while the total concentration of hydroxide remains constant. One experiment involves equal concentrations of cation in their hydroxide form; the other involves variable ratios (a factor of 10 both between Cs^+ and Rb^+ and between Rb^+ and K^+). Higher D_{Cs} values were expected in the experiments involving other cations due to the lower level of competition, and concomitant lower level of loading. The separation factors are approximately 12 for Cs^+ over Rb^+ and 110 for Cs^+ over K^+. These values increase (to 25 and 300 respectively) when all four cations are present in the same solution. It may be noted that the observed selectivity is comparable to that obtained with the calix[4]arene-crown-6 compounds. In all subsequent experiments described below, extractions were performed non-competitively with either cesium hydroxide or sodium hydroxide.

The first step of our study of substituent effects was aimed at comparing the effect due to an ortho benzyl group. We report in Table II the cesium distribution ratios and the cesium/sodium separation factors exhibited by 4-*sec*-butyl phenol (**Ph 1**), 4-*tert*-octyl phenol (**Ph 2**), and their respective ortho benzyl derivatives (**BPh 1** and **BPh 2**).

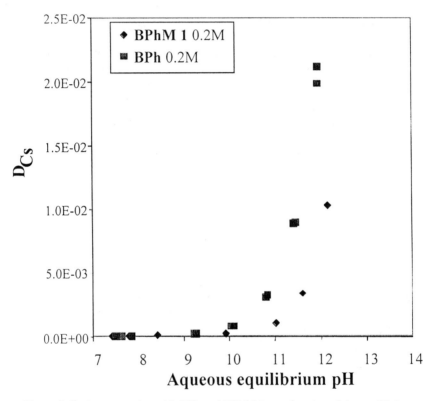

*Figure 6. Cesium extraction with **BPh** and **BPhM 1** as a function of the equilibrium pH. Organic phase: ligand 0.2 M in chloroform. Aqueous phase: mixture of cesium hydroxide and cesium chloride, total cesium concentration held constant at 0.1 M. O/A = 1. T = 25 °C.*

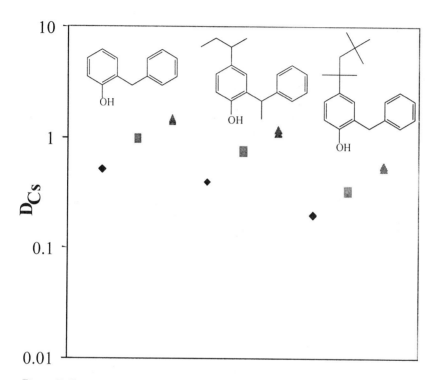

Figure 7. Cesium extraction with **BPh**, **BPhM 1** and **BPh 2**. Organic phase: ligand 0.75 M in chloroform. Aqueous phase: cesium hydroxide 0.1M or mixture of alkali cation hydroxide (♦:[CsOH] = 0.1 M, ■ : [NaOH] = [KOH] = [RbOH] = [CsOH] = 0.025 M, ▲:[NaOH] = 7.8×10^{-2} M, [KOH] = 2×10^{-2} M, [RbOH] = 2×10^{-3} M, [CsOH] = 2×10^{-4} M). O/A = 1. T = 25 °C.

Table II. Comparison Between Phenols and Ortho-Benzyl Phenol Derivatives.

Compound	Ph 1	BPh 1	Ph 2	BPh 2
D_{Cs}	0.24	0.42	0.64	0.22
D_{Na}	3.07×10^{-3}	1.12×10^{-4}	2.55×10^{-3}	8.15×10^{-4}
$SF_{Cs/Na}$	80	3800	25	270

Organic phases: ligand 0.75 M in chloroform.
Aqueous phases: cesium or sodium hydroxide 0.1 M.

The comparison shown in Table II clearly demonstrates the remarkable enhancement of selectivity upon benzyl substitution at the ortho position. All the distribution ratios are of the same order of magnitude, and no clear effect on D_{Cs} is obtained. However, a great increase in selectivity is achieved with the addition of the benzyl group. The "wrap around" effect featuring arene interactions seen in the crystal structures and studied by molecular modeling may be playing a major role in this case.

Table III. Comparison of the Substituents in the Para Position.

Compound	BPh	BPh 1	BPh2	BPh 3*
D_{Cs}	0.55	0.42	0.22	0.18
D_{Na}	2.21×10^{-4}	1.12×10^{-4}	8.15×10^{-4}	8×10^{-5}
$SF_{Cs/Na}$	2500	3800	270	2250

Organic phases: ligand 0.75 M in chloroform
Aqueous phases: cesium or sodium hydroxide 0.1 M
* This compound has also two methyl groups meta to the phenolic -OH. The influence of these groups is discussed below.

The influence of the substituent in the para position indicates that a bulky group lowers the extraction of Cs^+ among the benzyl phenols. Table III summarizes the results obtained with substituents –H, -CH_3, sec-butyl, and tert-octyl. Undoubtedly, the most efficient cesium extractant is the one that has the smallest group para to the –OH group, since a gradual decrease in the distribution ratios for Cs^+ is observed for increasing size of the para substituent. **BPh 3** bears only a methyl group, but also two other methyl groups meta to the phenol. This decrease in D_{Cs} across the series in Table 3 may be understood by noting that the arrangement of the ligands around Cs^+ (and K^+) in the crystal structures for **BPh** may be disrupted by substitution in the para and meta positions. The selectivities for Cs^+ over Na^+ for the ligands in Table III do not follow the same trend. Selectivity is apparently optimized by the presence of a sec-butyl group in the para position (**BPh 1**), yet it is at its worst when the sec-butyl group is replaced by a tert-octyl group (**BPh 2**). There does not seem to be an intuitively obvious explanation for these results, as we lack an understanding of the structural nature of the complexes formed between the phenols and Na^+. Apparently, alkyl substitution on the phenol may either help (for **BPh 2**) or hurt sodium extraction, possibly indicating that sodium extraction does not occur through a single mechanism.

Table IV. Influence of the Substitution on the Methylene Bridge

Compound	BPh 1	BPhM 1	BPh 2	BPhM 2
D_{Cs}	0.42	0.42	0.22	0.34
D_{Na}	1.12×10^{-4}	5.25×10^{-4}	8.15×10^{-4}	2.12×10^{-4}
$SF_{Cs/Na}$	3800	800	270	1600

Organic phases: ligand 0.75 M in chloroform
Aqueous phases: cesium or sodium hydroxide 0.1 M

The effects of substitution on the methylene bridge are shown in Table IV. The most effective extractant is **BPh 1**, both in terms of extraction strength (highest D_{Cs}) and selectivity (highest $SF_{Cs/Na}$). While addition of a methyl group to the methylene bridge of **BPh 1** (to give **BPhM 1**) does not seem to influence the value of the cesium distribution ratio, its presence clearly allows better extraction of sodium as reflected by the lowering of $SF_{Cs/Na}$ relative to **BPh 1**. On the other hand, addition of a methyl group to the methylene bridge of **BPh 2** (to give **BPhM 2**) results in a slight increase of cesium extraction and a dramatic *increase* in selectivity. Again, the sodium behavior may be reflecting more complicated equilibria.

As a final example, addition of a methyl group to **BPh 3** (to give **BPhM 3**, see Table V) dramatically lowers both D_{Cs} and $SF_{Cs/Na}$. Apparently the effect of methyl substitution on the methylene bridge is strongly dependent on the substitution pattern of the phenol. This particular point may need some extra attention in our upcoming studies.

Table V. Comparison of the Substituents in the Para Position

Compound	BPh	BPh 3	BPhM 3
D_{Cs}	0.55	0.18	0.018
D_{Na}	2.21×10^{-4}	8×10^{-5}	2.85×10^{-4}
$SF_{Cs/Na}$	2500	2250	63

Organic phases: ligand 0.75 M in chloroform
Aqueous phases: cesium or sodium hydroxide 0.1 M

Addition of three methyl groups to the phenol has a dramatic influence on the extraction of Cs^+. Alkyl substitution in the position meta to the –OH group, adjacent to the methylene bridge, was also suggested as being potentially beneficial by molecular mechanics. Unfortunately, substitution solely in this location is synthetically very difficult. Results presented in Table V compare **BPh** (totally unsubstituted) with **BPh 3** (methyl substituted in both meta positions and in the para position). As mentioned above, the effect may be explained as steric hindrance preventing an optimum "wrap" of the arene rings around the cesium cation. Obviously, this phenomenon has a similar impact on sodium extraction since the separation factor between Cs^+ and Na^+ is relatively unchanged between **BPh** and **BPh 3**. The effect due to the addition of the methyl on the methylene bridge in comparison of **BPh 3** with **BPhM 3** has a dramatic effect, as mentioned earlier, on both cesium extraction and cesium/sodium selectivity. The deleterious effect on cesium extraction can be related to the fact that the two methyl groups (meta to the phenol and on the methylene bridge) encounter steric interactions that prevent good complexation. The actual beneficial effect on D_{Na}, however, may imply a structure that does not develop this steric interaction, namely one that may not entail "wrap around".

An attempted double "wrap around" effect produced poor cesium extractants. We have investigated and thoroughly compared the effects of the different groups on benzyl phenol derivatives, described in the introduction as halves of calix[4]arenes. Additionally, we have investigated the extraction properties of dibenzyl phenol derivatives, namely where a second benzyl group is present at the other ortho position. These molecules can be considered as three fourths of a calix[4]arene. The results presented in Table VI show that these ligands are very poor cesium extractants, as indicated by the low D_{Cs} values. The addition of a second benzyl group apparently makes the complexing environment for Cs^+ very unfavorable. The cesium/sodium separation factors show that there is almost no selectivity exhibited by these compounds (SF between 1 and 10).

Table VI. Comparison of Dibenzyl phenol Derivatives

Compound	DBPhM 2	DBPh 3	DBPhM 3
D_{Cs}	1.2×10^{-3}	1.5×10^{-4}	3.4×10^{-3}
D_{Na}	2.4×10^{-4}	2.1×10^{-4}	1.1×10^{-3}
$SF_{Cs/Na}$	5	7	3

Organic phases: ligand 0.75 M in chloroform
Aqueous phases: cesium hydroxide 0.1 M

The disruption of the ligand arrangement about Cs^+ is predictable from the crystal structure and is undoubtedly the cause of the poor extraction.

Conclusion

This study confirms the high selectivity of benzyl phenols reported in the earlier literature and demonstrates the importance of certain alkyl substituent effects. Based on molecular modeling studies, alkylation at key sites on benzyl phenol was predicted to result in stronger binding affinities for alkali metal cations. The two sites identified include the methylene bridge between the phenol and the benzo unit, and the meta position to the –OH group (ortho to the benzyl). However, the predicted substitution effects were not observed in the extraction experiments. The cesium extraction results on various benzyl phenol derivatives exhibited a general trend depending on the position and the nature of the substituents. A bulky substituent in the para position to the phenol decreases the extraction efficiency, while a methyl group on the methylene bridge has at best a slightly beneficial influence. A methyl substitution at the meta position decreases D_{Cs}, even more so if there is a methyl on the methylene bridge. Since methyl substitutions did not yield the predicted increase in cesium extraction strength, it is concluded that self-assembly of the free ligands and their complexes is likely a stronger influence on extraction strength than conformational strain. This is supported by the effect of bulky substituents in the para position of the phenol. The interpretation of the selectivity for Cs^+ over Na^+ is much more challenging. In spite of multiple experiments, no real trend was found for this parameter, suggesting that the sodium complexes adopt structures of a different nature than the cesium complexes. If all these observations are extrapolated to calixarenes, it is reasonable to think that a bulky group para to the phenolic oxygen atoms or a substitution on the methylene bridges would not greatly improve the cesium extraction efficiency. It seems overall that the fewer groups attached to the benzyl or to the phenol unit, the greater is the cesium extraction. A prediction regarding selectivity does not seem possible.

Acknowledgments

This research was sponsored by the Division of Chemical Sciences, Office of Basic Energy Sciences, U. S. Department of Energy, under contract number DE-AC05-96OR22464 with Oak Ridge National Laboratory, managed by Lockheed Martin Energy Research Corp. This research was also sponsored in part by the Environmental Management Science Program under direction of the U.S. Department of Energy's Office of Basic Sciences, Office of Energy Research and the Office of Science and Technology, Office of Environmental Management under contract number DE-AC06-76RLO 1830 with Pacific Northwest National Laboratory, managed by Battelle Memorial Institute. In addition, this research was supported in part by an appointment (LHD) to the Oak Ridge National Laboratory Postdoctoral Research Associates Program administered jointly by the Oak Ridge Institute for Science and Education and Oak Ridge National Laboratory.

Literature Cited

‡ Benzyl phenol is the non-systematic terminology commonly used in solvent extraction literature to refer to 2-hydroxydiphenyl methane and derivatives that contain this moiety. This term is used throughout this manuscript as a matter of convenience.

1. Egan, B.Z.; Zingaro, R.A.; Benjamin, B.M. *Inorg. Chem.* **1965**, *4*, 1055
2. Zingaro, R.A.; Coleman, C.F. *J. Inorg. Nucl. Chem.* **1967**, *29*, 1287
3. Gutsche, C.D.; Muthukrishnan, R. *J. Org. Chem.* **1978**, *43* 4905
4. Casnati, A.; Pochini, A.; Ungaro, R.; Ugozzoli, F.; Arnaud, F.; Fanni, S.; Schwing, M.J.; Egberink, R.J.M.; de Jong, F.; Reinhoudt, D.N. *J. Am. Chem. Soc.* **1995**, *117*, 2767
5. Dozol, J.F.; Böhmer, V.; McKervey, A.; Lopez Calahorra, F.; Reinhoudt, D.; Schwing, M.J.; Ungaro, R.; Wipff, G.; *Nucl. Sci. Tech.* Report EUR 17615 **1996**
6. Hay, B. P.; Pacific Northwest National Laboratory, Richland, WA, Personal Communication **1998**
7. Lhotak, P.; Shinkai, S. *J. Phys. Org. Chem.* **1997** *10*, 273-285
8. Varnek, A.; Wipff, G.; *THEOCHEM* **1996** *363*, 67
9. Bryan, J. C.; Delmau, L. H.; Hay, B. P.; Moyer, B. A.; Rogers, L. M.; Rogers, R. D. *Struct. Chem.* **1999**, 10, 187
10. Martell, A. E.; Hancock, R. D. "Metal Complexes in Aqueous Solutions", Modern Inorganic Chemistry, Series Editor: J. P. Fackler, Jr.; Plenum Press, New York, 1996.
11. (a) Hancock, R. D. *Acc. Chem. Res.* **1990** *23*, 253. (b) Hancock, R. D.; Wade, P. W.; Ngwenya, M. P.; de Sousa, A. S.; Damu, K.V. *Inorg. Chem.* **1990** *29*, 1968. (c) Hancock, R. D. *J. Chem. Ed.* **1992** *69*, 615.
12. Hay, B. P.; Zhang, D.; Rustad, J. R. *Inorg. Chem.* **1996** *35*, 2650.
13. Hay, B. P.; Rustad, J. R. *Supramol. Chem.* **1996** *6*, 383.
14. Hay, B. P. in *Metal Ion Separation and Preconcentration: Progress and Opportunities*; Bond, A. H.; Dietz, M. L.; Rogers, R. D., Eds.; ACS Symposium Series 716, American Chemical Society: Washington, DC, 1998.
15. MM3(96) may be obtained from Tripos Associates, 1699 S. Hanley Road, St. Louis, MO 63144 for commercial users, and it may be obtained from the Quantum Chemistry Program Exchange, Mr. Richard Counts, QCPE, Indiana University, Bloomington, IN 47405, for non-commercial users.

Chapter 8

Calixarene-Containing Extraction Mixture for the Combined Extraction of Cs, Sr, Pu, and Am from Alkaline Radioactive Wastes

Igor V. Smirnov[1], Andrey Yu. Shadrin[1], Vasily A. Babain[1],
Mikhail V. Logunov[2], Marina K. Chmutova[3], and Vitaly I. Kal'tchenko[4]

[1]Khlopin Radium Institute, St.-Petersburg 194021, Russia
[2]"Mayak" PA, Ozersk, Russia
[3]Vernadsky Institute of Geochemistry and Analytical Chemistry,
Moscow 117975, Russia
[4]Institute of Organic Chemistry, Kiev 252660, Ukraine

The extraction mixture containing tert-butyl-calix[6]arene, 4-alkyl-2-di-(2-hydroxyethyl)aminomethylphenol and solubilizing additive in dodecane, was proposed for processing of alkaline high-level radioactive wastes. This extraction mixture extracts effectively cesium, strontium, rare-earth elements, americium and plutonium from basic solutions containing up to 8 M sodium salts and 0.1-4 M free alkaline. Acidic solutions could be used for the subsequent stripping of these elements from organic phase.
The extraction mixture was tested at "Mayak" PA (Chelyabinsk region). More than 99% of Cs and Sr and about 90% of gross alpha-activity were removed from actual alkaline HLW during the batch test.

In the "cold war" years, great amounts of radioactive wastes were accumulated at the American and Russian facilities for weapon plutonium production. High-level liquid alkaline wastes containing considerable quantities of long-lived radionuclides are most hazardous among them. The separation of these radionuclides from non-

---

* The greater part of the work reviewed in the present paper was financed by the European Commission in the frame-work of the Project INCO-COPERNICUS (Contract IC15-CT98-0208).

volative compounds would considerably reduce the volume of high-level glass to be disposed off. For processing of Hanford high-level waste (HLW) British Nuclear Fuels plc (BNFL) has proposed the multistage process involving strontium precipitation on a carrier, coprecipitation of actinides with iron, cesium removal by inorganic sorbent and technetium sorption on ion-exchange resin [1].

In our opinion, the more simple and cheap variant for processing of these wastes could be based on the combined extraction of all long-lived radionuclides in a single extraction cycle. It is known that cesium is readily extracted from alkaline media by calixarenes [2], while rare-earth elements and actinides are extracted by 4-alkyl-2-di-(2-hydroxyethyl)aminomethylphenol (DEAP) [3].

The object of our work was to check the possibilities for simultaneous recovery of the most long-lived radionuclides from alkaline media by one extractant.

Extraction from simulated solutions

Preliminary tests have shown that the mixtures of tert-butyl-calix[6]arene and DEAP extract Cs, Sr, RE and actinides from alkaline media. Some results are presented in Table I.

Table I. Extraction of Cs, Sr, Tc, Am from alkaline media by solution of 0.01 M tert-butyl-calix[6]arene and 0.01 M DEAP in 20% POR in dodecane

$NaNO_3$, M	NaOH, M	D_{Cs}	D_{Sr}	D_{Am}	D_{Tc}
	0.1	12.9			
0	1	11.3			
	2	6.5			
	0.1	5.8	3.1	0.4	0.5
1.0	1	4.0	3.3	2.4	
	2	3.8	1.7	0.5	2.3
	0.1	2.4			
2.0	1	2.2	3.4		2.3
	2	2.2			
	0.1	0.95			
4.0	1	0.85	2.4	1.3	
	2	0.92			
	0.1	0.30	2.4	0.49	
Saturated (~8)	1	0.39	4.3	0.55	5.6
	2	0.51	3.2	0.35	

It is seen from Table I that the proposed extraction mixture recovers all long-lived radionuclides over a wide range of concentration of nitrate and sodium hydroxide.

It should be noted that the main impediment to the use of tert-buryl-calix[6]arene as extractant is its low solubility in conventional solvents. The mixtures of paraffines and organophosphorus compounds like TBP and POR (different-radical phosphine oxide) turn out to be good solvents. An added advantage of POR as solubilizer of calixarene is that it extracts technetium, one of the most hazardous components of alkaline HLW, from alkaline media.

The good results obtained for simulated solutions have made possible to determine the conditions and to perform test on actual alkaline HLW of the "Mayak" PA.

Static test on actual HLW

The HLW used for the experiment was taken from the storage tank AD-5701/4 and contained about 3 M sodium hydroxide and 3 M sodium nitrate. The chemical and radiochemical composition of feed solution is given more fully in Tables II and III. The feed solution was used without any dilution and preparation. The extractant contained 0.02 M tert-butyl-calix[6]arene and 0.02 M DEAP in a mixture of 40%vol. POR and 60% dodecane.

The experiment involved four-fold successive treatment (contacting) of feed solution with fresh portions of extractant at slightly increasing volume ratio of phases O/A being equal to 3:3, 3:2.8, 3:2.6, 3:2.4, respectively. The first extract was contacted three times with fresh portions of the stripping agent containing 30 g/L of oxalic acid and 10 g/L of acetohydroxamic acid at constant phase ratio O/A=1:1. In all cases the phases were stirred for 15 min. Time of phase separation (settling) was 5-10 min.

Two to six independent analyses were performed for each component in all the samples. The main experimental data are presented in Tables II and III.

The analysis of these data shows that the proposed extraction mixture is highly efficient for recovery of cesium and strontium from alkaline HLW in 4 contacts. The purification coefficients of raffinate are 2-3 orders of magnitude for Cs and 1-2 orders of magnitude for Sr. Uranium and plutonium are extracted rather well (the purification coefficients lie within 14-27). As a whole, the alpha-emitters are removed by one order of magnitude, which is less than expected. The stripping efficiency of cesium, strontium and trivalent alpha-nuclides is rather high. Stripping of plutonium and uranium is not sufficient.

Since cesium-137 constitutes the major portion of radioactivity in feed solution, which may be a hindrance to the determination of other components, a portion of feed sample was treated by 0.12 M solution of chlorinated cobalt dicarbollyde (ChCoDiC) in metanitrobenzotrifluoride at phase ratio O/A=5:1. Under these conditions, ChCoDiC extracts more than 99% of cesium, without any practical extraction of other nuclides. The analysis of the treated and untreated feed solutions was carried out.

Figure 1 presents a schematic diagram of the test and conventional symbols used to represent the various solutions.

The γ-, α-spectral and chemical analyses were performed at the Central Plant Laboratory of "Mayak" PA by using the standard procedures.

Table II. Radiochemical composition of feed and resultant solutions

Component	Content of components in solutions, Bq/L							
	Feed	Raffinate	Strips			Extracts (Solvent)		Recycle solvent
			R_{11}	R_{12}	R_{13}	O_2	O_3	O_{13}
Gross β	3.5E10	1.8E7	3.5E10	1.2E8	4.1E6	3.4E9	2.1E8	1.8E6
Including:								
Cs-137	3.6E10	2.0E7	3.9E10	9.6E7	2.8E6	4.7E9	6.9E8	1.0E5
Cs-134	8.8E7	n.d.	5.7E7	3.5E5	n.d.	n.d.	n.d.	n.d.
Sr-90	1.7E8	≤7E6	5.6E7	≤1E6	≤1E6	-	-	-
U, mg/l	22.6	0.85	< 0.01	<0.02	<0.02	-	-	-
Pu, mg/l	3.26	0.24	1.5	< 0.5	< 0.5	-	-	-
Gross α	3.6E7	3.4E6	3.2E7	1.4E5	3.5E4	-	-	-
Including*:								
Am-241 (Pu-238, Th-228)	4.0E6	2.2E5	3.2E5	1.9E7	6.7E4	3.9E3	-	-
Pu-239 (Pu-240)	4.1E6	1.4E5	2.0E5	3.3E6	6.1E4	4.3E3	-	-
Am-243 (Po-210)	4.0E6	3.0E5	2.2E5	3.5E6	n.d.	6.1E3	-	-

Note: (*) Accompanying emitters contributing to appropriate peak of spectrum are indicated in parentheses. n.d. – not detected.

Table III. Chemical composition of feed and resultant solutions

Component	Content of components in solutions, mg/L				
	Feed	Raffinate	Strip		
			R_{11}	R_{12}	R_{13}
Sodium (total)	140000	100000	500	500	500
Sodium hydroxide	144000	-	-	-	-
Aluminium	6280	6500	200	2.5	5
Iron	< 300	100	50	45	100
Chromium	1890	1100	40	2.5	2.5
Nickel	300	-	25	10	10
Strontium	< 200	-	-	-	-
Molybdenum	100	50	0.5	0.5	0.5
Calcium	< 300	-	-	-	-
Magnesium	140	-	25	50	50
Cesium	50	-	25	-	-
Nitrate-ion	114000	-	-	-	-
Sulfate-ion	3600	-	-	-	-
Chloride-ion	1600	-	-	-	-
Fluoride-ion	200	-	-	-	-

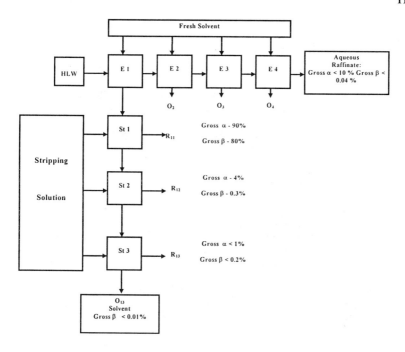

Figure 1. Batch contact flow diagram.

The spectral and mass-spectral analyses testify that the non-radioactive components in feed solution (except magnesium) are extracted very poorly. The strips contain considerable amounts of sodium and aluminium (the first strip), as a consequence of very high content of these elements in feed solution. A quantity of iron is likely to pass into the solvent and then is stripped. The cation content in the first strip containing most of the salts attains in aggregate 800 mg/l.

Thus, the proposed extraction mixture based on calixarene and DEAP enables to purify efficiently the alkaline HLW from all long-lived α, β and γ-radionuclides within a single extraction cycle.

References

1. Hanford Site TWRS, Programmatic environmental review report, July 1998, DOE/RL-98-54, Rev.0.
2. Izatt R.M., Lamb J.D., Hawkins R.T., J. Am. Chem. Soc. 105, 1782 (1983).
3. Karalova, Z.K.; Bukina, T.K.; Lavrinovich, E.A.; Myasoedov B.F., Radiochemistry 1989, vol.31, N 6, pp. 81-87 (In Russ.).

Chapter 9

Calix[4]arenes with a Novel Proton-Ionizable Group: Synthesis and Metal Ion Separations

Richard A. Bartsch, Hong-Sik Hwang, Vladimir S. Talanov, Chunkyung Park, and Galina G. Talanova

Department of Chemistry and Biochemistry, Texas Tech University, Lubbock, TX 79409–1061

Attachment of N-X-sulfonyl carboxamide functions, $C(O)NHSO_2X$, to the lower rim of calix[4]arene compounds provides a novel class of proton-ionizable metal ion extractants in which the acidity may be 'tuned' by variation of X. The syntheses of calix[4]arenes with one and two N-X-sulfonyl carboxamide groups are reported. The latter are efficient extractants for Pb(II), Hg(II), and Ag(I) ions. For calix[4]arene di(N-X-sulfonyl carboxamides), the efficiency and selectivity of Pb(II) extraction are found to be influenced by a change of the para substituent from *tert*-butyl to hydrogen. A fluorogenic reagent with highly selective recognition of Hg(II) is described.

In view of their efficient and selective binding of a variety of ionic and neutral guest species, calixarenes are an important class of host molecules (*1-5*). Calixarene compounds have applications in separations of organic and inorganic species, ion-selective electrodes, phase transfer catalysis, chromatography, etc. (*1*). Calixarenes with pendent proton-ionizable groups (carboxylic acid, hydroxyamic acid, phosphonic acid, and phosphonic acid monoalkyl ester) are of special interest for separations of polyvalent metal ions by solvent extraction and membrane transport processes (*6-11*) or sorption when immobilized on polymer matrices (*12-15*). The nature of the proton-ionizable group, particularly its acidity, controls the metal ion complexation properties of such compounds. For example, calixarene carboxylic acids were found to be much more efficient interphase carriers for alkaline earth metal cations than related unfunctionalized calixarenes with phenolic groups on the lower rim (*6*). The availability of new types of calixarenes with a wider variety of proton-ionizable groups should facilitate the development of new metal ion separation processes.

An important concept in the design of efficient and selective proton-ionizable ligands for metal ion separation processes is charge matching within the metal ion-ionized ligand complex. A match of the charge on the metal ion with an appropriate number of proton-ionizable groups in the ligand produces a neutral complex which facilitates metal-ion transfer into the organic phase in a solvent extraction process, as well as transport across an organic liquid membrane. This concept of charge matching is illustrated in Figure 1.

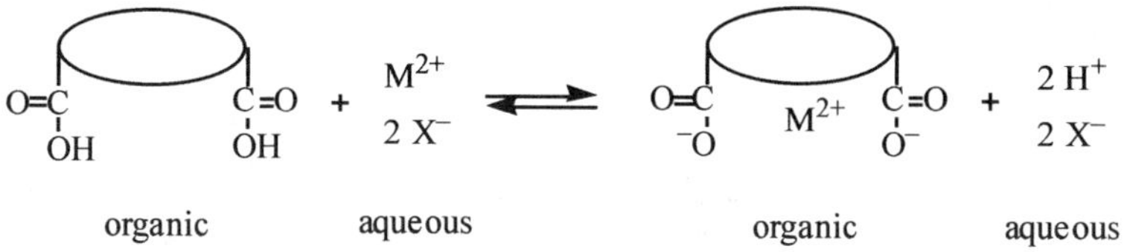

Figure 1. Matching of charges on the metal ion and ionized ligand to provide efficient extraction.

Thus, for use in separations of divalent metal ion species, calixarenes with two mono-ionizable groups or one di-ionizable group are needed. An outstanding example of the use of a calixarene with two proton-ionizable groups in a divalent metal ion separation is provided by Shinkai and coworkers (*16*). These researchers report exclusive extraction of Ca(II) from an aqueous solution containing four alkaline earth metal nitrate species at pH = 5.3 into chloroform by calix[4]arene dicarboxylic acid diamide **1** (Figure 2).

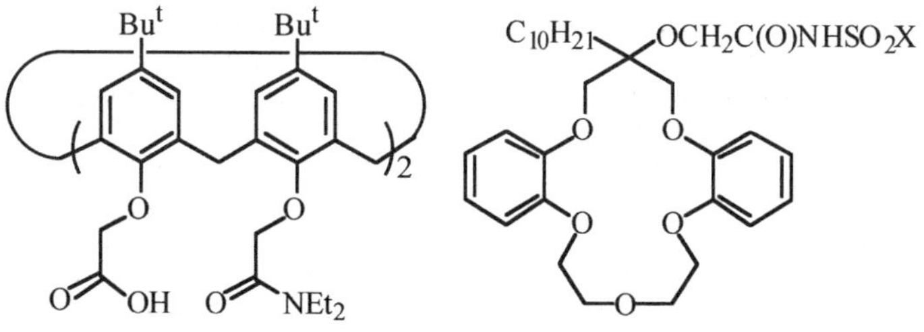

*Figure 2. Extractants **1** and **2** with high selectivities for Ca(II) and Na(I), respectively.*

Recently, we introduced the N-X-sulfonyl carboxamide function, $-C(O)NHSO_2X$, as a novel pendent proton-ionizable group for carbon-pivot lariat ethers (*17*). By variation of the electronic properties of X, the acidity of the macrocyclic ligand is 'tunable'. For competitive extractions of five alkali metal cation species from aqueous chloride and hydroxide solutions into chloroform by N-X-sulfonyl *sym*-(decyl)dibenzo-16-crown-5-oxyacetamides (**2**), transfer of Li(I), Rb(I), and Cs(I) ions into the organic diluent was undetectable. Ligand loading with Na(I) and K(I) cations was 100 %, based upon the formation of one-to-one complexes, and the Na(I)/K(I) selectivity was very high (nearly 50). Although the extraction selectivity was unaffected by variation of X, the acidity of the proton-ionizable lariat ethers, as judged from the extraction profiles, decreased in the order: X = CF_3 > $C_6H_4NO_2$-4 > Ph, Me. Proton-ionizable lariat ether 2 with X = CF_3 was an effective metal ion extractant from moderately acidic, as well as neutral and basic, aqueous solutions. For the other three X groups, alkali metal cations could be extracted only from basic aqueous solutions.

Although our studies of monovalent metal ion separations with such lariat ether N-X-sulfonyl carboxamides continue, we wanted to explore the utility of ligands containing this new proton-ionizable group in multivalent metal ion separations. Since the preparation of crown ether-type compounds with two or more carbon-pivot side arms is very challenging, attention was focused upon calix[4]arene compounds which provide an opportunity for attaching one, two, three, or four proton-ionizable functions to the lower rim. Herein we describe the initial results from our studies of calix[4]arenes which have one or two lower-rim N-X-sulfonyl carboxamide groups.

Calix[4]arene Di(N-X-Sulfonyl Carboxamides)

Synthesis

The *p-tert*-butylcalix[4]arene dicarboxylic acid **6** (Figure 3) was prepared in three steps from commercially available *p-tert*-butylcalix[4]arene (**3**). Reaction of **3** with methyl tosylate and potassium carbonate in acetone at reflux gave dimethoxy compound **4** (*18*) in 81% yield. From reaction of **4** with ethyl bromoacetate and potassium carbonate in acetone at reflux, a 71% yield of calix[4]arene diester **5** (*19*) was realized. To avoid trapping of metal cations, diester **5** was hydrolyzed with tetramethylammonium hydroxide in aqueous THF at reflux (*20*) to provide a 98% yield of calix[4]arene dicarboxylic acid **6** (*19*). Reaction of **6** with oxalyl chloride in benzene at reflux gave diacid chloride **7** which was used without purification. Treatment of **7** with a commercially available sulfonamide and potassium hydride in THF at room temperature produced 60-80% yields of *p-tert*-butylcalix[4]arene di(N-X-sulfonyl carboxamides) **8-11** with X = trifluoromethyl, 4-nitrophenyl, phenyl, and methyl, respectively (*21*). The positions and shapes of the ^{1}H NMR signals for **8-11** were found to vary with the identity of the solvent and the concentration, which demonstrated that the calixarenes exist in solution as mixtures of conformers and also associate by NH···O=C hydrogen bonding. The conformational and associative behavior was especially pronounced for **9**.

Extraction of Pb(II)

The new di-ionizable calixarenes **8-11** efficiently extracted Pb(II) from nitric acid solutions into chloroform (*21*) (Figure 4). Values of the pH for half-extraction were

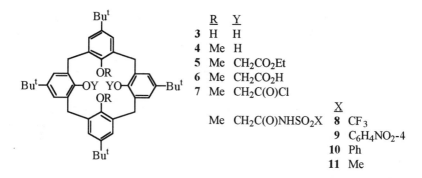

	R	Y		X
3	H	H		
4	Me	H		
5	Me	CH$_2$CO$_2$Et		
6	Me	CH$_2$CO$_2$H		
7	Me	CH$_2$C(O)Cl		
	Me	CH$_2$C(O)NHSO$_2$X	8	CF$_3$
			9	C$_6$H$_4$NO$_2$-4
			10	Ph
			11	Me

Figure 3. p-tert-Butylcalix[4]arene di(N-X-sulfonyl carboxamide) extractants 8-11 and their synthetic precursors.

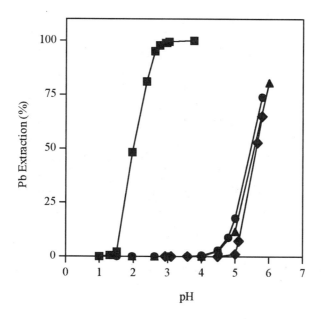

Figure 4. pH profiles for Pb(II) extraction from 0.50 mM aqueous Pb(NO$_3$)$_2$ with 1.00 mM solutions of calix[4]arere di(N-X-sulfonyl carboxamides) 8 (■), 9 (♦), 10 (●), and 11 (▲) in chloroform.

found to increase as X was varied in the order: trifluoromethyl (2.0) > phenyl (5.2) > 4-nitrophenyl (5.3) > methyl (5.5). Although the pH$_{1/2}$ values for **8**, **10**, and **11** fell in the expected ordering for the electron-withdrawing ability of X, that for **9** (X = 4-nitrophenyl) was considerably less acidic than anticipated. It was suggested that the reduced acidity of **9** arises from inter- or intra-molecular NH···O=C hydrogen bonding or from conformational peculiarities of the compound which hinder metal ion complexation.

The Pb(II) extraction selectivity of **8** (X = trifluoromethyl) at pH 4.3 from equimolar binary mixtures of Pb(II)-M, where M = Na(I), K(I), Cs(I), Ca(II), Sr(II), Ba(II), Ag(I), Cd(II), Co(II), Cu(II), Hg(II), Ni(II), Pd(II) and Zn(II), is shown in Figure 5. High Pb(II) selectivity is noted over K(I), Cs(I), Sr(II), Ba(II), and all of the transition metals except Ag(I), Hg(II), and Pd(II). Good extraction selectivity for Pb(II) over Na(I), Ca(II) and Pd(II) is also evident. Only for Hg(II) and Ag(I) is serious interference with Pb(II) extraction apparent.

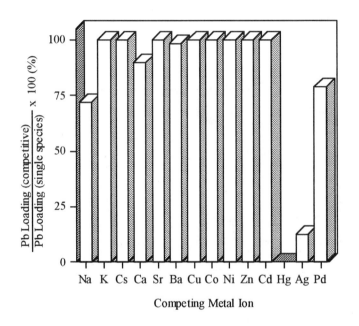

Figure 5. Competitive extraction of 0.50 mM Pb(II) and 0.50 mM competing metal ion from aqueous nitrate solutions at pH 4.3 with 5.00 mM solutions of ligand 8 in chloroform.

Effect of the Para Substituent on Extraction of Pb(II)

To probe the influence of the para substituent in the ligand upon the propensity for Pb(II) extraction, an analog of **8** was prepared in which the *p-tert*-butyl group has been replaced by hydrogen (Figure 6). Thus *p-tert*-butylcalix[4]arene (**3**) was dealkylated by the reported method (*22*) to produce calix[4]arene **12**. Dimethyoxy calix[4]arene **13** (*23*) was obtained in 94% yield by reaction of **12** with methyl tosylate and potassium carbonate in acetone at reflux. Reaction of **12** with ethyl bromoacetate and cesium carbonate in DMF at 70 °C was followed by hydrolysis of the resultant diester with tetramethylammonium hydroxide in aqueous THF at reflux to produce a 60% yield of calix[4]arene dicarboxylic acid **14**. Following treatment of **14** with oxalyl chloride in benzene at reflux, the crude diacid chloride was reacted with trifluoromethanesulfonamide and sodium hydride in THF at room temperature to provide calix[4]arene di(*N*-trifluoromethylsulfonyl carboxamide) **15** in 67% yield.

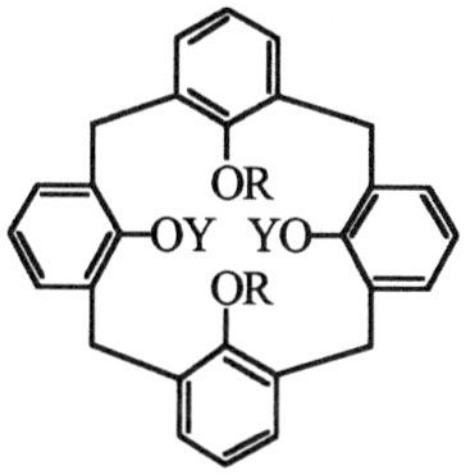

	R	Y
12	H	H
13	Me	H
14	Me	CH_2CO_2Et
15	Me	$CH_2C(O)NHSO_2CF_3$

Figure 6. Calix[4]arene di(N-trifluoromethylsulfonyl carboxamide) 15 and its synthetic precursors.

The abilities of *p-tert*-butylcalix[4]arene di(*N*-trifluoromethylsulfonyl carboxamide) **8** and calix[4]arene di(*N*-trifluoromethylsulfonyl carboxamide) **15** to extract Pb(II) from aqueous nitric acid solutions are compared in Figure 7. It is readily evident that removal of the *p-tert*-butyl group significantly decreases the Pb(II) extracting ability of the ligand. The competitive extraction abilities of ligands **8** and **15** were also compared for equimolar binary mixtures of Pb(II)-M. The Pb(II) extraction was unaffected by the presence of equimolar K(I), Cs(I), Mg(II), Sr(II), Ba(II), Cd(II), Co(II), Cu(II), Ni(II), and Zn(II). Data for the competitive extractions when M = Na(I), Ca(II), Hg(II), Ag(I), and Pd(II) are shown in Figure 8. These competing metal ions were those shown earlier to interfere to varying extents with Pb(II) extraction by **8** (Figure 5). Comparison of the data for calixarene extractants **8** and **15** reveals that the Pb(II) extraction selectivity in competitive extractions may also be influenced by the identity of the para substituent. Thus removal of the *p-tert*-butyl groups from **8** increases the Pb(II) extraction selectivity relative to Ca(II) and Ag(I), but decreases the Pb(II) selectivity relative to Na(I) and Pd(II). Additional experiments designed to identify the causative factor(s) for the observed influence of the para substituent upon extraction selectivity for calix[4]arene di(*N*-X-sulfonyl carboxamides) are in progress.

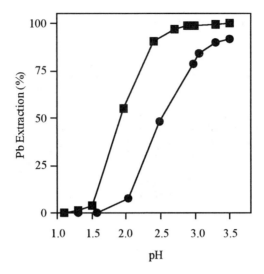

Figure 7. Effect of the para substituent for extraction of 0.50 mM aqueous $Pb(NO_3)_2$ with 1.00 mM solutions of 8 (■) and 15 (●) in chloroform.

Extraction of Soft Metal Cations

Unexpectedly, the *p-tert*-butylcalix[4]arene di(*N*-X-sulfonyl carboxamides) with hard donor groups were found to be efficient soft metal ion extractants (*24*). Thus, **8-11** efficiently extracted Hg(II) from nitric acid solutions into chloroform with excellent selectivity over alkali, alkaline earth, and many transition metal ions, including Pb(II), Ag(I), and Pd(II). Spectroscopic studies suggested a significant contribution to the complex stability by π-interactions between Hg(II) and the electron-rich aromatic rings of the calixarene framework. For a more complete description of soft metal ion extractions by these novel ligands, the Reader is directed to the chapter of this monograph entitled "Separations of Soft Heavy Metal Cations by Lower Rim-Functionalized Calixarenes".

A Fluorogenic Reagent for Hg(II) Recognition

We prepared the first calixarene-based fluorogenic Hg(II)-selective extractant **16** (Figure 9) by conversion of calix[4]arene dicarboxylic acid **6** to the corresponding diacid chloride **7** which was reacted with dansyl amide and potassium hydride in THF at room temperature to give the fluorogenic calix[4]arene di(*N*-X-sulfonyl carboxamide) **16**.

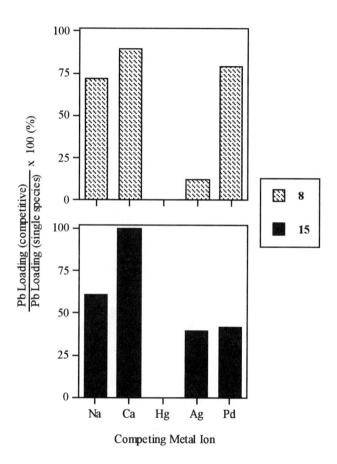

*Figure 8. Competitive extraction of 0.50 mM Pb(II) and 0.50 mM competing metal ion from aqueous solutions at pH 4.3 with 5.00 mM solutions of **8** and **15** in chloroform.*

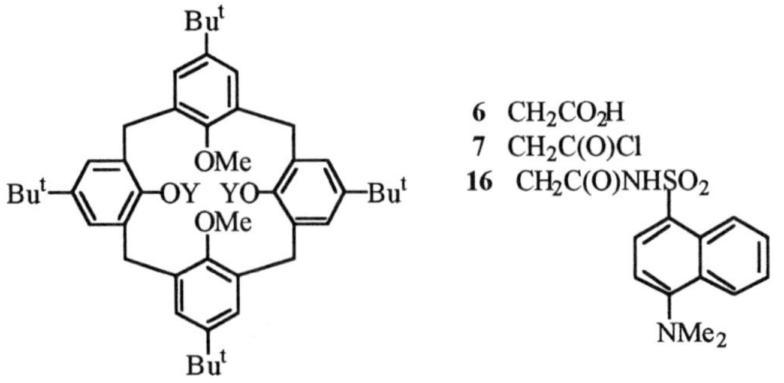

6	CH_2CO_2H
7	$CH_2C(O)Cl$
16	$CH_2C(O)NHSO_2$-

*Figure 9. Calixarene-based fluorogenic reagent **16** for selective recognition of Hg(II) and its synthetic precursors.*

In solvent extraction from aqueous nitric acid solutions, **16** exhibited excellent selectivity for Hg(II) over a wide range of alkali, alkaline earth, and transition metal cations. Quenching of its fluorescence due to Hg(II) coordination was unaffected by the presence of 100-fold excesses of alkali metal cations, alkaline earth metal cations, Ag(I), Tl(I), Cd(II), Co(II), Ni(II), Pb(II), Pd(II), Zn(II), and Fe(III).

Conformationally Locked Calix[4]arene Di(N-X-sulfonyl carboxamides)

^{1}H NMR studies of metal ion complexes of flexible *p-tert*-butylcalix[4]arene di(*N*-X-sulfonyl carboxamides) **8-11** show that some of the calixarenes are held in specific conformations (cone, partial cone, or 1,3-alternate). With the goal of attaining high selectivity in divalent metal ion recognition, we are preparing cone, partial cone, and 1,3-alternate conformations of *p-tert*-butylcalix[4]arene di(*N*-X-sulfonyl carboxamides) (Figure 10). Although there is good literature precedent for preparation of the requisite dicarboxylic acid precursor in the cone conformation (*25*), reported routes to similar 1,3-alternate and partial cone dicarboxylic acids involve either protection-deprotection steps (*26*) or isolation in low yield from mixtures of stereoisomers (*26-30*). Very recently, we have developed efficient synthetic routes to the cone *(24)*, 1,3-alternate, and partial cone (*31*) dicarboxylic acid precursors.

Preparation of the cone and 1,3-alternate isomers began by dialkylation of *p-tert*-butylcalix[4]arene with 1-bromobutane and potassium carbonate in DMF at 70 °C to give an 81% yield of dibutoxy compound **17** (Figure 11). Reaction of **17** with ethyl bromoacetate and sodium hydride in DMF at 80 °C provided diethyl ester **18** in the cone conformation in 75% yield. Hydrolysis of diester **18** with tetramethylammonium hydroxide in aqueous THF afforded a 97% yield of *p-tert*-butylcalix[4]arene dicarboxylic acid **19** in the cone conformation. On the other hand, reaction of dibutoxy compound **17** with methyl bromoacetate and potassium hydride in THF produced a 54% yield of dimethyl ester **20** in the 1,3-alternate conformation. Hydrolysis of this diester as before provided dicarboxylic acid **19** in the 1,3-alternate conformation. When purification of dimethyl ester **20** was omitted, the overall yield

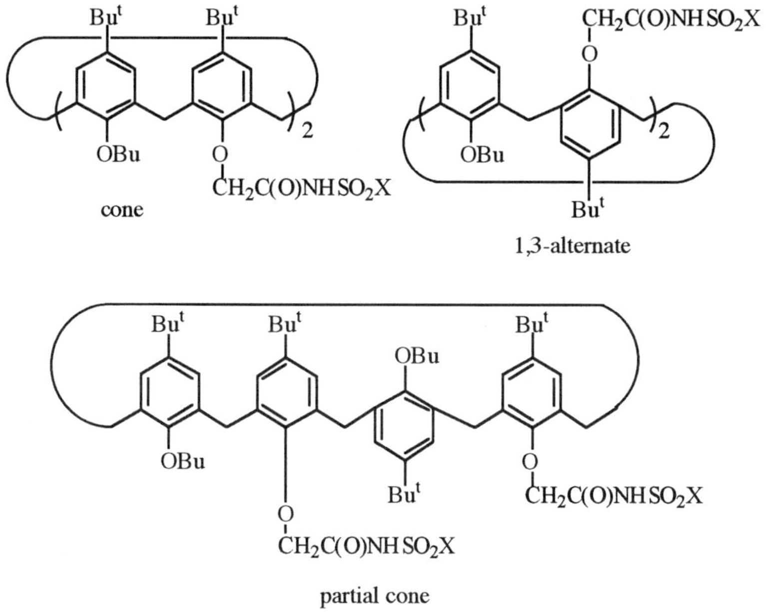

Figure 10. Isomeric p-tert-butylcalix[4]arene di(N-X-sulfonyl carboxamides).

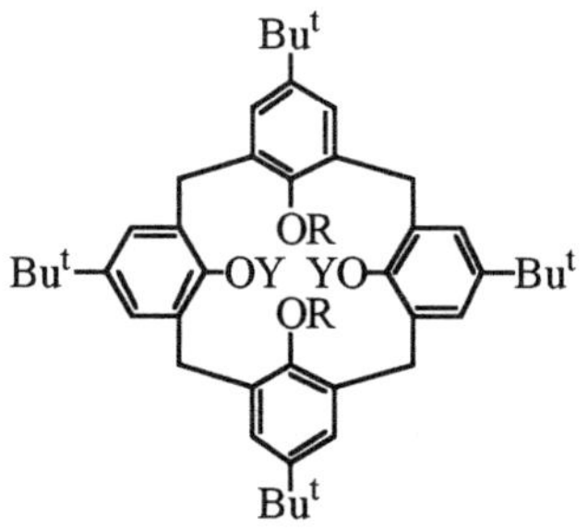

	R	Y
17	Bu	H
18	Bu	CH_2CO_2Et
19	Bu	CH_2CO_2H
20	Bu	CH_2CO_2Me
21	H	CH_2CO_2Me

Figure 11. Isomeric p-tert-butylcalix[4]arene dicarboxylic acids 19 and their synthetic precursors.

of the 1,3-alternate isomer of dicarboxylic acid **19** from dibutoxy compound **17** was 74% *(31)*.

For the partial cone dicarboxylic acid, the synthesis began by dialkylation of *p-tert*-butylcalix[4]arene with methyl bromoacetate and potassium carbonate in acetonitrile at 50-55 °C to produce a 76% yield of diester **21**. Reaction of **21** with 1-bromobutane and potassium hydride and THF at room temperature afforded the partial cone isomer of diester **20** which was hydrolyzed as before. Without purification of the dimethyl ester **20**, the overall yield of the partial cone isomer of dicarboxylic acid **19** from the compound **21** was 90% *(31)*.

Calix[4]arene Mono(*N*-X-Sulfonyl Carboxamides)

For exploration of their behavior as monovalent metal ion extractants, a series of *p-tert*-butylcalix[4]arenes containing one *N*-X-sulfonyl carboxamide group was envisioned (Figure 12). Trialkylation of *p-tert*-butylcalix[4]arene with 1-iodobutane, barium oxide, and barium hydroxide hydrate in DMF at room temperature *(32)* produced a 79 % yield of the tributoxy compound **22** in the cone conformation.

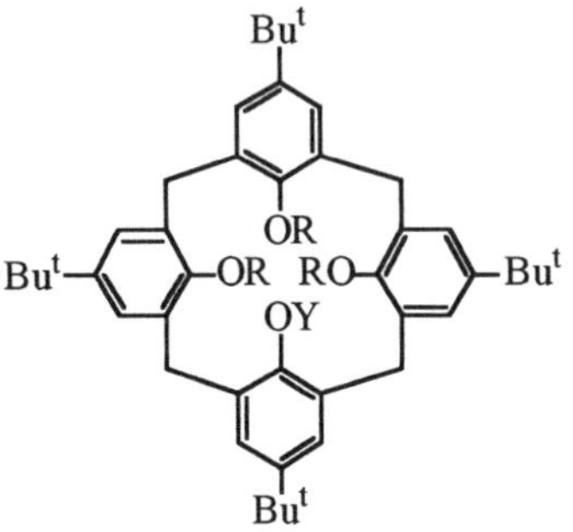

	R	Y		X
3	H	H		
22	Bu	H		
23	Bu	CH_2CO_2Et		
24	Bu	CH_2CO_2H		
25	Bu	$CH_2C(O)Cl$		
	Bu	$CH_2C(O)NHSO_2X$	26	$C_6H_4NO_2$-4
			27	Ph
			28	Me

Figure 12. Mono-ionizable p-tert-butylcalix[4]arene N-X-sulfonyl carboxamides 26-28 and their synthetic precursors.

Reaction of **22** with ethyl bromoacetate and sodium hydride in DMF at 80 °C gave a 93% yield of ester **23** in the cone conformation. Hydrolysis of the ester with tetramethylammonium hydroxide in aqueous THF at reflux produced the corresponding carboxylic acid **24** in 96% yield. Treatment of calix[4]arene carboxylic acid **24** with oxalyl chloride in benzene at reflux gave acid chloride **25** which was used without purification for the next step. Reactions of **25** with commercially available sulfonamides and potassium hydride in THF at room temperature afforded mono-ionizable calix[4]arene *N*-X-sulfonyl carboxamides **26-28** in 75-92% yields, respectively. When X = CF_3, the potassium salt of the desired product was isolated and purified as usual. However, treatment with 1 or 6 N hydrochloric acid failed to produce the desired calix[4]arene *N*-trifluoromethylsulfonyl carboxamide. The use of concentrated hydrochloric acid resulted in hydrolysis back to the precursor carboxylic acid **24**.

Assessment of the metal ion extracting abilities of the mono-ionizable calix[4]arene N-X-sulfonyl carboxamides is in progress.

Closing Remarks

Our initial investigations of calix[4]arenes with pendent N-X-sulfonyl carboxamide groups have yielded interesting and sometimes surprising results. Continued synthesis of such novel metal ion complexing agents and their evaluation in metal ion separation processes is definitely warranted. The future appears bright for further research in this area.

Acknowledgment

This research was supported by the Division of Chemical Sciences of the Office of Basic Energy Sciences of the U.S. Department of Energy (Grant DE-FG03-94ER14416).

Literature Cited

1. Calixarenes: A Versatile Class of Macrocyclic Compounds; Vicens, J.; Bohmer, V., Eds.; Kluwer: Boston, 1991.
2. McKervey, M. A.; Schwing-Weill, M. J.; Arnaud-Neu, F. In *Comprehensive Supramolecular Chemistry*, Gokel, G. W., Ed.; Elsevier: New York, 1996; Vol. 1, pp. 537-603.
3. Pochini, A.; Ungaro, R. In *Comprehensive Supramolecular Chemistry*, Vögtle, F., Ed.; Elsevier: New York, 1996; Vol. 2, pp. 103-142.
4. Ikeda, A.; Shinkai, S.; *Chem. Rev.* **1997**, *97*, 1713-1734.
5. Gutsche, C. D. *Calixarenes Revisited*, in Monographs in Supramolecular Chemistry; Stoddart, J. F., Ed.; The Royal Society of Chemistry: Cambridge, UK, 1998.
6. Ungaro, R.; Pochini, A.; Andretti, G. D. *J. Inclusion Phenom.* **1984**, *2*, 199-206.
7. Ohto, K.; Yano, M.; Inoue, K.; Yamamoto, T.; Goto, M.; Nakashio, F.; Shinkai, S.; Nagasaki, T. *Anal. Sci.* **1995**, *11*, 893-902.
8. Shinkai, S.; Shirahama, Y.; Satoh, H.; Manabe, O.; Arimura, T.; Fujimoto, K.; Matsuda, T. *J. Chem. Soc., Perkin Trans. 2* **1989**, 1167-1171.
9. Nagasaki, T.; Shinkai, S.; Matsuda, T. *J. Chem. Soc., Perkin Trans.1* **1990**, 2617-2618.
10. Nagasaki, T.; Arimura, T.; Shinkai, S.; *Bull. Chem. Soc. Jpn.* **1991**, *64*, 2575-2577.
11. Araki, K.; Hashimoto, N.; Otsuka, H.; Nagasaki, T.; Shinkai, S. *Chem. Lett.* **1993**, 829-832.

12. Shinkai, S.; Kawaguchi, H.; Manabe, O. *J. Polym. Sci., Polym. Lett.* **1988**, *26*, 391-396.
13. Hutchinson, S.; Keraney, G. A.; Horne, E.; Lynch, B.; Glennon, J. D.; McKervey, M. A.; Harris, S. J. *Anal. Chim. Acta* **1994**, *291*, 269-275.
14. Brindle, R.; Albert, K.; Harris, S. J.; Treltzsch, C.; Horne, E.; Glennon, J. D. *J. Chromatogr. A* **1996**, *731*, 41-46.
15. Ohto, K.; Tanaka, Y.; Inoue, K. *Chem. Lett.* **1997**, 647-648.
16. Ogata, M.; Fujimoto, K.; Shinkai, S. *J. Am. Chem. Soc.* **1994**, *116*, 4505-4506.
17. Huber, V. J.; Ivy, S. N.; Lu, J.; Bartsch, R. A. *J. Chem. Soc., Chem. Commun.* **1997**, *16*, 1499-1500.
18. Dijkstra, P. J.; Brunink, J. A.; Bugge, K.-E.; Reinhoudt, D. N.; Harkema, S.; Ungaro, R.; Ugozzoli, F.; Ghidini, E. *J. Am. Chem. Soc.* **1989**, *111*, 7567-7575.
19. Ostaszewski, R.; Stevens, T. W.; Verboom, W.; Reinhoudt, D. N. *Rec. Trav. Chim. Pays-Bas* **1991**, *110*, 294-298.
20. Chang, S. K.; Cho, I. *J. Chem. Soc, Perkin Trans. 1.* **1986**, 211-214.
21. Talanova, G. G. ; Hwang, H.-S.; Talanov, V. S.; Bartsch, R. A. *J Chem. Soc., Chem. Commun.* **1998**, 419-420.
22. Gutsche, C. D.; Levine, J. A. *J. Am. Chem. Soc.* **1982**, *104*, 2652-2653.
23. van Loon, J.-D.; Arduini, A.; Coppi, L.; Verboom, W.; Pochini, A.; Ungaro, R.; Harkema, S.; Reinhoudt, D. N. *J. Org. Chem.* **1990**, *55*, 5639-5646.
24. Talanova, G. G.; Hwang, H.-S.; Talanov, V. S.; Bartsch, R. A. *J. Chem. Soc., Chem. Commun.* **1998**, 1329-1330.
25. Arduini, A.; Pochini, A.; Reverberi, S.; Ungaro, R. *J. Chem. Soc., Chem. Commun.* **1984**, 981-982.
26. Iwamoto, K. Shinkai, S. *J. Org. Chem.* **1992**, *57*, 7066-7073.
27. Shinkai, S.; Fujimoto, K.; Otsuka, T.; Ammon, H. L. *J. Org. Chem.* **1992**, *57*, 1516-1523.
28. Beer, P. D.; Drew, M. G. B.; Gale, P. A.; Leeson, P. B.; Ogden, M. I. *J. Chem. Soc., Dalton Trans.* **1994**. 3479-3485.
29. Pitarch, M.; Browne, J. K., McKervey, M. A. *Tetrahedron* **1997**, *53*, 10503-10512.
30. Lugtenberg, R. J. W.; Egberink, R. J. M.; Engbersen, J. F. J.; Reinhoudt, D. N. *J. Chem. Soc., Perkin Trans. 2* **1997**, 1353-1357.
31. Talanov, V. S.; Bartsch, R. A. *J. Chem. Soc., Perkin Trans. 1* **1999**, in press.
32. Iwamoto, K.; Araki, K.; Shinkai, S. *J. Org. Chem.* **1991**, *56*, 4955-4962.

Chapter 10

Separations of Soft Heavy–Metal Cations by Lower Rim-Functionalized Calix[4]arenes

Galina G. Talanova, Vladimir S. Talanov, Hong-Sik Hwang,
Nazar S. A. Elkarim, and Richard A. Bartsch

Department of Chemistry and Biochemistry, Texas Tech University,
Lubbock, TX 79409-1061

Calix[4]arenes functionalized on the lower rim with various donor groups are used for recognition and separation of soft heavy metal cations, e. g., Ag(I), Au(III), Cd(II), Hg(II), Pd(II), and Pt(II). Recently, calix[4]arenes containing proton-ionizable N-X-sulfonyl carboxamide functionalities on the lower rim have been found to efficiently extract Hg(II), a soft metal ion, from acidic aqueous solutions with excellent selectivity over alkali, alkaline earth, and many transition and heavy metal cations, including Ag(I), Cd(II), Pb(II), Pd(II), and Pt(II). This unexpected favoring of Hg(II) complexation by ionophores containing hard donor groups has been used to obtain the first calixarene-based, mercury-selective fluorogenic reagent by incorporating a dansyl fluorophore moiety as a part of the N-X-sulfonyl carboxamide groups.

During the last decade, remarkable progress in the transition and heavy metal coordination chemistry of the calixarene ligands has been achieved (1-4). An important contributing factor was the introduction of softer donor groups with nitrogen, phosphorus, or sulfur atoms on the calixarene lower rim instead of hard oxygen-containing functions. This gave rise to new calixarene-based ionophores with enhanced affinity for d-metal ions, including the soft Pearson's acids of Ag(I), Au(III), Cd(II), Hg(II), and platinides(II), and diminished complexing ability toward hard alkali and alkaline earth metal cations. This feature of lower rim-functionalized calixarenes, particularly sulfur-containing derivatives (Figure 1), has been utilized in the separation of soft heavy metal ions.

Thus, calix[4]arene tetrathioamides **1** are efficient Ag(I) picrate extractants that show low extraction levels for Cd(II), Co(II), Cu(II), alkali and alkaline earth metal picrates (5). The calixarene-based bis(dithioamide) **2**, which is restricted to the 1,3-alternate conformation, possesses high selectivity for Cd(II) over Pb(II), Cu(II), Ca(II), and K(I) cations in plasticized PVC membranes of chemically modified field effect transistors (6). For another Cd(II) sensor based upon calix[4]arene tetrathio-amide **3** which is fixed in the cone conformation, Cu(II) interference is stronger (7). Ionophores **4** with four pendent dithiocarbamate moieties exhibit highly efficient sol-

© 2000 American Chemical Society

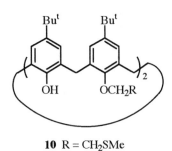

Figure 1. Structures of calix[4]arenes with soft sulfur-containing donor groups on the lower rim utilized in soft heavy metal ion separations.

vent extraction of Au(III) and Pd(II) with somewhat lower affinity for Hg(II) and Ag(I) (*8*). In contrast, structurally related calixarenes **4** are ineffective binders of soft Cd(II), Pt(II), and Pt(IV) ions, as well as harder Pb(II), Sn(II), Fe(II), Co(II), Ni(II), Zn(II), and Mn(II) cations.

Calix[4]arene **5** with four thiol groups on the lower rim shows a moderate extracting ability toward the soft metal ions of Ag(I), Au(III), Hg(II), and Pd(II). For the related tetrathioether **6a**, highly efficient Au(III) separation, moderate Ag(I) and Pd(II) extraction and very weak Hg(II) binding are observed (*8*). Ligand **6b** which does not contain *t*-Bu groups on the upper rim exhibits only low differentiation among Ag(I), Au(III), Hg(II), and Pd(II) (*8*).

Ionophores **7-9** that each possess four pendent thioether groups on the lower rim were utilized in silver ion-selective electrodes (*9*). Calixarenes **7** and **8** are restricted to the partial cone (paco) conformation and exhibit excellent selectivity for Ag(I) over a wide range of other metal ions, including Na(I), Hg(II), and Pb(II). Paco **8** provides more selective Ag(I) recognition than the cone isomer of the related calixarene **9**. Ligand **10** (*7*) is another example of a calixarene-based sensor for Ag(I).

Calix[4]arenes functionalized on the lower rim with harder oxygen-containing donor groups, *e. g.*, amide, phosphine oxide, ester, or ketone (Figure 2), also have been explored in separations of soft (mostly precious) metal ions. Thus, tetraamide **11** efficiently extracts Au(III) and Ag(I) ions from acidic nitrate solutions with discrimination over Pd(II) and Pt(II) (*10*). Tetraamide ligands **12** provide high levels of Ag(I) and Cd(II) picrate extraction from neutral aqueous solutions (*5*), but show poor selectivity for these metal salts over Na(I), K(I), Mg(II), Ba(II), and Pb(II) picrates. The calix[4]arene tetraketone **13** allows selective separation of Ag(I) present in small amounts from excess Pd(II) in highly acidic aqueous nitrate solution (*11*). Within the series of calix[4]arenes **14** which have various oxygen-containing donor groups, the ionophore with two pendent amide and two phosphine oxide groups appears to be the most efficient Ag(I) extractant and allows removal of trace amounts of Ag(I) in the presence of a large excess of Cu(II) (*12*).

Calixarenes **15** with four π-coordinating allyl groups on the lower rim were applied as carriers in silver ion sensors (*13*). Ionophore **15a** exhibits high Ag(I) selectivity while the ester group-containing analogue **15b** provides only poor Ag(I) recognition with severe Na(I) interference.

To the best of our knowledge, no report of Hg(II), Cd(II), Pd(II) or Pt(II)-selective calix[4]arenes functionalized on the lower rim with oxygen-containing donor groups has been published.

Recently, we prepared the series of calix[4]arene derivatives **16-19** (Figure 3) which contains on the lower rim two proton-ionizable *N*-X-sulfonyl carboxamide groups of "tunable" acidity (*14*) (see also the chapter of this monograph entitled "Calix[4]arenes with a Novel Proton-Ionizable Group: Synthesis and Metal Ion Separations"). These ionophores were found to efficiently extract Pb(II) from acidic aqueous nitrate solutions with good-to-excellent selectivity over many transition, alkali, and alkaline earth metal cations. However, the soft heavy metal ions of Ag(I), Pd(II) and especially Hg(II) produced significant interference with Pb(II) extraction by the calix[4]arene *N*-X-sulfonyl carboxamides. This unexpected favoring of soft metal cations over harder metal ions by ligands containing hard donor groups encouraged us to investigate the solvent extraction of Ag(I), Cd(II), Hg(II), Pd(II), and Pt(II) by ionophores **16-19**.

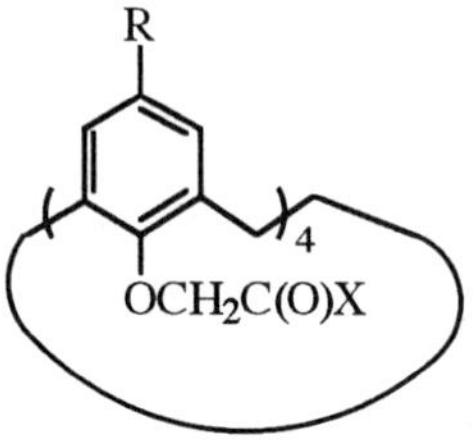

11 R = C(CH$_3$)$_2$C(CH$_3$)$_3$; X = NEt$_2$

12 R = Bu^t; X = NEt$_2$, N⟨ ⟩

13 R = C(CH$_3$)$_2$C(CH$_3$)$_3$; X = Me

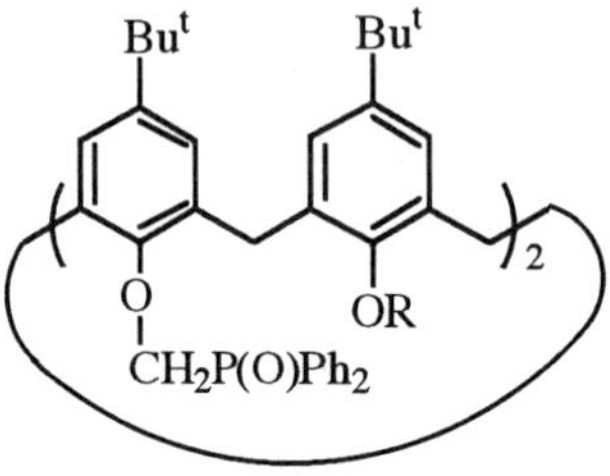

14 R = CH$_2$C(O)NEt$_2$, CH$_2$P(O)Ph$_2$, CH$_2$CO$_2$Et or Me

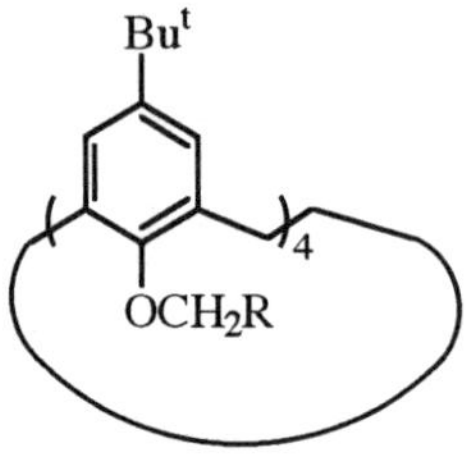

15
R
15a CH=CH$_2$
15b CO$_2$CH$_2$CH=CH$_2$

Figure 2. Structures of calix[4]arenes with hard oxygen-containing donor groups on the lower rim utilized in soft heavy metal ion separations.

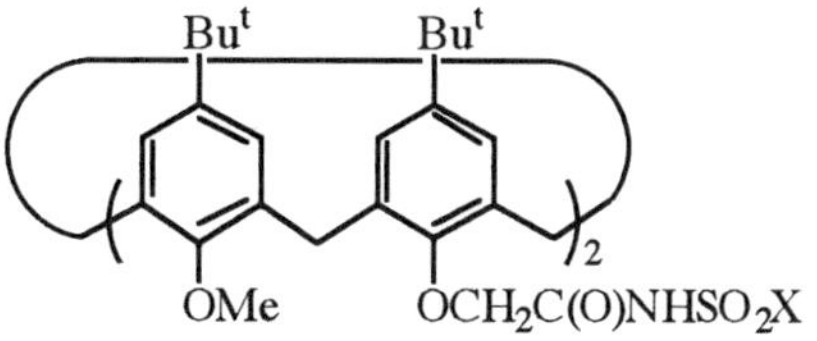

	X
16	CF$_3$
17	Me
18	Ph
19	C$_6$H$_4$NO$_2$-4

Figure 3. Structures of calix[4]arene di(N-X-sulfonyl carboxamides).

Separations of Soft Heavy Metal Ions with Calix[4]arene N-X-Sulfonyl Carboxamides

Solvent Extraction of Silver(I)

Extractions of Ag(I) from acidic and neutral aqueous nitrate solutions into chloroform by calix[4]arene di(N-X-sulfonyl carboxamides) **16-19** (*15*) showed that their propensity for Ag(I) binding is controlled by their acidities in accordance with the electron-withdrawing ability of the substituents X, *e.g.*, **16** (CF$_3$) >> **19** (C$_6$H$_4$NO$_2$-4) > **18** (Ph) > **17** (Me). A similar trend was noted for solvent extractions of Pb(II) (*14*) and of alkali and alkaline earth metal cations by **16-19**. Ligand **16** efficiently separates Ag(I) from dilute nitric acid (pH < 3), while for the weaker NH-acids **17-19**, a higher pH is required.

Calixarene **16** was found to interact with Ag(I) in a water-chloroform extraction system to yield complexes with a 2:1 metal-to-ligand stoichiometry, as described by following equation:

$$2Ag^+_{aq} + H_2L_{org} = Ag_2L_{org} + 2H^+_{aq}$$

where H$_2$L is the di(proton-ionizable) ligand **16** and the subscripts aq and org denote species in the aqueous and organic phases, respectively. The log of the equilibrium constant for this reaction (*e. g.*, the extraction constant) is 2.81.

To probe for the interference of Ag(I) extraction by other metal cations, particularly Na(I), Cu(II), Pb(II), Pd(II), Hg(II), and Tl(I), competitive extractions of Ag(I) from aqueous equimolar mixtures of two metal nitrates at pH 2.5 into chloroform by **16** was studied. Relative to the single species Ag(I) extraction under otherwise identical conditions, the level of Ag(I) extraction was unaffected by the presence of Na(I), Cu(II), Pb(II), and Pd(II), while with Hg(II) and Tl(I) it decreased by 20-25%. Thus, calix[4]arene di(N-trifluoromethylsulfonyl carboxamide) **16** was shown to efficiently and selectively separate Ag(I) from aqueous acidic nitrate solutions.

Extraction of Palladium(II) and Platinum(II)

Calixarenes **16-19** were also tested in single species extractions of Pd(II) and Pt(II) nitrates from acidic and neutral aqueous solutions. These ionophores were found to be much less efficient complexants for Pd(II) than for Ag(I). Only the most acidic ligand **16** was found to be capable of extracting Pd(II) at a moderate level into chloroform from aqueous solutions with pH < 6. Under otherwise identical conditions, the Pd(II) loading of **16** was about three times lower than that of Ag(I). The percentage of Pd(II) extraction by **16** further decreased when the aqueous phase anion was changed from nitrate to chloride. This was attributed to formation in the aqueous phase of the anionic complex species $PdCl_3^-$ and $PdCl_4^{2-}$.

None of the calix[4]arene di(*N*-X-sulfonyl carboxamides) **16-19** gave significant extraction of Pt(II) from aqueous solutions into chloroform.

Extraction of Hg(II) and Cd(II)

Although calixarenes **16-19** were found to be unable to extract Cd(II) from aqueous nitrate solutions into chloroform, all of them exhibited strong affinity for Hg(II) ions (*15*). The pH profiles for extractions of mercuric nitrate from aqueous solutions into chloroform by **16-19** (Figure 4) demonstrate that the calix[4]arene di(*N*-

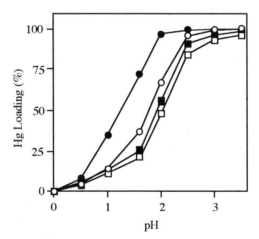

Figure 4. pH profiles for Hg(II) extraction from aqueous 0.50 mM mercuric nitrate solution into the equal volume of chloroform by 0.50 mM calix[4]arene di(N-X-sulfonyl carboxamides) **16** *(●),* **17** *(○),* **18** *(■), and* **19** *(□).*

X-sulfonyl carboxamides) provide efficient Hg(II) separation from acidic aqueous media. Unlike the extractions of soft Ag(I) and Pd(II) described above and those of harder metal ions reported earlier, Hg(II) loadings of these ligands vary only slightly when the identity of the sulfonyl group substituent X is changed. Moreover, the propensities of this calixarene series for Hg(II) extraction decrease in the order: **16** (CF_3) > **17** (Me) > **18** (Ph) > **19** ($C_6H_4NO_2$-4). This ordering differs appreciably from that observed for other metal ions and indicates that the size of X, rather than its electron-withdrawing ability, is important for Hg(II) extraction.

Due to this unique feature of the calix[4]arene di(N-X-sulfonyl carboxamides), the lower acidity ligands **17-19** provide extremely selective Hg(II) separation from acidic aqueous solutions since they completely extract Hg(II) at pH < 4 with negligible loadings of other metal cations, e.g., alkali, alkaline earth, and many transition and heavy metal ions, including Ag(I), Cd(II), Cu(II), Pd(II), and Pt(II). To the best of our knowledge, compounds **16-19** are the first representatives of mercury-selective calixarene-type ionophores which contain no soft donor functions.

Calixarenes **16-19** have been found to interact with Hg(II) ions forming complexes of either 1:1 or 2:1 metal-to-ligand stoichiometry. In particular, the complexation reaction in an extraction system with an excess of Hg(II) over the ionophore proceeds in accordance with the following equation:

$$2Hg^{2+}_{aq} + H_2L_{org} + 4NO_3^-{}_{aq} = [(HgNO_3)_2L]_{org} + 2H^+_{aq}$$

Evaluated extraction constants for such complexes exceed 13.5 log units and decrease in the order: **16** > **17** > **18** > **19**.

To provide insight into the mode of Hg(II) coordination with the calix[4]arene di(N-X-sulfonyl carboxamides), UV spectra of **16-19** in chloroform solutions before and after mercuric nitrate extraction were studied (*15*). Hg(II) coordination caused changes in the spectra of all four of the calixarenes. The absorption bands for the substituted benzene rings of the ligands at 270-279 nm showed hypsochromic shifts of 17-23 nm in the spectra measured after Hg(II) extraction. Analogous spectral changes were not observed after lead nitrate extraction or after contact of the calixarene solutions in chloroform with 1.0 M aqueous NaOH.

The results presented above suggest a significant contribution of the π-electron-rich aromatic units in **16-19** to the coordination of Hg(II). Although π-cation interactions were previously observed in some of the calixarene complexes with soft heavy metal ions (see references *2* and *16* for examples), they have not been advocated before in the Hg(II) complexes formed by calixarene-based ionophores utilized for Hg(II) separations. Evidently in coordination of Hg(II) by calix[4]arenes functionalized in the lower rim with soft sulfur-containing donor groups for which strong sulfur-mercury(II) bonding plays a dominant role, the weaker interaction of the π-donor aromatic units with the metal cation is unimportant. However for calixarenes that do not contain soft donor groups, participation of the π-electron-rich calixarene cavity in the coordination arrangement for Hg(II) ion may increase dramatically.

A Fluorogenic Calix[4]arene N-X-Sulfonyl Carboxamide for Selective Hg(II) Recognition

Optical chemosensors for the determination of heavy metal ions are receiving ever-increasing attention (*17, 18*). A convenient approach to the preparation of fluorogenic reagents for selective metal ion recognition consists of attaching fluorophore moieties to the platform of a macrocyclic or chelating complexant. This methodology was applied to obtain the recently reported optical sensors for the detection of Hg(II) (*19-21*). However, calixarene platforms have not been utilized previously in the synthesis of fluorogenic reagents for Hg(II) recognition.

We incorporated a dansyl moiety as a part of the *N*-X-sulfonyl carboxamide groups in calix[4]arene derivative **20** (Figure 5) and explored the properties of this ligand as a Hg(II) extractant (*22*).

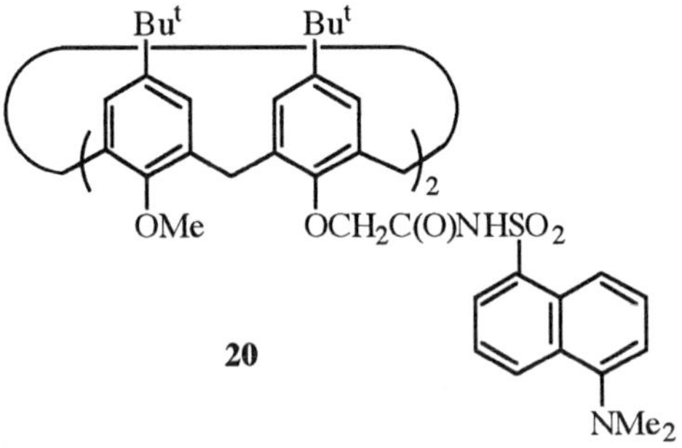

Figure 5. Structure of fluorogenic calix[4]arene di(N-X-sulfonyl carboxamide).

Similarly to its structural analogs **16-19**, ionophore **20** efficiently extracted Hg(II) from acidic aqueous mercuric nitrate solutions into chloroform. As shown in Figure 6, Hg(II) uptake by **20** was accompanied by quenching of its fluorescence emission at 520 nm. The relative emission intensity I/I_0 (where I_0 and I are emission intensities observed in the spectrum of **20** before and after the Hg(II) extraction, respectively) diminished as the Hg(II) loading increased. With a 50.0 μM chloroform solution of the fluorogenic calixarene, the detection limit of Hg(II) in the aqueous phase at pH 2.5 was determined to be 5.00 μM.

Ionophore **20** exhibited excellent extraction selectivity for Hg(II) over a wide variety of alkali, alkaline earth, transition, and heavy metal ions. Its fluorescence quenching due to the Hg(II) coordination was unaffected by the presence of 100-fold excesses of Na(I), Ag(I), Tl(I), Ca(II), Cd(II), Co(II), Cu(II), Ni(II), Pb(II), Pd(II), Zn(II), or Fe(III) (*22*). Ligand **20** is the first example of a calixarene-based fluorogenic sensor for selective Hg(II) recognition.

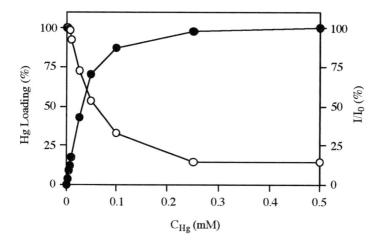

Figure 6. Dependence of the mercury loading (●) and relative emission intensity (○) of fluorogenic calixarene 20 in a 50.0 μM chloroform solution after Hg(II) extraction on the initial aqueous phase mercuric nitrate concentration (pH 2.5).

Acknowledgment

This research was supported by the Division of Chemical Sciences of the Office of Basic Energy Sciences of the U.S. Department of Energy (Grant DE-FG03-94ER14416).

Literature Cited

1. Gutsche, C. D. *Calixarenes Revisited*; in Monographs in Supramolecular Chemistry; Stoddart, J. F., Ed.; The Royal Society of Chemistry: Cambridge, UK, 1998.
2. Wieser, C.; Dieleman, C. B.; Matt, D. *Coord. Chem. Rev.* **1997**, *165*, 93-161.
3. McKervey, M. A.; Scwing-Weill, M.-J.; Arnaud-Neu, F. In *Molecular Recognition: Receptors for Cationic Guests*; Gokel, G. W., Ed.; Comprehensive Supramolecular Chemistry; Elsevier: New York, 1996; Vol. 1, pp. 537-603.
4. Roundhill, D. M. In *Progress in Inorganic Chemistry*; Karlin, K. D., Ed.; Wiley: New York, 1995; Vol. 43, pp. 533-591.
5. Arnaud-Neu, F.; Barrett, G.; Corry, D.; Cremin, S.; Ferguson, G.; Gallagher, J. F.; Harris, S. J.; McKervey, M. A.; Schwing-Weill, M.-J. *J. Chem. Soc., Perkin Trans. 2* **1997**, 575-579.

6. Lugtenberg, R. J. W.; Egberink, R. J. M.; Engbersen, J. F. J.; Reinhoudt, D. N. *J. Chem. Soc., Perkin Trans. 2* **1997**, 1353-1357.
7. Cobben, P. L. H. M.; Egberink, R. J. M.; Bomer, J. G.; Bergveld, P.; Verboom, W.; Reinhoudt, D. N. *J. Am. Chem. Soc.* **1992**, *114*, 10573-10582.
8. Yordanov, A. T.; Whittlesey, B. R.; Roundhill, D. M. *Inorg. Chem.* **1998**, *37*, 3526-3531.
9. O'Connor, K. M.; Henderson, W.; O'Neil, E.; Arrigan, D. W. M.; Harris, S. J.; McKervey, M. A.; Svehla, G. *Electroanalysis* **1997**, *9*, 311-315.
10. Ohto, K.; Yamaga, H.; Murakami, E.; Inoue, K. *Talanta* **1997**, *44*, 1123-1130.
11. Ohto, K.; Murakami, E.; Shinohara, T.; Shiratsuchi, K.; Inoue, K.; Iwasaki, M. *Anal. Chim. Acta* **1997**, *341*, 275-283.
12. Yaftian, M. R.; Burgard, M.; El Bachiri, A.; Matt, D.; Wieser, C.; Dieleman, C. B. *J. Incl. Phenom.* **1997**, *29*, 137-151.
13. Kimura, K.; Tatsumi, K.; Yajima, S.; Miyake, S.; Sakamoto, H.; Yokoyama, M. *Chem. Lett.* **1998**, 833-834.
14. Talanova, G. G.; Hwang, H.-S.; Talanov, V. S.; Bartsch, R. A. *J. Chem. Soc., Chem. Commun.* **1998**, 419-420.
15. Talanova, G. G.; Hwang, H.-S.; Talanov, V. S.; Bartsch, R. A. *J. Chem. Soc., Chem. Commun.* **1998**, 1329-1330.
16. Ikeda, A.; Shinkai, S. *Chem. Rev.* **1997**, *97*, 1713-1734.
17. Czarnik, A. W. *Acc. Chem. Res.* **1994**, *27*, 302-308.
18. Fabrizzi, L.; Poggi, A. *Chem. Soc. Rev.* **1995**, *24*, 197-202.
19. Yoon, J.; Ohler, N. E.; Vance, D. H.; Aumiller, W. D.; Czarnik, A. W. In *Chemosensors for Ion and Molecule Recognition*; Desvergne, J. P.; Czarnik, A. W., Eds.; Kluwer Academic Publishers: Boston, 1997; pp. 189-194.
20. Vaidya, B.; Zak, J.; Bastiaans, G. J.; Porter, M. D.; Hallman, J. L.; Nabulsi, N. A. R.; Utterback, M. D.; Strzelbicka, B.; Bartsch, R. A. *Anal. Chem.* **1995**, *67*, 4101-4111.
21. Sasaki, D. Y.; Padilla, B. E. *J. Chem. Soc., Chem. Commun.* **1998**, 1581-1582.
22. Talanova, G. G.; Elkarim, N. S. A.; Talanov, V. S.; Bartsch, R. A. *Anal. Chem.* **1999**, in press.

Chapter 11

CMPO-Substituted Calixarenes

Volker Böhmer

Fachbereich Chemie und Pharmazie, Johannes Gutenberg-Universität Mainz, Duesbergweg 10-14, D-55099 Mainz, Germany

The synthesis of calixarenes substituted by carbamoylmethylphosphine oxide groups either on their wide rim or on their narrow rim is reviewed. Some typical results of extraction, membrane transport, and complexation studies are reported. In comparison to CMPO all CMPO-substituted calixarenes show a drastically improved extraction efficiency and also a strongly increased selectivity within the lanthanides and between actinides and lanthanides.

Carbamoylmethylphosphine oxides (**1**) are excellent extractants for actinides. Indeed, (N,N-di-isobutylcarbamoylmethyl)octylphenylphosphine oxide (**1a**, referred to as CMPO in this article) is used in the TRUEX process (*1*). For trivalent actinides, such as Am(III), the species extracted is believed to contain three CMPO molecules per actinide cation. Thus, it seems reasonable to attach three (or more) functional groups of the CMPO type in a suitable way on an appropriate skeleton, which would allow their simultaneous, "coordinated" action in the complexation of the metal cation.

Calixarenes represent a class of cyclic oligomers which are not only easily available in larger quantities, but offer also nearly unlimited possibilities of chemical modification (*2,3*). Calix[4]arenes have especially been used in numerous ways as such a basic scaffold on which various ligating groups have been attached.

Looking for improved extractants for actinides, it was therefore logical to think of calixarenes substituted by CMPO-like functions. This article gives a short survey of the results obtained so far, concentrating mainly on the synthesis. A brief discussion of typical results for extraction, complexation, and structure of the complexes formed is also presented. For a more detailed discussion of these latter topics, the reader is referred to the contributions from F. Arnaud-Neu and J. Desreux.

© 2000 American Chemical Society

Wide Rim CMPO-Calixarenes

Synthesis

While many calixarene based ionophores were known, in which ligating functions are attached to the narrow rim (*4*) (caused probably by the fact, that the phenolic hydroxyl groups are especially prone to chemical modification), our first attempts concentrated on the attachment of CMPO-functions via amide bonds to the wide rim (*5*). The synthetic strategy is outlined in the following schemes.

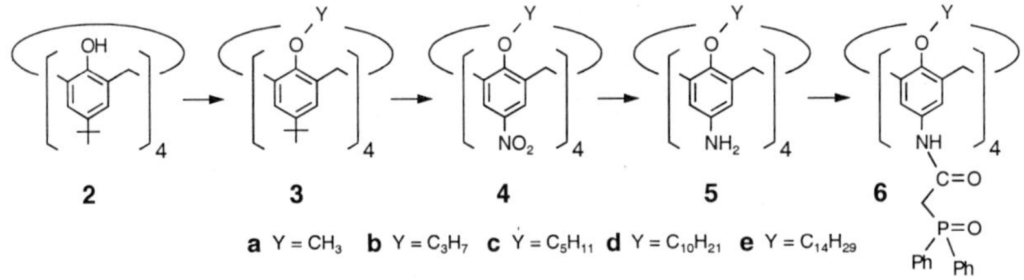

a Y = CH$_3$ **b** Y = C$_3$H$_7$ **c** Y = C$_5$H$_{11}$ **d** Y = C$_{10}$H$_{21}$ **e** Y = C$_{14}$H$_{29}$

t-Butylcalix[4]arene is easily converted to tetraethers, in which the cone-conformation is fixed, if the residues attached to the phenolic oxygen are at least as large as a propyl group. Such tetraethers can be efficiently ipso-nitrated (room temperature, HNO$_3$, dichloromethane/acetic acid) in high yields (*6,7*) The tetranitro derivatives may then be reduced to the tetraamino derivatives in numerous ways. (For standard preparations of sufficiently soluble compounds we normally use catalytic hydrogenation with Raney-Ni at room temperature (*7*).)

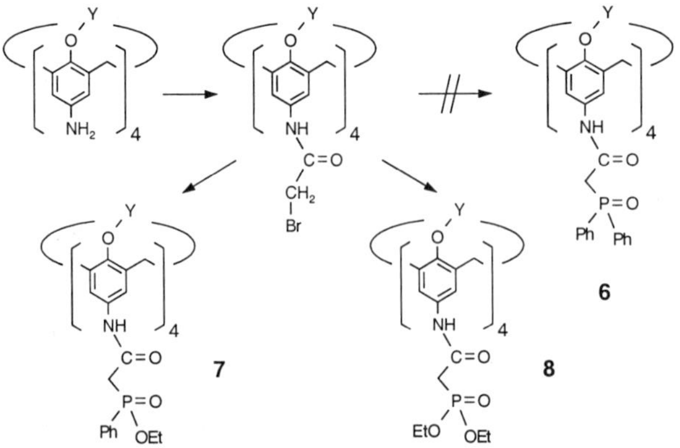

For the next steps we initially planned the acylation with chloro- or bromoacetyl chloride followed by Arbouzov reaction with various esters of trivalent phosphorus.

Although this reaction sequence has been successfully used for the synthesis of phosphinates **7** and phosphonates **8** (see below), it failed for the preparation of phosphine oxides **6**.

The introduction of the desired phosphine oxide functions was possible, however, using an active ester, easily available from bromoethylacetate via Arbouzov reaction with the isopropyl ester of diphenylphosphinous acid followed by hydrolysis and esterification with p-nitrophenol (5).

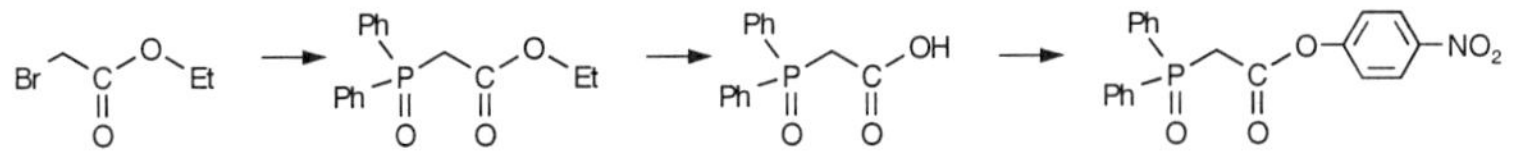

The (diphenylphosphoryl)acetic acid **9** and the active ester **10** have been structurally characterized by X-ray diffraction (Fig. 1) (8). This active ester **10** can be used as a stable and storable, crystalline reagent, to attach carbonylmethyldiphenylphosphine oxide functions via amide bonds to various amines, provided their nucleophilicity is sufficiently high. This comprises of course aliphatic amines, while kinetic studies show (9) that the reactivity of simple aromatic amines is too low, if they are not "activated" by alkoxy groups.

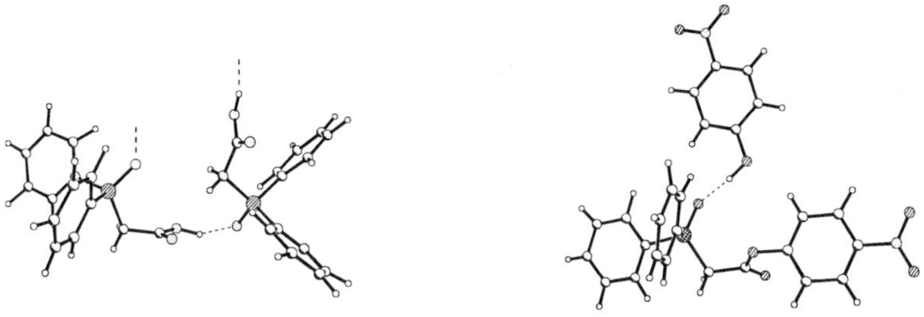

*Figure 1. Single crystal X-ray structure of (diphenylphosporyl)acetic acid **9** (two molecules of a centrosymmetric tetrameric arrangement) and its p-nitrophenyl ester **10** (co-crystallized with p-nitrophenol) (8).*

Ionophoric Properties

Ligands **6** were compared in extraction and ion transport experiments with typical "standard" extractants like CMPO or TOPO (trioctylphosphine oxide) (5).

Using dichloromethane as the diluent, and an organic-to-aqueous volume ratio of 1 %E (percent extracted) values >50 are reached for the extraction of Th(IV) nitrate (from 1 M HNO_3) with a ligand concentration as low as 10^{-4} M (equal to the initial aqueous-phase concentration of metal ions). These values are not obtained by CMPO even at a

100-fold concentration. For the extraction of Eu(III) nitrate, the concentration of CMPO must be 250-fold, to get the same %E of about 70 reached by a 10^{-3} M solution of **6**.

Extraction under conditions closer to the technical requirements (aqueous phase 1 M HNO_3, 4 M $NaNO_3$; organic phase o-nitrophenyl hexyl ether, $c_L = 10^{-3}$ M) gave extraction values of ≥99% for ^{152}Eu, ^{239}Pu, and ^{241}Am, values which were not reached by CMPO with a 10-fold higher concentration.

Some of the wide rim CMPOs were checked in transport studies using supported liquid membranes (o-nitrophenyl hexyl ether, $c_L = 10^{-3}$ M). The results are shown in Table I. Again CMPO **1a** must be applied with a 10-fold concentration to reach permeabilities which still are generally lower (^{237}Np, ^{241}Am) or at least comparable (^{239}Pu).

Table I. Transport through supported liquid membranes (NPHE)

Carrier	Y	c_L (M)	Permeabilities P (cm h^{-1})		
			^{237}Np	^{239}Pu	^{241}Am
6c	C_5H_{11}	10^{-3}	<0.2	3.6	3.4
6d	$C_{10}H_{21}$	10^{-3}	3.0	3.6	5.0
6e	$C_{14}H_{29}$	10^{-3}	1.2	2.9	5.3
11	$C_{14}H_{29}$	10^{-3}	1.6	1.8	3.0
12b'	$CH_2CH(C_2H_5)C_4H_9$ [a]	10^{-3}	1.6	4.2	2.9
12b''	$C_{10}H_{21}$ [a]	10^{-3}	1.6	1.8	3.0
12b'''	$C_{18}H_{37}$ [a]	10^{-3}	1.3	7.0	3.2
1a CMPO		10^{-2}	0.74	3.44	0.17

[a] ligands in 1,3-position alternating with methyl in 2,4-position, compare compound **12b**

Selectivities

Compound **6c** (Y = C_5H_{11}) was chosen to study the extraction efficiencies of wide rim CMPOs towards nine lanthanides (La, Pm, Sm, Eu, Gd, Tb, Ho, Er, Yb) and two actinides (Am, Cm) (*10*). CMPO **1a** in a concentration appropriate to obtain comparable distribution coefficients was again used for comparison. The extraction conditions and the results are represented in Fig. 2. A strong decrease of the distribution coefficients along the lanthanide series, from 140 for lanthanum to 0.19 for ytterbium was found which corresponds to a separation factor (= ratio of the distribution coefficients) of nearly three orders of magnitude. CMPO showed only a weak and broad maximum of D for cations with an intermediate ion radius.

The observed size selectivity is strongly decreased, if the phenyl groups in the phosphine oxide structures are replaced by hexyl, an observation made also for the much lower selectivity of simple CMPOs. The selectivity is kept, however, for the diphenylphosphine oxide derivative **6c** for the extraction from a strongly acidic (3 M HNO$_3$) medium, where essentially the same trend is observed as shown in Fig. 2. A separation factor $D_{Am}/D_{Eu} = 10.2$ is found under these conditions which is probably one of the highest ever observed for this acidity (*10*).

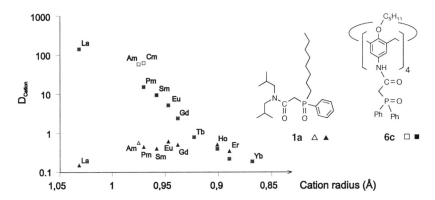

*Figure 2. Comparison of the distribution coefficients obtained for various lanthanides and actinides with **1a** and **6c** (Y = C$_5$H$_{11}$) (10). Aqueous phase: c(NaNO$_3$) = 4 M, c(HNO$_3$) = 0.01 M; c(Ln^{3+}) = 10^{-6} M (Pm and actinides at trace levels); organic phase: c(**1a**) = 0.2 M or c(**6c**) = 10^{-3} M in CHCl$_3$.*

Composition and Structure of Complexes

Intuitively, it was assumed that the cation would be surrounded by the four pendant arms of the ligand **6** forming a 1:1 complex. However, there are various indications that the complexation mode is not so simple. Extraction of Th(IV)-nitrate by **6** from an aqueous phase (1 M HNO$_3$) into dichloromethane led to a slope of about 1 in plots of log D versus log c(L), consistent with a 1:1 complex (*5*). Similar experiments with Eu(III)-nitrate, however, gave log/log-plots with a slope close to 2 suggesting a 1:2 stoichiometry of the extracted species. (Saturation of the organic phase by offering an excess of cations in the aqueous phase is complete again at a ratio 1:1.)

Complexes of **6** with the diamagnetic Th^{4+} and paramagnetic cations like Yb^{3+} were studied also by NMR. Meaningful spectra were obtained only when perchlorates were used in slight excess under strictly anhydrous conditions in acetonitrile (see Fig. 3) (*11*). The results are in agreement with a 1:1 complex (in the case of Yb^{3+} at least as the main species) with an unusual, up to then never observed C$_{2v}$-geometry where the two symmetry planes intersect opposite methylene bridges. This is indicated for instance by two pairs of Ar-CH$_2$-Ar doublets in the Th-complex and two signals (correlated in the COSY spectrum) for the aromatic protons of the calixarene, showing that the methylene groups

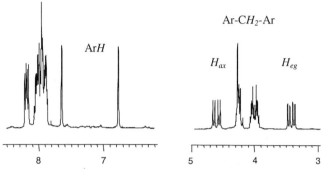

Figure 3. Section of the ^{1}H-NMR-spectrum of the complex obtained from **6b** with anhydrous Th(ClO$_4$)$_4$ in CD$_3$CN (11).

are pairwise different, while the four equivalent aromatic units have two different Ar-H protons.

It is not entirely clear, if this apparent C$_{2v}$-symmetry imposed by the coordination requirements of the cation is connected with a diamond-like distortion of the calixarene shape (Fig. 4a) with different diagonal distances between opposite methylene groups or just by an appropriate arrangement of the ligating NH-C(O)-CH$_2$-P(O)-Ph$_2$ groups around the cation (Fig. 4b). It may be that both effects illustrated as an alternative in Fig. 4 are superimposed.

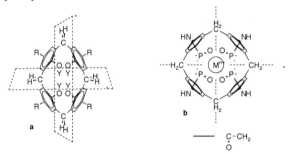

Figure 4. Schematic representation of two alternative explanations for the apparent C$_{2v}$-symmetry of lanthanide complexes with **6**; a) distortion of the calixarene shape (seen from the narrow rim); b) arrangement of ligating functions, P=O as an example (seen from the wide rim) (11).

A further corroboration is given, however, by two derivatives having C$_{2v}$-symmetry by constitution, due to the substitution of the phenolic oxygens (compare sections below). While the 1,3-mixed ether **12b** (Y^1/Y^3 = Pr, Y^2/Y^4 = Me) with symmetry planes through the phenolic units has identical methylene groups, the methylene groups are different in the bis-crown **14a** where the symmetry planes intersect the bridges. Thus, a single complex exists for **12b** while two isomeric forms are found for **14a** in an equilibrium slow on the NMR-time scale (*12*).

The spectra are less clear if the metal ions are added in a 1:1 stoichiometry, conditions under which relaxivity measurements suggest the presence of oligomeric species (*11*). It must be emphasized also, that well-resolved NMR spectra for paramagnetic cations like Yb^{3+} (spread over a range of 30 to -40 ppm) disappear upon addition of the stronger co-ordinating nitrate anions or of water, indicating averaging of signals by rapid exchange between various structures.

Structural Modification of Wide Rim CMPO-Calixarenes

The general structure presented so far can be varied in numerous ways. Some of these variations, undertaken with the aim to increase the extraction efficiency and especially to create or to improve selectivities for certain cations are described in the following sections.

Calix[5]arenes

An obvious variation is the ring size of the calixarene skeleton. Calix[5]arenes functionalized with CMPO-groups at the wide rim can be obtained exactly as described above. To obtain a conformationally pure derivative in the cone-conformation without complications, the ether residues should be octyl or larger. Shorter chains led to mixtures of conformers in our hands (*13*). Most probably the synthetic strategy holds also for calix[n]arenes with n>5. However, since only *t*-butylcalix[5]arene can be fixed in the cone conformation by ether groups, we have not studied larger calixarenes so far.

Extraction results with calix[5]arenes **11** are in principle similar to calix[4]arenes, although an interesting selectivity for Np was observed (*5*). Its distribution coefficient (1 M HNO_3, 4 M $NaNO_3$ / NPHE) is distinctly higher than for analogous calix[4]arenes **6**, while the opposite trend is seen for Am. This may be due to higher oxidation states (IV - VI) possible for Np.

Flexible Calix[4]arenes

The basic conformations of calix[4]arenes can be completely fixed by O-alkyl groups larger than ethyl, while methyl groups can pass the annulus. Therefore mixed (*syn*)alkyl/methyl ethers of calix[4]arenes do not necessarily assume exclusively the cone conformation. And even if complexation requires the cone conformation they should show slight differences in their conformational preferences, which might be used to "fine tune" the complexation properties of wide rim CMPO calix[4]arenes **6**. With this idea in mind we prepared a series of CMPO-derivatives bearing on their narrow rim all possible combinations of O-propyl and O-methyl groups (*14*).

Their synthesis starts with the partial O-alkylation with propylbromide (or any other alkylating agent, if desired) followed by exhaustive methylation. This sequence is crucial to ensure the *syn* arrangement of the propoxy groups. The subsequent steps are those, described already above.

Among the various compounds the 1,2-dipropyl-3,4-dimethylether **12c** shows the highest (73%), the tetramethylether **6a** the lowest (35%) extraction values for Eu^{3+} from 1 M HNO_3 into dichloromethane at a 1:1 o/a ratio. This corresponds to a factor of five for the distribution coefficients. Differences are less pronounced for Th^{4+}, where all compounds **12a-d** are slightly better (66-70%) than **6b** (Y = Pr) (*14*).

In Fig. 5 the extraction abilities of the different calixarenes **12** for various actinides and lanthanides are compared for the extraction from 3 M HNO_3 to o-nitrophenyl hexyl ether (extraction from highly acidic solution being technically most interesting). Again **12c** shows the best results (expressed here by the distribution coefficient).

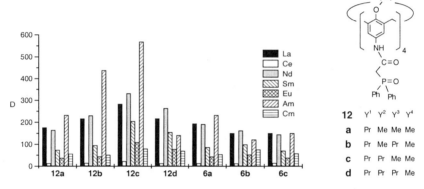

*Figure 5. Extraction of different cations, expressed by the distribution coefficient, from 3 M HNO_3 to NPHE by different CMPO-calixarenes **12** and **6** ($c_L = 10^{-3}$ M) (14).*

It is difficult to explain in detail the subtle differences observed in this series of conformationally mobile CMPO calixarenes, since they will be influenced also by differences in the lipophilicity. It seems reasonable, however, to assume that the diamond like C_{2v} distortion (compare above) is easiest achieved by the 1,2/3,4 substitution pattern of the ether groups present in **12c**.

Linear Oligomers

The most flexible alternative that can be obtained within the general structural features of wide-rim CMPO-calixarenes are their non cyclic analogues in which of course any preorganization of the CMPO-like functions is more or less lost. On the other hand this offers additional possibilities to empirically understand the requirements for an efficient extractant.

Synthesis starts with p-nitrophenol or its dimer (obtained by condensation with formaldehyde) from which linear oligomers are built up by alternative bis-bromomethylation and condensation with excess of p-nitrophenol. Thus, two steps convert a "n-mer" into the corresponding "(n+2)-mer" a principle that could be continued still to higher oligomers. After complete O-alkylation the further synthetic steps are identical to the synthesis of **6**.

Two remarkable results should be mentioned here: a) The extraction ability increases from the simple phenol derivative **13a** to the tetramer **13d**, but is lower again for the pentamer **13e**. b) For **13d** the extraction values for Eu(III) are similar to those of calix[4]arenes **6** and for Th(IV) they are at least in the same order of magnitude (*14*).

Rigidified Calix[4]arenes

In line with the above considerations, it seemed reasonable also to study the effect of a more rigid skeleton on which to assemble CMPO-functions (*15*). It has been shown (*16*) that the connection of adjacent oxygens in calix[4]arenes by 3-oxapentano bridges leads to a calix[4]arene skeleton fixed in a nearly ideal cone conformation with almost fourfold symmetry (*17*) (disregarding the C_{2v}-symmetrical crown ether part) which lends itself for the attachment of CMPO-functions.

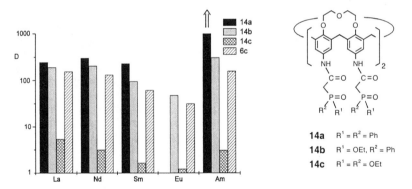

*Figure 6. Extraction of different cations from 3 M HNO_3 to NPHE ($c_L = 10^{-3}M$) calixarenes **14** in comparison to **6c** ($Y = C_5H_{11}$) (21).*

Table II. Extraction of ^{241}Am from aqueous nitric acid into NPHE

Ligand ($c_L = 10^{-3}$ M)	Distribution coefficients D		
	0.1 M HNO$_3$	1 M HNO$_3$	3 M HNO$_3$
14a	397	>1000	>1000
14b	25	247	308
7 (Y = C$_5$H$_{11}$)	0.45	12.1	40.4
14c	0.25	1.1	3.1

The most difficult step in the synthesis of CMPO-derivatives **14** was the ipso-nitration (*18*) which could be finally achieved in boiling acetic acid (*19*). Due to the low solubility the reduction to the amino groups was carried out with hydrazine hydrate, while phosphine oxide, phosphinate and phosphonate groups were introduced as described above.

Compounds **14** showed higher extraction efficiencies (distribution coefficient) in comparison to the corresponding tetraalkyl ether compounds **6-8** (Fig. 6, Table II) (*20*). Unfortunately however, no improved selectivity could be observed. The size selectivity shown in Fig. 2 was found more or less unchanged (*21*).

Partial CMPO-Derivatives

Mainly in order to get a better understanding of the complexation mode we also prepared a series of calix[4]arenes partially substituted by CMPO-functions on the wide rim. The synthesis of these compounds was based upon partial (ipso)-nitration or on calixarene synthesis by fragment condensation (*22*).

As expected, the extraction abilities of these calixarenes, in which the remaining p-positions were substituted by *t*-butyl, varied in the order mono-CMPO<<1,3-di-CMPO<1,2-di-CMPO<<tri-CMPO. All of them showed much lower extraction abilities than the reference CMPO-calixarene **6a**, but the tri-CMPO was even distinctly less effective than the linear trimer **13c**. For example, under conditions described in Fig. 2, with $c_L = 10^{-2}$ M distribution coefficients of 2.2 (tri-CMPO), 70 (**13c**) and 653 (**6c**) were found for Eu(III) to give one example (*20*).

Since this might be due to the hydrophobic nature of the *t*-butyl groups in the remaining p-positions or to steric reasons due to their bulkiness, we studied also calixarene derivatives, where these p-positions are unsubstituted. Again their extraction ability is very low. However, the structure of the 1,3-di-CMPO-calixarene could be confirmed by single crystal X-ray analysis, as a first example of a wide rim CMPO (*23*) Fig. 7 shows the result and gives an impression of the molecular proportions. The conformation is mainly determined by two intramolecular NH···O=P hydrogen bonds occurring between the opposite CMPO-groups.

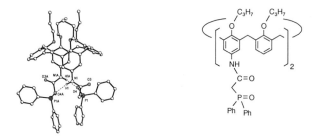

Figure 7. Single crystal X-ray analysis of a 1,3-di-CMPO-calixarene (23).

We recently found conditions for the easy protection of one, two adjacent, or three amino groups in a tetraamino calix[4]arenes **5** by the BOC-group (*24*). This will open up the way, to prepare calix[4]arenes bearing different CMPO-like functional groups (see below) or similar groups like malonamide structures at the wide rim.

Variation of the Phosphorus Function

As already indicated, the two phenyl groups of the phosphine oxide structures may be replaced by two alkyl groups. The extraction ability of these dialkyl phosphine oxide derivatives for medium sized lanthanide ions is similar, but lower for e.g. La(III) and higher for e.g. Yb(III). This loss of selectivity makes them obviously less attractive.

As mentioned above, phosphinates (**7**) and phosphonates (**8**) are accessible via Arbouzov reaction via the bromoacetamides (*19*). The extraction ability of these compounds decreases (in agreement with literature data on simple CMPOs (*25*)) in the order phosphine oxide > phosphinate > phosphonate, as shown in Table II for the rigid compounds **14**.

Phosphinates and phosphonates may be hydrolyzed, however, to the corresponding acids via the silyl-derivatives (*19*). In combination with phosphine oxide groups the incorporation of these potentially ionizable groups may lead to elaborate ligands able to compensate the cationic charge (completely or in part) by anionic groups, which may be favorable for the extraction.

The combination of CMPO-functions with malonic acid functions attached via amide bonds (calix-NH-C(O)-CH$_2$-COOH) is another option (26).

Narrow Rim CMPO-Calix[4]arenes

Functional groups attached to the wide rim of a calix[4]arene fixed in the cone conformation are divergently oriented, and this could well be one reason for the complicated situation with respect to their interaction with metal cations. The situation may change, if such groups are attached to the narrow rim, having therefore primarily a more convergent orientation.

Syntheses

The preparation of calix[4]arenes bearing CMPO-functions at the narrow rim (27) requires again suitable tetraamines, which in the last step can be acylated by the active ester **10** (28). The shortest inert and stable connection to the narrow rim seems to be an ether linkage (-O-(CH$_2$)$_n$-NH$_2$) of at least two carbon atoms length. Such aminoether derivatives **15** have been partly described before (29). In our hands O-alkylation of *t*-butylcalix[4]arene by N-(ω-bromoalkyl)-phthalimide followed by deprotection of the aminogroups with hydrazine gave the best results for n > 2, while for n = 2 the known ethyleneoxy tetratosylate (available in three steps from *t*-butylcalix[4]arene) was substituted by azide and the product reduced to the amine **15a** (n=2) without isolation.

Extraction

First extraction studies with ligands **16a-c** (n = 2-4) were done again from 1 M HNO$_3$ to dichloromethane. All the narrow rim CMPO-calixarenes are better extractants for Th(IV) as compared to **6c**, while for the extraction of lanthanides only **16c**, surprisingly the compound with the longest tether, can compete with **6c**. However, the %E-values reveal, that the selectivity for the light over the heavy lanthanides is not so pronounced as for **6c** (27).

Fig. 8 shows in comparison typical results for the extraction of selected cations with wide and narrow rim CMPO-calixarenes as a function of the concentration of HNO_3 in the source phase. In contrast to **6c** no decrease of the extraction ability is observed for higher concentrations in the case of **16c**. Although the reason for this remarkable difference is not known yet, this may be well a potential advantage of narrow rim CMPOs over wide rim CMPOs.

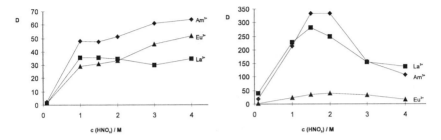

*Figure 8. Distribution coefficient as function of c(HNO_3) for the extraction with **16c** (left) and **6c** (right). Organic phase NPHE, $c_L = 10^{-3}$ M (27).*

Conclusions and Outlook

Calix[4]arenes bearing –NH-C(O)-CH_2-P(O)-Ph_2 functions, either at their narrow or at their wide rim, are excellent extractants for lanthanides and actinides. Extraction levels comparable or even superior to those of the "classical" CMPO **1a** are reached with ligand concentrations of 10^{-3} M or less under conditions relevant for the treatment of technical nuclear waste streams. This high efficiency would compensate also a higher price for their synthesis. In addition, wide rim CMPO-calixarenes **6** show an interesting size selectivity within the lanthanides and between actinides and lanthanides.

At present it must be stated, however, that these excellent extraction properties cannot yet be convincingly explained by structural features of the complexes formed and that even the composition of the extracted species is not entirely clear. The structural modifications which are possible with these ligands, especially the construction of calixarene derivatives with mixed ligating functions, may be used to fine tune their properties. Simultaneously this will lead also to a better understanding of these properties and to a detailed description of the interaction between cations and ligand.

Acknowledgement: These studies were supported by the European Community (Contract FI4W-CT96-0022). I am grateful to all the colleagues and coworkers who contributed to these results. Their names are mentioned in the respective references.

References

1. Horwitz, E. P.; Kalina, D. G.; Diamond, H.; Vandegrift, D. G.; Schultz, W. W. *Solv. Extr. Ion Exch.* **1985**, *3*, 75.
2. Böhmer, V. *Angew. Chem.* **1995**, *107*, 785; *Angew. Chem. Int. Ed. Engl.* **1995**, *34*, 713.
3. Gutsche, C. D. *Calixarenes Revisited* in *Monographs in Supramolecular Chemistry*, ed. F. Stoddart, F., Ed.; The Royal Chemical Society, Cambridge, 1998.
4. For reviews on calixarene based ionophores see: a) McKervey, M. A.; Arnaud-Neu, F.; Schwing-Weill, M.-J. *Cation binding by calixarenes* in *Comprehensive Supramolecular Chemistry*, Vol. 1, p. 537, Gokel, G. W., Ed.; Pergamon, Oxford, **1996** b) Ikeda, A.; Shinkai, S. *Chem. Rev.* **1997**, *97*, 1713.
5. Arnaud-Neu, F.; Böhmer, V.; Dozol, J.-F.; Grüttner, C.; Jakobi, R. A.; Kraft, D.; Mauprivez, O.; Rouquette, H.; Schwing-Weill, M.-J.; Simon, N.; Vogt, W. *J. Chem. Soc., Perkin Trans. 2*, **1996**, 1175.
6. Verboom, W.; Durie, A.; Egberink, R. J. M.; Asfari, Z.; Reinhoudt, D. N. *J. Org. Chem.* **1992**, *57*, 1313.
7. Jakobi, R. A.; Böhmer, V.; Grüttner, C.; Kraft, D.; Vogt, W. *New J. Chem.* **1996**, *20*, 493.
8. Ugozzoli, F.; Böhmer, V.; et al., unpublished results.
9. Klüh, U.; Vogt, W.; Böhmer, V., unpublished results.
10. Delmau, L. H.; Simon, N.; Schwing-Weill, M.-J.; Arnaud-Neu, F.; Dozol, J.-F.; Eymard, S.; Tournois, B.; Böhmer, V.; Grüttner, C.; Musigmann, C.; Tunayar, A. *Chem. Commun.* **1998**, 1627.
11. Lambert, B.; Jacques, V.; Shivanyuk, A.; Matthews, S. E.; Tunayar, A.; Baaden, M.; Wipff, G.; Böhmer, V.; Desreux, J. F. *Inorg. Chem.*, submitted.
12. Desreux, J. F.; Böhmer, V.; et al., unpublished results.
13. Musigmann, C.; Böhmer, V., unpublished results.
14. Matthews, S. E.; Saadioui, M.; Böhmer, V.; Barboso, S.; Arnaud-Neu, F.; Schwing-Weill, M.-J.; Garcia Carrera, A.; Dozol, J.-F. *J. prakt. Chem.* **1999**, *341*, 264.
15. For the use of cavitands as molecular scaffold for CMPO-like functions see: a) Boerrigter, H.; Verboom, W.; Reinhoudt, D. N. *J. Org. Chem.*, **1997**, *62*, 7155; b) Boerrigter, H.; Verboom, W.; Reinhoudt, D. N. *Liebigs Ann./Recueil*, **1997**, 2247; c) Boerrigter, H.; Verboom, W.; De Jong, F.; Reinhoudt, D. N. *Radiochim. Acta*, **1998**, *81*, 39.
16. Arduini, A.; Fanni, S.; Manfredi, G.; Pochini, A.; Ungaro, R.; Sicuri, A. R.; Ugozzoli, F. *J. Org. Chem.* **1995**, *60*, 1454.
17. For an X-ray structure see: Arduini, A.; McGregor, W. M.; Paganuzzi, D.; Pochini, A.; Secchi, A.; Ugozzoli, F.; Ungaro, R. *J. Chem. Soc., Perkin Trans. 2*, **1996**, 839.
18. For the nitration see: Arduini, A.; Mirone, L.; Paganuzzi, D.; Pinalli, A.; Pochini, A.; Secchi, A.; Ungaro, R. *Tetrahedron* **1996**, *52*, 6011.
19. Shivanyuk, A.; Böhmer, V., unpublished results.
20. Delmau, L. H.; Dozol, J.-F; Böhmer, V.; et al., unpublished results.

21. Garcia Carrera, A.; Dozol, J.-F.; Böhmer, V.; et al., unpublished results.
22. Böhmer, V. *Liebigs Ann./Recueil* **1997**, 2019.
23. Paulus, E. F.; Shivanyuk, A.; Böhmer, V., unpublished results.
24. Saadioui, M.; Shivanyuk, A.; Böhmer, V.; Vogt, W. *J. Org. Chem.* **1999**, *64*, 3774.
25. Kalina, D. G.; Horwitz, E. P.; Kaplan, L.; Muscatello, A. C. *Sep. Sci. Techn.* **1981**, *16*, 1127.
26. Shivanyuk, A.; Tunayar, A.; Böhmer, V., unpublished results.
27. Barboso, S.; Garcia Carrera, A.; Matthews, S. E.; Arnaud-Neu, F.; Böhmer, V.; Dozol, J.-F.; Rouquette, H.; Schwing-Weill, M.-J. *J. Chem. Soc., Perkin Trans. 2*, **1999**, 719.
28. See also: Lambert, T. N.; Jarvinen, G. D.; Gopolan, A. S. *Tetrahedron Lett.* **1999**, *40*, 1613.
29. e.g. Scheerder, J.; Fochi, M.; Engbersen, J. F. J.; Reinhoudt, D. N. *J. Org. Chem.*, **1994**, **59**, 7815.

Chapter 12

Binding of Lanthanides(III) and Thorium(IV) by Phosphorylated Calixarenes

F. Arnaud-Neu, S. Barboso, D. Byrne, L. J. Charbonnière,
M. J. Schwing-Weill, and G. Ulrich

Laboratoire de Chimie-Physique, UMR 7512 du CNRS, ULP-ECPM,
25, rue Becquerel, 67087 Strasbourg Cedex 2, France

Liquid-liquid extraction of lanthanide and thorium nitrates from acidic aqueous solutions into dichloromethane by different series of phosphorylated calixarenes are reported. The nature and the stability constants of the complexes formed in methanol are also presented. These receptors extract and complex these cations more efficiently than CMPO or TOPO, and some of them exhibit interesting selectivities.

The nuclear industry is currently confronted with the major problem of the concentration of generated radioactive wastes for further disposal or decontamination (1). Their treatment requires at some stage the extraction of lanthanides and actinides, together or separately, from effluents of high salinity and acidity. Much effort has already been devoted to their simultaneous extraction, leading in particular to the development of the TRUEX process, based on the use of the (N,N'-diisobutylcarbamoylmethyl)octylphenylphosphine oxide (CMPO), which has been thus far, along with TOPO (trioctylphosphine oxide), one of the best extractants of these kinds of metal ions (2). The extracted complex of Am(III) involves three CMPO molecules, three co-extracted HNO_3 molecules and three nitrate counterions (3). However, it is still debatable whether both the phosphoryl and the carbonyl (4) or only the phosphoryl groups (5) participate in the complexation. On the basis of the remarkable abilities of calixarene derivatives for the selective binding of metal ions (6), several groups have introduced CMPO and other phosphorylated residues on calixarenic structures (7-17). Some of these derivatives have demonstrated to be effective extractants for lanthanide and thorium ions. Although much is known about extraction with these ligands and related compounds including CMPO, there is a need to acquire a better understanding of the thermodynamic data of complexation, in particular in the case of calixarenes to understand the role played by the macrocyclic skeleton.

We report here extraction data for calixarene derivatives substituted by phosphine oxides or carbamoyl phosphine oxides as well as complexation data in a homogeneous medium. These data are very important for a better knowledge of the complexation and the further improvement in the performances of these compounds.

The results are compared with those obtained with acyclic analogues as well as CMPO and TOPO. Different factors affecting the binding abilities of the ligands are examined such as (i) the nature and the positionning of the functional groups; (ii) the conformation and flexibility of the ligands; (iii) the substituents at the lower or upper rim and at the functional groups; (iv) the nature of the metal ion and of the counterion.

Upper Rim CMPO-Calixarenes

A series of p-CMPO calix[4]arene tetraalkoxy (R = CH_3, C_3H_7, C_5H_{11}, $C_{10}H_{21}$, $C_{12}H_{25}$, $C_{14}H_{27}$, $C_{16}H_{33}$ and $C_{18}H_{37}$) ("homo-calixarenes") and related "mixed" derivatives combining two different alkoxy substituents (R = CH_3 and C_3H_7) as well as some of the corresponding acyclic compounds (Figure 1) have been studied (7, 11). Except the flexible methoxy derivative, all the "homo" calixarenes are in the cone conformation.

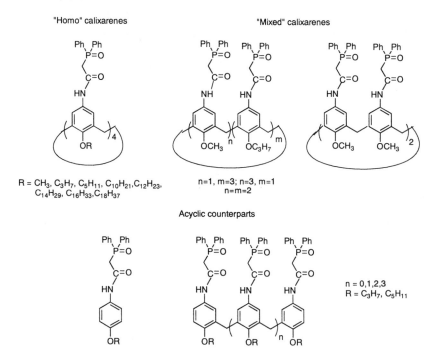

Figure 1. Upper rim CMPO calix[4]arenes and acyclic counterparts

Extraction studies

Extent of extraction

Extraction experiments from 1M HNO_3 aqueous solution into dichloromethane were performed with some lanthanides and thorium nitrates (C_M = 10^{-4} M) (7, 11). Selected results are given in Table 1 for thorium and europium. They show the remarkable efficiency of all the "homo" calixarenes for thorium as low calixarene concentrations (C_L = 10^{-4} M) allow percentage extraction ranging from 50 to 63% (at an organic-to-aqueous phase ratio of 1). These compounds are also good extractants of europium, although ligand concentrations of 10^{-3} M (10 times higher than for thorium) must be used to reach a percentage extraction ranging from 35 to 72 %. They are much more efficient than CMPO and TOPO, which must be used at least at a concentration of 0.025 M (for thorium) or 0.25 M (for europium) to reach similar extraction levels.

Table 1. Percentage extraction of europium and thorium nitrates (C_M = 10^{-4} M) from aqueous HNO_3 1M solutions into dichloromethane[a] by "homo" upper rim CMPO calix[4]arenes[b] and related acyclic compounds[c]

Calixarenes (R)	Thorium (C_L = 10^{-4}M)	Europium (C_L = 10^{-3}M)	Acyclics (R)	Thorium (C_L = 10^{-4}M)	Europium (C_L = 10^{-3}M)
CH_3[c]	61	35	Monomer (C_3H_7)	6[d]	4[e]
C_3H_7[c]	62	64	Dimer	18[d]	24[e]
C_5H_{11}	60	58	Trimer	6	23
$C_{10}H_{21}$	53	68	Tetramer	35	57
$C_{12}H_{23}$	63	68	Pentamer	31	38
$C_{14}H_{29}$	54	72			
$C_{16}H_{33}$	52	70	Monomer (C_5H_{11})	4[d]	6[e]
$C_{18}H_{37}$	50	59	Dimer	22[d]	18[e]
			Trimer	4.3	15
CMPO	70[f]	18[g]	Tetramer	15	43
TOPO	64[f]	70[g]	Pentamer	22	17

a) Organic-to-aqueous phase ratio = 1; b) reference 5; c) reference 11; d) C_L = 10^{-3} M; e) C_L = 0.01 M; f) C_L = 0.025 M; g) C_L = 0.25 M

Importance of the calixarenic structure

Data in Table 1 also show that extraction of thorium and europium increases on going from the acyclic monomers to the tetramers (R = C_3H_7 and R = C_5H_{11}), but tends to decrease for the pentamers. It is important to note that these acyclic compounds, although better than CMPO, are all less efficient than the corresponding calix[4]arenes, especially for thorium. For instance, the percentage extraction of this cation is only 15 % with the tetramer (R = C_5H_{11}) whereas it is 60 % with the corresponding calixarene.

Influence of the calixarene flexibility

As seen in Table 1, only a slight influence on the extraction of europium and thorium nitrates is observed by varying the alkyl chain length of the lower rim substituents, except in the case of the methoxy compound with europium, which leads to a percentage extraction significantly lower (%E = 35 %) than those for the other derivatives (average %E = 65±7 %).

The mixed calixarenes studied correspond to the progressive replacement of the four CH_3 groups at the lower rim of the "homo" calixarene by bulkier C_3H_7 groups. This substitution, which results in rigidification of the calixarene structure, has little influence on the extraction of thorium, which covers the range 60-70% for $C_L = 10^{-4}M$ (Figure 2). In contrast, the extraction levels of europium increase significantly with the number of propyl groups. However, the best extractant is the 1,2-dimethoxy-3,4-dipropoxy derivative (%E = 73). No satisfactory interpretation of these results has been found so far.

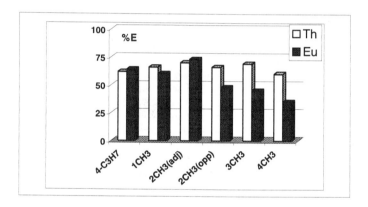

Figure 2. Extraction percentage of thorium and europium nitrates from 1M HNO_3 into dichloromethane by two "homo" p-CMPO calix[4]arenes (R = CH_3 and R = C_3H_7) and their corresponding "mixed" derivatives ($C_M = C_L = 10^{-4}M$; organic-to-aqueous phase ratio = 1)

Intra-lanthanide series selectivities

With the propoxy calixarene derivative, the extraction level decreases in the lanthanide series from 98% for La^{3+} to 64% for Eu^{3+} and 6% for Yb^{3+} in the above mentioned conditions (Figure 3). Such a decrease is not displayed by CMPO nor by the corresponding acyclic tetramer which extract europium better than lanthanum or ytterbium. Similar trends are observed with these compounds in different experimental conditions (extraction from aqueous HNO_3 0.01M/$NaNO_3$ 4M into $CHCl_3$ (8)). This emphasizes the importance of the calixarenic structure in the demonstration of intra-series selectivity.

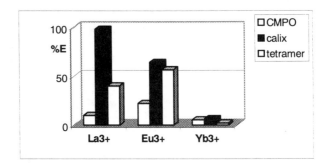

Figure 3. Comparison of percentage extraction of some lanthanide nitrates from 1M HNO_3 into dichloromethane with CMPO ($C_L = 0.2$ M), the p-CMPO calix[4]arene (R = C_3H_7) and its acyclic counterpart ($C_L = 10^{-3}$ M) ($C_M = 10^{-4}$ M, organic-to-aqueous phase ratio = 1)

Stoichiometry of the extracted species

The slopes of the log D – log[L] plots for the extraction of thorium by some of the "homo" calixarenes are near 1, as expected for the formation of 1:1 complexes (7). In contrast, slopes of 2 have been found with europium suggesting that the stoichiometry of the extracted complexes is 1:2 under these conditions, i.e. in the presence of an excess of ligand. However determination of the loading capacity of the organic phase indicated the formation of a 1:1 complex (15). The results show the possible formation of two kinds of complexes during the extraction process, with either 1:1 or 1:2 stoichiometries depending on the respective metal/ligand concentrations. With the mixed derivative 1,2-dimethoxy-3,4 dipropoxy a slope of 2 has also been found in the case of europium, suggesting again 1:2 stoichiometry for the extracted complex (11).

Complexation studies

Complexation data in methanol for some lanthanides(III) and thorium(IV) have been established with two "homo" calixarenes (R = CH_3 and C_5H_{11}), two "mixed" calixarenes (the 1,2-dimethoxy-3,4-dipropoxy and the 1,3-dimethoxy-2,4-dipropoxy derivatives) as well as with some related acyclic compounds and CMPO (18). The nature of the complexes formed and their stability constants were determined in methanol, in the presence of $NaNO_3$ as inert salt, by absorption spectrophotometry. Direct titrations of the ligands were performed in cases of significant spectral changes. When this was not possible, a competitive spectrophotometric method using the auxiliary coloured ligand PAN (1-(2-pyridylazo)-2-naphtol) was implemented (19). In these experiments, typical PAN concentration was 10^{-4} M and C_L ranged from 10^{-2} to 10^{-3} M. The wavelength range monitored was 400 to 700 nm. Table 2 lists the results obtained for europium and thorium.

Rigid calixarene : p-CMPO calix[4]arene (R = C_5H_{11})

The results obtained by competitive absorption spectrophotometry with the rigid pentoxy derivative, fixed in the cone conformation, show the formation of 1:1 complexes with all lanthanides studied. An interesting feature is the simultaneous formation of 1:2 complexes with cations of the middle of the series (Pr^{3+}, Eu^{3+}, Tb^{3+}, Er^{3+}), which is in agreement with the assumed stoichiometry of the extracted europium complex. Distribution curves show that the two types of complexes coexist for $C_L/C_M = 2$. The electrospray mass spectrum of a solution of ligand and europium in this ratio presents peaks which can be attributed to the mono- and bi-ligand complexes thus confirming the spectrophotometric results. With thorium, only the 1:1 species is found, a result which is again in agreement with the extraction data.

Table 2. Overall stability constants (log β_{xy})[a] of europium and thorium complexes of "homo" *p*-CMPO calix[4]arenes and related compounds in methanol (T = 25 °C, I = 0.05 M ($NaNO_3$))

Ligands	Europium		Thorium	
	log β_{11}	log β_{12}	log β_{11}	log β_{12}
"Homo" calix (R = C_5H_{11})	6.2	11.1	6.4	
Acyclic monomer (R = C_5H_{11})	4.5	7.4		
Acyclic dimer (R = C_5H_{11})	5.0	9.0		
Acyclic trimer (R = C_5H_{11})	5.5	10.3		
Acyclic tetramer (R = C_5H_{11})	5.6	9.9		
Acyclic pentamer (R = C_5H_{11})	6.2	10.1		
CMPO	3.6	5.5	5.1	9.4

a : corresponding to the equilibrium : $xEu^{3+} + yL \leftrightarrow Eu_xL_y^{3x+}$

The stability constants of the 1:1 complexes of lanthanides are all close to 6 log units and hence no particular selectivity is observed along the series, although the lanthanum complex is slightly less stable (log β_{11} = 5.0) (Figure 4). The stability of the 1:2 complexes reveals a slight maximum for terbium.

The 1:1 and 1:2 complexes also formed with CMPO in nitrate medium are much less stable than the corresponding complexes with the calixarene derivative (eg. Δ log β_{11} = 2.6 for europium). In addition there is a slight increase in stability along the series for the 1:2 complexes. It is also worth noting that there is evidence for a third 1:3 complex in the presence of chlorides. In this medium, the complexes are more stable and their increase in stability along the series more pronounced than in the presence of nitrates. This strongly suggests coordination of nitrate anions.

In order to gain further insight into the influence of the calixarenic structure and the mode of coordination of the CMPO moieties, a series of acyclic analogues – from the monomer to the pentamer - has been studied with europium (Table 2) (18). The results show the formation of 1:1 and 1:2 species as with the calixarene derivative.

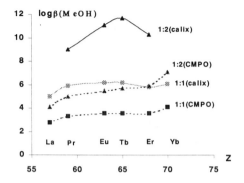

Figure 4. Stability constants of lanthanide complexes with CMPO and the upper rim CMPO calix[4]arene (R = C_5H_{11}) as a function of the atomic number Z of the cations.

The stability of the 1:1 complexes slightly increases when a supplementary unit is included in the ligand (Δlog β_{11} c.a. 0.4). These small differences, which can be accounted for only by statistical reasons, suggest that only one CMPO arm is involved in the complexation. Otherwise, larger increases would be expected. Regarding the 1:2 complexes, a higher stabilisation of the complexes is noticed on going from the monomer to the dimer and to the trimer. This could be explained by a second order effect, e.g. anion complexation. The uncomplexed arms of the compounds could stabilize the complexes by hydrogen bonding between the amide protons and nitrate anions. In the case of the tetra- and pentamers, the additional arms are useless and therefore a plateau is observed for the stepwise stability constants log K_2. The complexes of the calixarene appear to be only slightly more stable than those of the tetramer. Also in the calixarene a single arm may be involved in the complexation whereas the other three would stabilize the complexes via nitrate coordination. Further studies must be undertaken to confirm these assumptions.

Flexible calixarenes : "homo" p-CMPO calix[4]arene (R = CH_3) and "mixed" calix[4]arenes

Data for the complexation of europium and thorium by the flexible calixarenes (the "homo" methoxy and the two dimethoxy-dipropoxy derivatives) show the formation of 1:1 and 2:1 complexes with both cations (Table 3). A 1:2 species is also found with thorium and the 1,3-dimethoxy-2,4-dipropoxy calixarene.

The existence of binuclear complexes may be explained by the conformational mobility of the receptors. For instance, the 1,2-dimethoxy-3,4-dipropoxy derivative is likely to adopt the 1,2 alternate conformation, which would enable the formation of binuclear species. An alternative explanation would be a possible opening of the cone conformation owing to the presence of rather small substituents, which would enable the coordination of two cations as shown by the structure of the lanthanum complex with the tetrapropoxy derivative in the solid state (20).

Table 3. Overall stability constants (log β_{xy})[a] of europium complexes of "homo" and "mixed" p-CMPO calix[4]arenes in methanol
(T = 25 °C, = 0.05 M (NaNO$_3$))

Calixarenes	Europium		Thorium		
	log β_{11}	log β_{21}	log β_{11}	log β_{21}	log β_{12}
"Homo" calix (R = CH$_3$)	4.4	8.4	6.6	12.0	-
1,2-dimethoxy-3,4-dipropoxy	6.4	11.4	6.8	12.4	-
1,3-dimethoxy-2,4-dipropoxy	5.7	10.9	6.8	11.8	13.0

a) corresponding to the equilibrium : $xEu^{3+} + yL \leftrightarrow Eu_xL_y^{3x+}$

The 1,2-methoxy-3,4-propoxy derivative forms the most stable complexes with europium. It is interesting to note that the corresponding complexes with the tetramethoxy calixarene are much less stable (eg. $\Delta \log \beta_{11}$ = 2) and that the 1,3-dimethoxy-2,4-dipropoxy complexes have intermediate stability. These results perfectly mirror the extraction abilities of these compounds. Again, in agreement with extraction results, such differences are not found with thorium, whose complexes have roughly the same stability.

Lower Rim CMPO-Calixarenes

Another way to substitute calixarenes is to graft the CMPO residues on the lower rim, as previously carried out with other functional groups (6). A series of such derivatives differing in the number of CH$_2$ spacers between the phenolic oxygens and the functional groups as well as the nature of the *para*-substituents (H or *tert*-butyl) have been studied (Figure 5)(21).

Extraction studies

Figure 6 shows the extraction levels of Th^{4+}, La^{3+}, Eu^{3+} and Yb^{3+} nitrates (C$_M$ = 10^{-4}M) from a 1M nitric acid solution into dichloromethane (21).

All these compounds are highly efficient for Th^{4+} (%E = 80-100 for C$_L$ = 10^{-4} M), even more so than their upper rim counterparts. The lanthanides, however, are extracted to a much lesser degree. Maximum extraction of europium is achieved by the four CH$_2$ derivative (%E = 68% for C$_L$ = 10^{-3} M). For all cations, the extraction levels depend upon the length of the alkyl chain linking the functional groups : four appears to be the optimum number of CH$_2$ in terms of efficiency (Figure 6). This probably results from different competing factors such as, the size of the cavity defined by the donor sites, the flexibility of the receptors and their lipophilicity (21). For each ligand, the extraction level is similar for La^{3+} and Eu^{3+} and then decreases for Yb^{3+}.

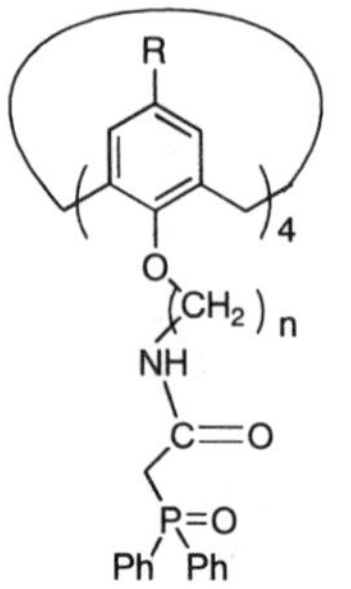

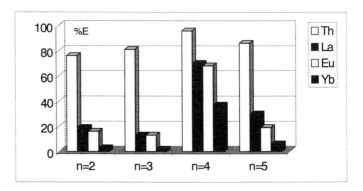

n = 2, 3, 4, 5; R = *tert*-butyl
n = 2, 3, 4; R = H,

Figure 5. Lower rim CMPO calix [4]arenes

An important feature of these lower rim CMPO derivatives is their remarkable selectivity for thorium over lanthanides, much higher than that displayed by their upper-rim counterparts (Figure 6). The most Th^{4+}/Eu^{3+} selective extractant is the three CH_2 derivative (n = 3). This selectivity, expressed as the ratio %E(Th)/%E(Eu), is increased upon dealkylation of the compounds, eg. from 4.75 to 14 when n = 2.

Figure 6. Percentage extraction of thorium and lanthanide nitrates by lower rim CMPO calixarenes (R = tert-butyl) as a function of the number n of CH_2 spacers. ($C_M = 10^{-4}$ M; $C_L = 10^{-4}$ M (thorium); $C_L = 10^{-3}$ M (europium); organic-to-aqueous phase ratio = 1).

The log D vs log [L] plot analysis for the extraction of europium by two *p-tert*-butyl derivatives (n = 3 and 4) shows slopes close to 1 instead of 2 for their upper rim counterparts (Figure 7). This suggests the 1:1 stoichiometry for the extracted complexes.

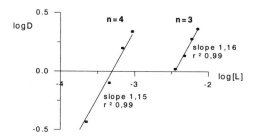

Figure 7. log D - log [L] plots for the extraction of europium nitrate by lower rim CMPO calixarenes (R = tert-butyl, n = 3 and n = 4).

Complexation studies

The complexation results in methanol for three ligands (n = 2, 3 and 4 ; R = *tert*-butyl), obtained by competitive spectrophotometry, show the formation of 1:1 complexes with the lanthanides and thorium (18), which is consistent with the extraction results (21) (Table 4). In some cases, the formation of a mixed complex involving one calixarene and one PAN molecule must be considered.

Table 4. Stability constants (log β_{11}) of the 1:1 complexes of lanthanide and thorium complexes with lower rim CMPO calix[4]arenes in methanol (I = 0.05 M in NaNO$_3$, T = 25 °C)

Cations	Complexes	n = 2	n = 3	n = 4
La^{3+}	1:1		7.0	
Pr^{3+}	1:1	7.0	6.8	6.5
Eu^{3+}	1:1	6.8	7.1	6.9
Tb^{3+}	1:1		7.1	
Er^{3+}	1:1	6.9	7.1	6.8
Yb^{3+}	1:1		6.8	
Th^{4+}	1:1	5.5	5.3	5.4

There is no evidence for the formation of 1:2 or 2:1 complexes like those which are formed with the upper rim derivatives. The 1:1 lanthanide complexes are more stable than the corresponding complexes with the *p*-CMPO calix[4]arene (R = C$_5$H$_{11}$) (0.6 $\leq \Delta$log $\beta_{11} \leq$ 2). No real selectivity is observed with these ligands. In addition their complexing power is similar and the trends observed in extraction are thus not reflected in complexation.

All the three ligands exhibit the same affinity for thorium. The stability of the complexes, ranging between 5.3 and 5.5 log units, is lower than that of the corresponding complexes with the upper rim derivatives. In contrast to the expectations based on extraction results, thorium complexes are less stable than lanthanide complexes in methanol.

Calix[n]arene Phosphine Oxides

Simple diphenylphosphine oxides groups have been attached to the lower rim of a series of *p*-H and *p-tert*-butyl calix[n]arenes (n = 4, 6, 8) with the functional groups separated by two CH_2 groups from the phenolic oxygens. Three other derivatives with different chain lengths have also been obtained (Figure 8) (10-13).

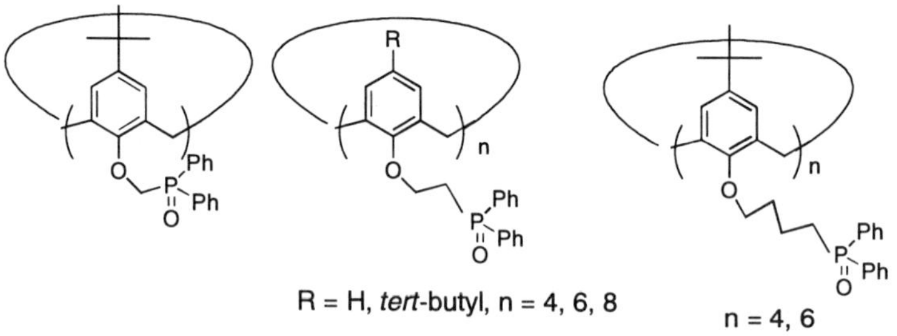

R = H, *tert*-butyl, n = 4, 6, 8   n = 4, 6

Figure 8. Structures of calix[n]arene phosphine oxides

Extraction studies

All these compounds are stronger extractants for europium and thorium nitrates than TOPO and CMPO. The only exception concerns the *p-tert*-butyl octamer, which does not extract europium at $C_L = 2.5 \times 10^{-2}$M. As seen on Figure 9 illustrating their performances, the dealkylated series is particularly efficient, with %E values for thorium close to or even higher than 80% for $C_L = 10^{-3}$ M. Thorium is always better extracted than europium. With the latter, much higher ligand concentrations ($C_L \geq 2.5 \times 10^{-2}$M) are needed to reach similar extraction levels.

The replacement of the phenyl groups on the phosphine oxide functions by n-butyl groups leads to a complete loss of extraction efficiency, since there is no extraction of thorium at $C_L = 10^{-3}$M.

The extraction of thorium by the *p-tert*-butyl tetra- and hexamers has also been shown to depend on the length of the alkyl chain between the phenolic oxygen and the functional groups. The efficiency decreases as this distance increases. In contrast to the observation made with the ethoxy derivatives (Figure 9), the butoxy tetramer is better than its hexameric counterpart (%E = 50% instead of 20% for $C_L = 10^{-2}$ M).

Nevertheless the *p*-H tetramer with two CH_2 spacers remains a better extractant than the "short" methoxy *p-tert*-butyl derivative with just one CH_2 group. For europium, the shortening of the spacer leads to an even more spectacular increase in the extraction level, from 0% with the ethoxy *p*-tert-butylcalix[4]arene derivative to 86% with its methoxy counterpart for $C_L = 2.5 \times 10^{-2}$ M.

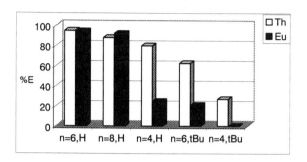

Figure 9. Extraction of thorium and europium nitrates from 1M HNO_3 into dichloromethane by calix[n]arene phosphine oxides ($C_M = 10^{-4}M$; $C_L = 10^{-4}M$ (thorium); $C_L = 10^{-3}$ M (europium); organic-to-aqueous phase ratio = 1).

The slopes of the log D – log [L] plots for the extraction of thorium by the three *p*-H calix[n]arenes are near to 2, suggesting that two ligand molecules may be involved in the extracted complex (12). This has been confirmed by saturation experiments of the organic phase (14). Slopes ranging between 1.8 and 2.6 have been found with the p-tert-butyl derivatives suggesting coextraction of complexes of different stoichiometries, although coextraction of nitric acid or activity coefficient effects must not be excluded (22). For the extraction of europium, the slopes range between 2.0 and 2.5, except with the *p*-H hexamer for which the slope is near 3 as for TOPO (23).

Complexation studies

Complexation studies are in progress. The first results concern the two ethoxy tetramers. 1:1 species are formed with the *p-tert*-butyl derivative and the cations Pr^{3+}, Eu^{3+}, and Er^{3+} (24). The corresponding stability constants are : 4.8, 4.9 and 5.1 log units. A 1:1 complex is also found with the *p*-H derivative and europium, which is 0.7 log units more stable than its *p-tert*-butyl homologue. This result is consistent with the extraction data. With the other cations, there is evidence for 1:2 complexes only, with similar stability (log $\beta_{12} = 9.2$).

Conclusions

All the phosphorylated calixarenes studied are better extractants for thorium and lanthanides than the classical CMPO and TOPO. However their efficiency depends on the nature of the phosphorylated functional groups and on their positionning on the calixarenic structure. Figure 10, which summarizes the results for the most effective compounds, shows that the best extraction of thorium is achieved with the lower rim substituted CMPO calixarenes.

High extraction of europium is achieved by both series of calixarenes although higher ligand concentrations are required to reach the same extent of extraction as that of thorium. An important result is the remarkable Th^{4+}/Eu^{3+} selectivity of the lower rim CMPO calixarenes (in particular with n = 3), which is still enhanced upon *p*-dealkylation of the compounds.

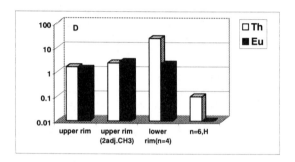

Figure 10. Best extractants for thorium ($C_L = 10^{-4}$ M) and europium ($C_L = 10^{-3}$ M)

The influence of some features on the extraction efficiency has been pointed out in each series of compounds.

- With the *upper rim CMPO calix[4]arenes*, an increase in the ligand flexibility leads to a decrease in the extraction level of europium and to no significant changes in the extraction of thorium. The flexibility also has a great influence on the stability of the complexes formed in methanol.

In general there is agreement between the stoichiometry of the complexes formed in methanol and the extracted species. However, whereas extraction selectivity is observed in the lanthanide series, there is only little discrimination in complexation. In the other hand, if there is an obvious calixarene effect in extraction (an increase on going from the acyclic oligomers to the calixarene), complexation results suggest that not all the CMPO arms may be involved in the complexation and that stabilisation of the complexes may occur via nitrate coordination.

- With the *lower rim CMPO calix[4]arenes*, the extraction efficiency depends on the length of the spacer separating the functional groups and the phenolic oxygens as well as the nature of the *p*-substituents. For instance, the presence of four CH_2 spacers leads to optimal efficiency and of three spacers to the best selectivities. With these compounds, the extraction results mirror the complexation data.

- The *lower rim phosphine oxide calixarenes*, although good extractants, are slightly less effective than the CMPO derivatives, certainly because of the lack of the carbonyl oxygen donors. Their efficacy also depends on the length of the spacers between the phenolic oxygen and the functional groups. But here, the shorter derivatives are the more efficient certainly because they enable the participation of the phenolic oxygens to the complexation. This is not the case for the lower rim CMPO derivatives for which the participation of the phenolic oxygen is not necessary to achieve effective complexation, because of the presence of the carbonyls from the carbamoyl groups.

Ackowledgements

These studies were mainly supported by the commission of the European Communities (Contract F12W-CT96-0022). The authors thank Dr. V. Böhmer, Pr. M.A. McKervey and R. Ungaro for providing the ligands and Dr. A. Van Dorsselaer and Dr. E. Leize for the interpretation of the ES/MS spectra.

Literature Cited

1. Grossi, G.; Cecille, L. in *"New Separation Chemistry Techniques for Radioactive Waste and Other Specific Applications"*, (Cecille, L.; Casarci, M.; Pietrelli, L. Eds.; Elsevier: London, **1991**, p 11.
2. Horwitz E. P.; Kalina, D.G.; Diamond, H.; Vandegrift, G.F.; Schulz, W.W. *Solvent Extr. Ion Exch.*, **1985**, 3, 75.
3. Chamberlain, D.B.; Leonard, L.A.; Hoh, J.C.; Gay, E.C.; Kalina, D.G.; Vandegrift, G.F. Truex Hot Demonstration : Final report, Report ANL-89/37, Argonne, Illinois, April 1990.
4. Nakamura, T.; Yoshimura, T.; Nakatani, A.; Miyake, C. *J. Alloys and Compounds.*, 1993, 192, 302.
5. Schulz, W.W.; Horwitz, E.P.; *Sep. Sci. Technol.*, **1988**, 23, 1191.
6. McKervey, MA.; Schwing-Weill, M.J.; Arnaud-Neu, F. *Comprehensive Supramolecular Chemistry*; Lehn, J.M.; Gockel, G.W. Eds.; **1996**, 1, p 534.
7. Arnaud-Neu, F.; Böhmer, V.; Dozol, J.F.; Grüttner, C.; Jakobi, R.A.; Kraft, D.; Mauprivez, O.; Rouquette, H.; Schwing-Weill, M.J.; Simon, N.; Vogt, W. *J. Chem. Soc., Perkin Trans. 2*, **1996**, 1175.
8. Delmau, L.H.; Simon, N.; Schwing-Weill, M.J.;Arnaud-Neu, F.; Dozol, J.F.; Eymard, S.; Tournois, B.; Böhmer, V.; Grüttner, C.; Musigmann, C.; Tunayar, A. *Chem. Commun.*, **1998**, 1627.
9. Delmau,L.H.; Simon, N.; Schwing-Weill, M.J.; Arnaud-Neu, F.; Dozol, J.F.; Eymard, S.; Tournois, B.; Gruttner, C.; Musigmann, C.; Tunayar, A.; Bohmer, V. *Sep. Sci. Technol.*, **1999**, 34, 863.
10. Delmau, L.H. PhD. Thesis, Universite Louis Pasteur, Strasbourg, France, 1997.

11. Matthews, S.E.; Saadioui, M.; Böhmer, V.; Barboso, S.; Arnaud-Neu, F.; Schwing-Weill, M.J., Garcia-Carrera, A.; Dozol, J.F.; *J. Prakt. Chem.*, 1999, 341, 264.
12. Malone, J.F.; Marrs, D.J.; McKervey, M.A.; O'Hagan, P.; Thompson, N.; Walker, A.; Arnaud-Neu, F.; Mauprivez, O.; Schwing-Weill, M.J.; Dozol, J.F.; Rouquette, H.; Simon, N. *J. Chem. Soc., Chem. Commun.*, **1995**, 2151.
13. Schwing-Weill, M.J.; Arnaud-Neu, F. *Gazz. Chim. Ital.*, **1997**, 127, 11, 687.
14. Arnaud-Neu, F.; Browne, J.K.; Byrne, D.; Marrs, D.J.; McKervey, M.A.; O'Hagan, P.; Schwing-Weill, M.J.; Walker, A. *Chem. Eur. J.*, **1999**, 5, 175.
15. Dozol, J.F.; Böhmer, V.; McKervey, M.A.; Lopez-Calahorra, F.; Reinhoudt, D., Schwing, M.J., Ungaro, R.; Wipff, G. European Commission, Nuclear Science and Technology, Report EUR 1761SEN, **1997**.
16. Yaftian, M.R.; Burgard, M.; Matt, D.; Dielemann, C.B.; Rastegar, F. *Solvent Extr. Ion Exch.*, **1997**, 15, 975.
17. Wieser-Jeunesse, C.; Matt, D.; Yaftian, M.R.; Burgard, M.; Harrowfield, J. *C. R. Acad. Sci. Paris*, **1998**, 1, 479.
18. Arnaud-Neu, F. ; Charbonnière, L.J. ; Schwing-Weill, M.J. ; Ulrich, G. to be published .
19. For competitive spectrophotometric methods, see for instance : Thompson, S W.; Byrne, R.H. *Anal. Chem.*, **1988**, 60, 19.
20. Cherfa, S. PhD. Thesis, Université de Paris XI Orsay, France, 1999.
21. S. Barboso, S. ; Garcia Carrera, A. ; Matthews, S.E. ; Arnaud-Neu, F. ; Böhmer, V. ; Dozol, J.F. ; Rouquette, H. ; Schwing-Weill, M.J. *J. Chem. Soc. Perkin Trans.2*, **1999**, 719.
22. Diamond, H.; Horwitz, E.P.; Danesi, P.R. *Solvent Extr. Ion Exch.*, **1986**, 4, 1009.
23. Horwitz, E.P.; Diamond, H.; Martin, K.A. *Solvent Extr. Ion Exch.* **1987**, 5, 447.
24. Arnaud-Neu, F. ; Byrne, D. ; Schwing-Weill, M.J. unpublished results.

Chapter 13

Lanthanide Calix[4]arene Complexes Investigated by NMR

B. Lambert, V. Jacques, and J. F. Desreux[1]

Coordination and Radiochemistry, University of Liège, Sart Tilman (B6), B–4000 Liège, Belgium

> The solution behavior of calix[4]arene lanthanide perchlorate complexes in anhydrous acetonitrile depends on the location of coordinating substituents such as carbamoylmethylphosphine oxide groups. Relaxation titrations and NMR dispersion curves indicate that oligomeric species are formed with ligands bearing substituents on the wide rim while stable monomeric derivatives are obtained if substituents are located on the narrow rim. The structure of the 1:1 complexes is deduced from 1D and 2D NMR spectra of dia- and paramagnetic complexes. The high selectivity of some of these ligands is not directly related to their steric requirements.

Classically, the analysis of extraction processes is performed by slope analysis and by saturation experiments. Log-log plots of distribution coefficients of metal ions vs. the concentrations of ligands or of anions yield information on the stoichiometry of the extracted compounds. However, this approach is hampered by the non-ideal behavior of the species involved in the extraction processes and it can lead to spurious results. Moreover, no insight is gained into the solution structure of the extracted complexes and it is difficult to carry out a separate analysis of the role played by each factor affecting the metal extraction. On the other hand, spectroscopic techniques give access to information that cannot be obtained from the measurements of distribution coefficients. A variety of techniques have been used for this purpose, the choice of the most appropriate approach being directed by the properties of the extracted metal ions. Nuclear magnetic resonance spectroscopy (NMR) is particularly suitable for the study of paramagnetic metal ions such as lanthanides because of the large shifts induced by some of these ions and because of their pronounced influence on the relaxation times of their neighbouring nuclei, whether they are located in ligands or in solvent molecules(*1*). NMR spectroscopy is applied in the present paper to a detailed study of the

[1]Corresponding author.

complexation of lanthanide ions by calix[4]arenes. These ligands were selected because some of them recently proved to be very selective extraction agents of the lanthanides(2). Moreover, the steric requirements of the aromatic calix[4] units of these ligands should impart rigidity to their lanthanide complexes, a prerequisite for effectively using NMR spectroscopy(1).

Analyzing Lanthanide Calix[4]arene Complexes by Nuclear Magnetic Resonance

The present study starts with systems as simple as possible: mixtures of anhydrous lanthanide perchlorate salts(3) and of ligands in an aprotic solvent. Water or coordinating anions are subsequently added so as to approach the conditions met in practice in solvent extraction. Perchlorate salts are used because of the poor coordinating ability of the ClO_4^- anion(4) despite the explosive nature of these compounds when they are in contact with organic matter. Triflate lanthanide salts can be handled more safely but less well resolved NMR spectra are recorded in the presence of calix[4]arenes, presumably because of the somewhat stronger coordinating properties of $CF_3SO_3^-$ at the concentrations used here(5).

CAUTION: Anhydrous lanthanide perchlorate salts are known to become explosive when in contact with organic materials. Only small amounts of these salts should be handled at a time in a glove-box with proper eye protection.

All the calix[4]arenes under investigation in the present study feature bidentate donor groups that are good complexing agents of the lanthanides. Carbamoylmethylphosphine oxide groups are the core of extractants such as CMPO that are used on the industrial scale in the TRUEX process(6). Amides, malonamides and carbamoylmethyl esters are other mono- or bidentate functions that have been tested as extractants and that have been covalently linked to calix[4]arenes in hope of increasing both the selectivity and the efficacy of the extraction of metal ions (7). The synthesis of many of these ligands has been reported by Böhmer et al.(8).

The structure of some of the calix[4]arenes studied in the present report are presented in the scheme below. These ligands can be divided in two classes depending on whether they are substituted at the narrow (lower) or at the wide (upper) rim of the calix ring.

Experimental

NMR spectra were acquired on a Avance DRX-400 Bruker spectrometer. Mixing times of 20 ms and 500 ms were used to record the COSY and ROESY spectra respectively. Relaxivity data were collected at 20 MHz on a Bruker Minispec 120 and on a field cycling relaxometer as reported elsewhere(*5*).

Relaxivity Studies

The solution behavior of the calix[4]arene lanthanide complexes is strongly dependent on the position of the substituents as will be shown here by measurements of the longitudinal relaxation times T_1 of anhydrous acetonitrile solutions of gadolinium complexes. Gadolinium(III) is an S state ion with an electronic relaxation time τ_s that is relatively long compared to that of the other paramagnetic lanthanides. This ion is thus able to significantly enhance the relaxation rates of nuclei in its immediate vicinity by through space dipolar interactions(*1*). The Solomon-Bloembergen equation accounts for the reduction of T_{1is}, the experimental relaxation time of solvent molecules exchanging between the metal coordination inner sphere and the bulk of the solution:

$$\left[\frac{1}{T_{1is}}\right] = \frac{P_M \, q_{solvent}}{[solvent]} \frac{1}{T_{1M} + \tau_m} \quad (1)$$

$$\frac{1}{T_{1M}} = \frac{2}{15} \frac{\gamma_H^2 \, g^2 \, S(S+1) \, \mu_B^2}{r^6} \left[\frac{7\tau_c}{1 + \omega_S^2 \tau_c^2} + \frac{3\tau_c}{1 + \omega_H^2 \tau_c^2}\right] \quad (2)$$

where ω_s and ω_H are the Larmor frequencies of the electron and of the proton, r is the distance between Gd^{3+} and the proton under study, P_M and $q_{solvent}$ are the mole fraction of metal ion and the number of inner sphere solvent molecules and where the other factors have their usual meaning. The correlation time τ_c is given by

$$\frac{1}{\tau_c} = \frac{1}{\tau_s} + \frac{1}{\tau_m} + \frac{1}{\tau_r} \quad (3)$$

where τ_m and τ_r are the correlation times for the exchange of solvent molecules and for the molecular rotation respectively. τ_c primarily depends on the smallest of the correlation times in eq. 3. For small chelates, τ_r is the dominant factor (typically $\tau_r = 10^{-10}\text{-}10^{-11}$ s in water) while τ_s predominates for slowly tumbling molecules(*9*). Relaxation rate vs. frequency plots, also called NMR dispersion or relaxivity curves, show the effect of Gd^{3+} complexes on the longitudinal relaxation of water at frequencies ranging from 0.01 to 50-100 MHz. The shapes of these curves depend on the relative values of the correlation times and have been the subject of intense scrutiny as kinetically inert Gd^{3+} chelates very effectively improve the contrast of magnetic resonance images of the human body(*1,10,11*). However, NMR dispersion studies are not limited to water solutions and can be performed with organic solvents although, to our knowledge, such studies have not been reported so far. In the present work, ligands **1** and **2** were selected as models of calixarenes as they lend themselves to a complete analysis of both the relaxivity data and the NMR spectra.

As shown in Figure 1, the longitudinal relaxation rate of the methyl protons of acetonitrile measured at 20 MHz regularly decreases when calix[4]arene **1** is added to anhydrous Gd^{3+} perchlorate until a 1:1 ligand/metal ratio is reached. The plateau observed for higher ratios clearly indicates that a stable 1:1: Gd^{3+}.**1** complex is fully formed. When the complexation takes place, solvent molecules are removed from the inner coordination sphere of the metal ion and are relaxing more slowly as they are no longer close to a paramagnetic center, hence the decrease in relaxivity. By contrast, progressively forming the Gd^{3+}.**2** complex leads to a relaxivity maximum for an approximately equimolecular solution of metal and ligand. This maximum is followed by a steady decrease in relaxivity until a plateau is reached for ligand/metal ratios higher than 2.5. The very unusual behavior of tetra-CMPO ligand **2** is ascribed to the formation of oligomeric structures that are tumbling sufficiently slowly to bring about an increase in relaxivity despite the release of solvent molecules into the bulk of the solution due to the encapsulation of the metal ions. The stoichiometry of the complex formed in the presence of a large excess of ligand cannot be assessed with certainty but the low relaxivity found in this case indicates that the metal ions are poorly solvated by acetonitrile.

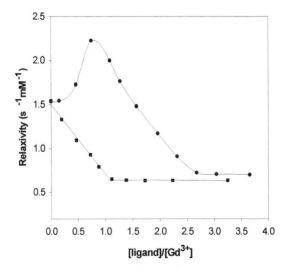

Figure 1. Longitudinal relaxation rates of 1 mM Gd^{3+} solutions in anhydrous acetonitrile at 25°C titrated by ligands 1 (■) and 2 (●). Relaxation rates $1/T_1$ are expressed per mM of Gd^{3+} (relaxivity in s^{-1} mM^{-1})

The non-ideal behavior of ligand **2** when compared to ligand **1** is clearly seen in the NMR dispersion curves of stoichiometric Gd^{3+} solutions of these two ligands (Figure 2). Calix[4]arene **1** yields a S-shaped relaxivity curve with an inflection point at about 6 MHz. Such curves are classically obtained for rapidly rotating small Gd^{3+} chelates in water(*12*). In contrast, the relaxivity curve of a 1:1 solution of Gd^{3+} and ligand **2** features a plateau followed by a maximum at about 50 MHz. This maximum clearly proves the formation of oligomeric species. It has indeed been amply demonstrated (*12*) that relaxivity maxima are obtained at high frequencies when studying aqueous solutions of high molecular weight Gd^{3+} complexes because the rotational correlation time τ_r of these species is no longer the smallest correlation time in equation 3. The frequency dependence of the electronic relaxation time τ_s then becomes the dominant factor in equation 3 and leads to a bell-shape curve at high frequencies.

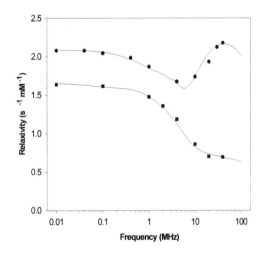

Figure 2. NMR dispersion curves of 1 mM stoichiometric mixtures of Gd^{3+} and 1 (■) or 2 (●) in anhydrous acetonitrile at 25°C. The solid lines through the data points result from a least-squares fit(13).

A best fit treatment of the experimental data in Figure 2 using a computer program(*13*) that takes into account zero-field splittings yields correlation times τ_r of 36, 190 ps and around 4000 ps for uncomplexed Gd^{3+}, Gd^{3+}.**1**, and Gd^{3+}.**2**, respectively. A τ_r value of 120 ps is deduced for La^{3+}.**1** from R_1^{DD}, the dipolar part of the ^{13}C relaxation rate of the calix CH groups at 400 MHz by applying the following equation(*14*)

$$\tau_r = \frac{R_1^{DD} \, r^6}{2 \, \gamma_C^2 \, \gamma_H^2 \, \hbar^2} \qquad (4)$$

This approach cannot be used for La^{3+}.**2** because τ_r is too large and the extreme narrowing limit conditions are no longer met as indicated by very small NOE effects(*14*). The hypothesis of an oligomerisation of the complexes with **2** is thus well supported by different relaxation studies. As a relaxivity maximum in titration curve of **2** is reached at an approximately stoichiometric ligand:Gd^{3+} ratio (Figure 1), it is assumed that the major components in these conditions are linear polymers in which metal ions alternate with calixarenes and are encapsulated between CMPO groups belonging to different ligands. Polymeric arrays appear to be formed in the presence of a large excess of metal ions as indicated by an increase in relaxivity for small Gd^{3+}:**2** ratios in Figure 1.

The formation of oligomeric species seems to be a characteristic of all Gd^{3+} calix[4]arene complexes substituted on the wide rim. Maxima are found in the relaxation rate titration curves of ligands substituted by four CMPO groups as in **2** but also by amide-ester and malonamide functions. In addition, the nature of the substituents on the narrow rim of **2** appears to have only a minor influence as maxima are found in the titration curves of complexes with **2** bearing either methyl groups or pentyl chains on the phenolic functions. Contrasting with this, a classical titration curve with a break at a 1:1 ligand/metal ratio is obtained not only for ligand **1** but also for its analogue substituted by four -$(CH_2)_3$-CMPO groups. It thus appears that whatever their exact chemical structure, coordinating groups at the wide rim of calix[4]arenes are prone to take part into the formation of oligomeric lanthanide complexes. Despite their flexibility, substituents on the narrow opening of these ligands are not involved in aggregation processes, presumably because of the steric requirements of the most rigid part of the calix ring. It is noteworthy that the relaxivity measurements presented in Figures 1 and 2 are in keeping with the as yet unreported solid state structure of a ligand **2** complex in which two calixarenes are coordinated to five $Eu(NO_3)_2^+$ or $Eu(NO_3)_3$ units. In this structure, one bridging Eu^{3+} ion is coordinated with two CMPO arms belonging to different calixarenes(*15*).

Spectral Analyses

NMR spectroscopy is not only a useful tool for analyzing the solution behavior of lanthanide calixarene complexes but it is also a powerful technique for unraveling the solution structure of these compounds. The Yb^{3+} ion is well known to induce large paramagnetic shifts that are essentially due to through space dipolar interactions between the unpaired electronic spins of the metal and the nuclei of the surrounding ligands and solvent molecules. These shifts depend on the structure of the complex according to the equation

$$\delta_i = -D \left\langle \frac{3\cos^2\theta_i - 1}{r_i^3} \right\rangle - D' \left\langle \frac{\sin^2\theta_i \cos 2\varphi_i}{r_i^3} \right\rangle \tag{5}$$

where D and D' are magnetic susceptibility factors and where r_i, θ_i and φ_i are the spherical coordinates of the nucleus i under investigation with respect to the Yb^{3+} ion at the origin of the set of axes of the magnetic susceptibility tensor(*1*). Provided the Yb^{3+} complex features at least a twofold symmetry, the principal magnetic axis of the system is the main symmetry axis. If the metal complex is axially symmetric (C_3 or above), there is no longer an anisotropy in the xy plane and equation 5 simplifies to (D'=0):

$$\delta_i = \frac{1}{3N} (\chi_{zz} - \chi_{xx}) \left\langle \frac{3\cos^2\theta_i - 1}{r_i^3} \right\rangle \qquad (6)$$

Equations 5 and 6 have been extensively used in the past to interpret the NMR spectra of substrates interacting with paramagnetic lanthanide β-diketonates called shift reagents. However, these equations fell out of favor with spectroscopists because very few lanthanide chelates are sufficiently symmetric and rigid to allow a reliable quantitative interpretation of the induced paramagnetic shifts(*1*). On the other hand, small macrocyclic ligands most often have very strict steric requirements and are thus able to impose their symmetry and their rigidity to their lanthanide chelates despite the lack of directionality of the *f* orbitals. Equations 5 and 6 proved most useful for elucidating the solution conformation of lanthanide macrocyclic polyaminopolyacetic chelates of high symmetry(*16,17*). These equations will be used to the same aim in the case of calix[4]arenes which themselves are potentially able to form rigid complexes of fourfold symmetry. The oligomerisation of the chelates with ligand **2** will obviously be a limitation in these studies. Central to our analyses is the recourse to two-dimensional NMR techniques that are constantly used with diamagnetic compounds but that are much more difficult to implement with paramagnetic metal ions because they drastically reduce the proton relaxation times so that cross peak intensities cannot be accumulated very effectively. It is only recently that 2D spectra of paramagnetic lanthanide chelates have been reported thanks to the very short pulse repetition times currently available on NMR spectrometers(*18-20*).

The various resonances in the spectrum of La^{3+}.**1** in anhydrous acetonitrile are easily assigned from their relative peak areas and splitting patterns. As expected for a fourfold symmetry, a single resonance peak is observed for the aromatic groups and the protons of the bridging CH_2 moieties in the calix unit appear as two doublets because they are either in an axial or an equatorial position. The proton spectrum of Yb^{3+}.**1** is presented in Figure 3 together with the corresponding COSY patterns. The Yb^{3+} ion induces paramagnetic shifts that cover a 25 ppm range and it also broadens the NMR peaks to the point that ^1H-^1H couplings are no longer discernible in the 1D spectra. However, most peaks can be reliably assigned from the COSY and EXSY spectra and as expected, the number of peaks and their relative areas indicate that Yb^{3+}.**1** adopts a geometry of fourfold symmetry. Applying equation 6 requires a model of Yb^{3+}.**1**. This model can be simulated by a molecular mechanics approach suggested by Hay(*21*) and Cundari *et al.*(*22*). With the force field parameters they proposed, these authors were able to reproduce accurately the structure of more than 60 inorganic and organic complexes. The geometry optimization of the Yb^{3+}.**1** structure leads to a C_4 symmetrical arrangement of the ligand with a nearly exact square antiprismatic geometry of the coordination sphere around the metal ion. The predicted structure is in excellent agreement with a preliminary crystallographic analysis of the solid state structure of the complex shown in Figure 4. The geometric factors deduced from these structures for each proton are in good agreement(*1*) with the experimental shifts

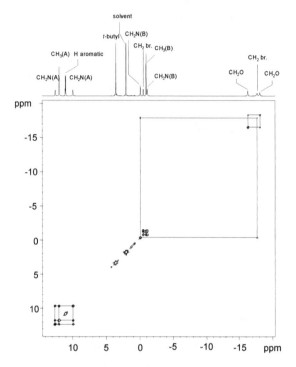

Figure 3. COSY spectrum of the $Yb^{3+}.1$ perchlorate complex in anhydrous acetonitrile at 400 MHz and 25°C.

Figure 4. Crystallographic structure of the $Yb^{3+}.1$ complex.

as shown in Figure 5 (R = 10.4 %) and it can be assumed that Yb^{3+} forms only a 1:1 complex with ligand **1**, the structure of which is identical in the solid state and in solution in contrast with the well-known lability of more usual lanthanide complexes.

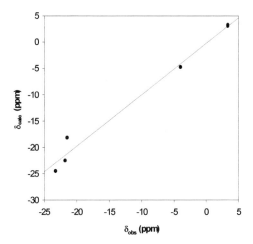

*Figure 5. Calculated vs. experimental paramagnetic shifts of the Yb^{3+}.**1** perchlorate complex in anhydrous acetonitrile.*

A structural analysis of the complexes with ligand **2** will inevitably be thwarted by the oligomerisation revealed by the relaxivity measurements mentioned above. The presence of oligomers in 1:1 mixtures of lanthanides and ligand **2** is borne out by the ^1H NMR spectra that feature considerably broadened peaks whether the metal ions are dia- or paramagnetic. The NMR peaks of La^{3+}.**2** remain broad at all temperatures (Figure 6) and their large bandwidth is assigned to the slow tumbling rates of oligomers that cause a shortening of the longitudinal relaxation times. However, a 20% excess of La^{3+} suffices to reduce drastically the proton peak widths. The same phenomenon is observed in the case of Yb^{3+} and it is assumed that the excess of metal ions favors the formation of a 1:1 species. However, the relaxivity data collected in Figure 1 clearly show that oligomeric species cannot be entirely neglected if a 50 % excess of metal ions is used for recording the NMR spectra and a structural investigation based on the dipolar equations will remain on a semi-quantitative level at best.

The protons of the aromatic groups in the calix unit of La^{3+}.**2** appear as two doublets (J =3 Hz) in the 1D and COSY spectra of La^{3+}.**2** and the bridging CH_2 protons give rise to two AB forms but singlets are found for the amide protons and the phosphine oxide groups in the ^1H and ^{31}P spectra respectively. Hence, the calix ring is in a conformation of twofold symmetry and the coordinating part of the ligand adopts a geometry of fourfold symmetry whether it is static or dynamically averaged. The molecular geometry of the La^{3+} complex with **2** is thus less symmetric than that with ligand **1**.

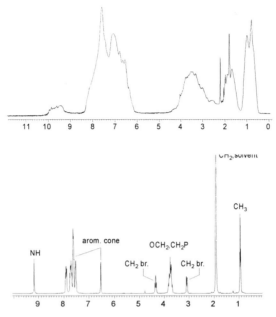

Figure 6. ^1H NMR spectrum of $La^{3+}.2$ perchlorate in anhydrous acetonitrile at 400 MHz. Top: stoichiometric metal/ligand mixture; bottom: 25% excess of metal added.

The same conclusion is reached in the case of $Yb^{3+}.2$, the 1D NMR spectrum of which is shown in Figure 7. This spectrum features 19 NMR peaks shifted toward high and low fields and covering a 70 ppm range. A total of 46 relative area units is obtained for 92 protons (the amide protons are not considered in this analysis as they give only very broad peaks at low temperatures). Hence, a twofold symmetry is also ascribed to $Yb^{3+}.2$. Assigning the NMR peaks in Figure 7 requires a detailed analysis of COSY and EXSY patterns recorded between -45 and 70°C because several peaks are overlapping in a broad temperature range and because exchange cross peaks are only observed at about 70°C. The most notable features of these spectra are that EXSY cross-peaks are observed between the two phenyl substituents of the phosphine oxide functions, between the two types of aromatic protons in the calix ring and between protons of the bridging CH_2 groups. In the latter case, the COSY spectra clearly indicate that the exchange takes place between protons located on the same carbon atom. The full assignment of the spectrum of $Yb^{3+}.2$ is given in Figure 7. It should be noted here that assigning the featureless NMR peaks of Yb^{3+} complexes as large as the calix[4]arenes has rarely been possible because the very short relaxation times of this ion most often do not permit the recording of cross-peaks for both coupling and exchange processes for all protons.

The molecular force field approach used in the case of ligand **1** cannot be applied to the derivatives of ligand **2** as parameters for the phosphine oxide groups are lacking. Molecular mechanics and dynamic simulations performed with the AMBER 4.1 software by Wipff *et al.(23)* on a 1:1 $Yb^{3+}.2$ complex yield conformations that can be compared with the NMR observations.

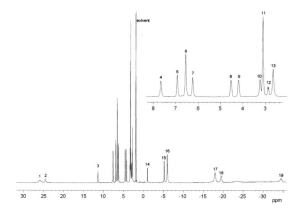

Figure 7. ^1H NMR spectrum of Yb3+.2 perchlorate in anhydrous acetonitrile at 400 MHz. Assignments from the COSY and EXSY spectra as follows. Calix phenyl H: 1,15; calix bridging CH_2: 2, 3, 12 ,14; C(O)-C$\underline{H}_2$-P(O): 18,19; P-phenyl H: 5,6,10,13,16,17.

As illustrated schematically in Figure 8, these conformations feature a series of parallel lines and planes containing atoms in symmetrically identical locations. In some minimized structures, the calix ring is elongated by around 0.5 Å along a line joining two bridging CH_2 groups facing each other. Because of this elongation, the aromatic groups are tilted and feature two types of protons and the cycle contains two types of bridging methylenes that give rise to four proton peaks in keeping with the NMR spectra. Computing the induced paramagnetic shifts with the full dipolar equation 5 leads to a very poor agreement factor of 46% and it can only be concluded that the conformation of $Yb^{3+}.2$ is probably close to the structure reproduced in Figure 8 as it accounts for the number and relative areas of NMR peaks. Other possible sources of the twofold symmetry of the metal complexes with **2** can be put forward. For instance, the dissymmetry does not stem from an elongation of the calix ring but from the position of the phosphine oxide functions or from specific hydrogen bonds between perchlorate anions and the amide groups. Such structures have been found by minimization by Wipff *et al.*(*23*) and they are also built from a succession of parallel planes as illustrated in Figure 8. These structures do not give a better agreement between the calculated and the experimental paramagnetic shifts. It thus seems that rapid exchanges between oligomeric structures and 1:1 and possibly even 2:1 complexes of Yb^{3+} with ligand **2** preclude a quantitative conformational analysis. The overall symmetry of the major solution species and an approximate geometrical arrangement of the ligand are the only information that can be gained by NMR in the present case.

As expected, the same difficulties are encountered with ligand **3**, an analogue of **2** with two ether bridges substituting the narrow rim. However, it is noteworthy that all the NMR peaks of $La^{3+}.2$ and $Yb^{3+}.2$ are split in two when the ether bridges are added. The twofold symmetry is kept in the lanthanide chelates of **3** but two conformers in slow exchange give rise to separate peaks in the NMR spectra. The relative populations of these conformers decrease from 1:1 in the case of $La^{3+}.3$ to 1:3 for $Lu^{3+}.3$. An elongation in the calix unit towards the ether bridges or in the opposite direction would be in agreement with these observations but other possibilities are also plausible as mentioned above.

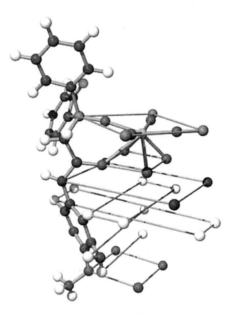

Figure 8. Schematic model of the conformation of $Yb^{3+}.2$

Effect of Water and Nitrate Ions: Getting Closer to Real Extraction Conditions

Spectroscopic measurements carried out in the absence of water and of coordinating anions facilitate the study of the solution behavior of calix[4]arene lanthanide complexes because this approach limits the number of parameters to be taken into account. However, an anhydrous environment with no complexing inorganic anions is a situation far removed from the conditions met in practice in solvent extraction separations for which water and complexing inorganic anions are always present. The effect of H_2O or NO_3^- ions on the NMR spectra is readily investigated by adding water or tetramethylammonium nitrate to anhydrous acetonitrile solutions of the metal complexes. The spectra of $La^{3+}.1$ and $Yb^{3+}.1$ remain unchanged after the addition of 1-2 equivalents of water or of $Me_4N^+NO_3^-$. It thus appears that the inner coordination sphere of the metal ions is not modified by these agents. On the other hand, the spectra of the metal complexes with ligand **2** are drastically altered: all peaks become extremely broad and cover a small shift range of 0-11 ppm in the case of Yb^{3+} and some of the NMR peaks of the La^{3+} complex are shifted by 0.5 ppm (spectra recorded in the presence of an excess of metals as above). It thus appears that water and nitrate ions can enter the inner coordination sphere of the metals and take part in the formation of oligomers as found by crystallography by Nierlich *et al*(*15*). The differences in rigidity observed here between the complexes of **1** and **2** are entirely in keeping with the conclusions drawn from the relaxivity data.

Conclusions

The complexation of lanthanide ions and their extraction by calix[4]arenes are much more complicated processes that one could have anticipated by relying exclusively on slope analyses. NMR spectroscopy and particularly relaxivity measurements cast light on different aspects of the properties of these compounds, for instance on striking differences between ligands substituted on the narrow and wide rims. The very high selectivity of calix[4]arene **1** for lanthanides(2) is not directly related to the structure of this ligand.

Acknowledgments

We gratefully acknowledge the financial support of the Fonds National de la Recherche Scientifique and the Institut Interuniversitaire des Sciences Nucléaires of Belgium. The authors also express their thanks for a grant from the European Commission in the framework of the research program "Nuclear Fission Safety". Professors V. Böhmer (Mainz) and G. Wipff (Strasbourg) are sincerely thanked for many helpful discussions.

References

(1) Peters, J.A.; Huskens, J.; Raber, D. J. *Prog. Nucl. Magn. Reson. Spectrosc.* **1996**, *28*, 283-350.

(2) Delmau, L. H.; Simon, N.; Schwing-Weill, M.-J.; Arnaud-Neu, F.; Dozol, J.-F.; Eymard, S.; Tournois, B.; Böhmer, V.; Grüttner, C.; Musigmann, C.; Tunayar, A. *J. Chem. Soc.,Chem. Commun.* **1998**, 1627-1628.

(3) Pascal, J. L.; Potier, J.; Zhang, C. S. *J. Chem. Soc.,Dalton Trans.* **1985**, 297-305.

(4) Bünzli, J.-C. G.; Yersin, J. R.; Mabillard, C. *Inorg. Chem.* **1982**, *21*, 1471-1476.

(5) Lambert, B.; Jacques, V.; Shivanyuk, A;; Matthews, S.E.; Tunayar, A.; Bohmer, V.; Baaden, M.; Wipff, G.; Desreux, J. F. to be submitted for publication.

(6) Horwitz, E. P.; Schultz, W. W. T. In *New separation chemistry techniques for radioactive waste and other specific applications*; Cecille, L.; Casarci, M.; Pietrelli, L., Eds. Elsevier Applied Science: London, 1991; pp 21-29.

(7) Arnaud-Neu, F.; Böhmer, V.; Dozol, J.-F.; Grüttner, C.; Jakobi, R.A.; Kraft, D.; Mauprivez, O.; Rouquette, H.; Schwing-Weill, M.-J.; Simon, N.; Vogt, W. *J. Chem. Soc., Perkin Trans. II* **1996**, *2*, 1175-1182.

(8) Böhmer, V. *Angew. Chem. Int. Ed. Engl.* **1995**, *34*, 713-745.
(9) Banci, L.; Bertini, I.; Luchinat, C. *Nuclear and electron relaxation*; VCH: Weinheim, 1991; pp 1-208.
(10) Elster, A.D. *Magnetic resonance imaging*; Mosby: St Louis, 1994; pp 1-278.
(11) Powell, D.H.; Ni Dhubhghaill, O. N.; Pubanz, D.; Helm, L.; Lebedev, Y. S.; Schlaepfer, W.; Merbach, A. E. *J. Am. Chem. Soc.* **1996**, *118*, 9333-9346.
(12) Lauffer, R. B. *Chem. Rev.* **1987**, *87*, 901-927.
(13) Bertini, I.; Galas, O.; Luchinat, C.; Parigi, G. *J. Magn. Resonance* **1995**, *113*, 151-158.
(14) Neuhaus, D.; Williamson, M. P. *The nuclear Overhauser effect in structural and conformational analysis*, VCH: Weinheim, 1989; pp. 31-38.
(15) S. Sherfa, Ph thesis, University of Paris Sud, France.
(16) Desreux, J. F.; Loncin, M. F. *Inorg. Chem.* **1986**, *25*, 69-74.
(17) Desreux, J. F. *Inorg. Chem.* **1980**, *19*, 1319-1324.
(18) Jenkins, B.G.; Lauffer, R. B. *J. Magn. Resonance* **1988**, *80*, 328-336.
(19) Jacques, V.; Desreux, J. F. *Inorg. Chem.* **1994**, *33*, 4048-4053.
(20) Aime, S.; Botta, M.; Fasano, M.; Marques, M. P. M.; Geraldes, C. F. G. C.; Pubanz, D.; Merbach, A. E. *Inorg. Chem.* **1997**, *36*, 2059-2068.
(21) Hay, B. P. *Inorg. Chem.* **1991**, *30*, 2876-2884.
(22) Cundari, T. R.; Moody, E. W.; Sommerer, S. O. *Inorg. Chem.* **1995**, *34*, 5989-5999.
(23) Wipff, G. *personal communication*.

Chapter 14
Bimetallic Lanthanide Supramolecular Edifices with Calixarenes

Jean-Claude G. Bünzli[1], Frédéric Besançon, and Frédéric Ihringer

Institute of Inorganic and Analytical Chemistry, University of Lausanne, BCH 1402, CH-1015 Lausanne, Switzerland

The structural, kinetic and photophysical properties of bimetallic lanthanide complexes with simple calix[n]arenes (n = 5 and 8) are discussed. The intensity of the f-f transitions of the Eu(III)-containing edifices strongly depends on the mixing between 4f and ligand-to-metal charge-transfer (LMCT) wave functions. In particular, the $^5D_0 \leftarrow {}^7F_0$ transition displays oscillator strengths in the range 0.1–0.5×10^{-6} (ε in the range 1-5 l·mol^{-1}·cm^{-1}). This allows one to follow the 2-step formation of the 2:1 bimetallic assemblies by monitoring this transition. On the other hand, Eu(III) luminescence is severely quenched by a low lying LMCT state. The ligand-to-metal energy transfer process ("antenna effect") can be adjusted by changing the substituent in the para position of the phenol ring, leading to easy tuning of the Eu(III) and/or Tb(III) sensitization processes. As a consequence, UV-Vis absorption and emission measurements prove to be a useful tool to study complex equilibria similar to those found in extraction processes. Moreover, the short Ln-Ln distances evidenced in the bimetallic complexes (365-390 pm) lead to a weak anti-ferromagnetic interaction in the Gd(III) complexes ($J = -0.06$ to -0.07 cm^{-1}).

Background

Trivalent lanthanide ions Ln(III) (here Ln represents La-Lu, except Pm) are hard acids and therefore interact preferentially with hard bases, typical examples of which are oxygen-containing molecules or anions such as water, alcohols, carboxylates, catecholates, or β-diketonates (*1*). At the end of the 1980's, the availability of numerous substituted calixarenes offered the possibility of controlling aryloxide/Ln(III) interaction and of using the macrocyclic effect to produce trivalent

[1]Corresponding author (E-mail: Jean-Claude.Bunzli@icma.unil.ch).

lanthanide edifices with a good command of their structural and photophysical properties. The first Ln(III) complexes with parent calixarenes have been synthesized by J. M. Harrowfield and coworkers (*2*) who isolated series of monometallic and bimetallic compounds with *p-tert*-butylcalix[n]arenes (n = 4, 6, 8) between 1987 and 1991 (*3*), and who studied their solid state and solution structure. Since calixarenes may be readily derivatized both on the upper and lower rims, the potential of such edifices was soon taken advantage of and work was initiated with three main goals: (i) investigation of Ln(III) ion supramolecular chemistry (*3*), (ii) development of efficient extraction and separation processes (*4,5*), and (iii) design of highly luminescent stains (*vide infra*).

In this paper, we first present the requirements to be met by lanthanide-containing macrocyclic edifices if they are to act as informative luminescent probes. We then present achievements in this area by other researchers before reviewing our work on bimetallic complexes with calix[n]arenes (n = 5, 8), including some new data. In particular, we show how sensitization of the Ln^{III} luminescence is easily tuned by changing the *para* substituent on the upper rim of the calixarene. We also demonstrate how optical properties may be used to unravel information both on the kinetics of formation of the complexes and on the number and nature of the species in solution. Such data could be of great importance for a thorough understanding of extraction processes in which several differently complexed species often form with different rate constants.

Design of Lanthanide Containing Luminescent Probes (*6*)

The inner-shell nature of 4f electrons results in lanthanide trivalent ions having remarkable spectroscopic properties, with narrow and easily recognizable f-f transitions; in addition, some of the ions have long-lived excited states which endow them with intense metal-centered luminescence. Numerous practical applications stem from these properties (*7*), *e.g.* lasers, phosphors for fluorescent lamps and cathodic or electroluminescent color displays, IR-imaging devices, and optical amplifiers. Uses in biology and medicine are also developing at a fast pace, especially in the field of time-resolved fluoroimmunoassays and protein labelling (*8*). The more popular ions used in these processes are Eu(III), a $4f^6$ ion, and Tb(III) with $4f^8$ configuration (Figure 1). The former displays a strong red luminescence from the $^5D_0$ excited state (main transitions: $^5D_0 \rightarrow {}^7F_1$ at 590 nm and $^5D_0 \rightarrow {}^7F_2$ at 620 nm) while the latter has green emission from its $^5D_4$ state (main transition : $^5D_4 \rightarrow {}^7F_5$, 550 nm).

Since f-f transitions are forbidden and therefore very weak, excitation usually relies on energy transfer from the host matrix or from the ligands surrounding the lanthanide ion, an effect discovered in 1942 by S.I. Weismann (*9*), presently termed "antenna effect" (*10*) and sketched in Figure 2. In addition to efficient sensitization, the design of lanthanide containing luminescent stains has to overcome another difficulty arising from the easy de-excitation of the Ln(III) excited states through high-energy vibrations (*11*) or, for Eu(III), low-lying ligand-to-metal charge-transfer (LMCT) states (*12*), while Tb(III) is amenable to back transfer processes. As a consequence, there are several stringent requirements for the ligand to produce efficient luminescent probes.

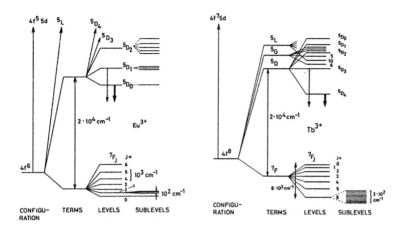

Figure 1. Partial energy diagram of trivalent Eu (left) and Tb (right) ions.

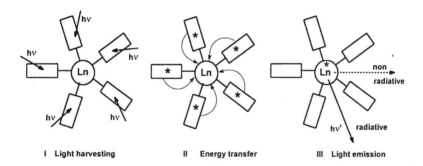

Figure 2. Antenna effect in sensitized Ln(III) complexes.

The main path of energy transfer between the excited ligand molecule and the long-lived Ln(III) states (with lifetimes in the ms range) involves the ligand-centered $^3\pi\pi^*$ state (*13*). Therefore the overall quantum yield of the metal-centered luminescence is influenced by the population rate of the $^3\pi\pi^*$ state, the rate of ligand-to-metal energy transfer, the energy gap between the Ln(III) excited and ground states, the extent of the 4f mixing with ligand and LMCT states, and the presence of high energy vibrations (Figure 3). The relationship between the observed quantum yields for series of similar compounds and the energy difference between the $^3\pi\pi^*$

transition and the luminescent Ln(III) state has been rationalized (*14,15*), leading to the following experimental rules: the energy gap ΔE between the ligand $^3\pi\pi^*$ state (taken as the 0-0 transition) and the excited ^5D_J level must be at least 3,500 cm^{-1} for an efficient transfer to take place. For Tb(III) probes, phonon-assisted back transfer ($^5D_4 \rightarrow {}^3\pi\pi^*$) occurs easily, especially when $\Delta E < 1,500$ cm^{-1}.

One strategy to solve these problems relies on the encapsulation of the Ln(III) ions in supramolecular edifices providing a rigid and protective environment for the metal ion. This is done either by using preorganized macrocyclic ligands (*16*), multidentate podands (*17*), or self-assembly processes (*18*). In this context, versatile ligands like calixarenes (*19*) are very good candidates and several sensitizer-modified calixarenes have been proposed.

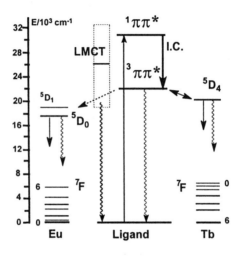

Figure 3. Energy migration processes in sensitized Ln(III) complexes.

Luminescent Lanthanide Complexes With Calixarenes

The first example stemmed from the work of N. Sabbatini and coworkers, who showed in 1990 that *p-tert*-butylcalix[4]arene substituted at the lower rim by four amide functions is a good sensitizer for Tb (quantum yield Φ_w in water : 20 %) and a fair one for Eu ($\Phi_w = 0.2$ %) (*20,21*). Sato and Shinkai (*22*) further improved the antenna effect of calix[4]arene by grafting one phenacyl group or diphenacylcarbonyl group besides the remaining three amide functions, achieving very good sensitization for Tb(III) with a quantum yield of 27% in acetonitrile, and reasonable energy transfer to Eu(III) ($\Phi_{MeCN} = 6\%$). The same authors synthesized water soluble calix[n]arenes (n = 4, 6, 8) by introducing sulfonate groups on the upper rim, leading to quantum yields for Tb(III) as large as 20% (23). On the other hand, D. M. Roundhill and collaborators have proposed a series of 2-aminoethoxy and carbamoyloxy substituted calix[4]arenes and calix[6]arenes which display good antenna effects for Tb(III) in methanol while Eu(III) is only weakly sensitized (24). In 1995, Reinhoudt and coworkers described a series of calix[4]arenes with three different sensitizing chromophores attached to the lower rim via a short spacer. In the Eu(III) and Tb(III) complexes of these calixarenes, photoexcitation of the antenna induces sizable lanthanide emission via intramolecular energy transfer. The best antenna for Eu(III) was found to be triphenylene (14). Neutral Ln(III) complexes with calix[4]arene triacids have also been isolated (25). More recently, Ziessel and coworkers have attached bipyridine units on the lower rim of *p-tert*-butylcalix[4]arene and obtained a good transfer to Eu(III) in acetonitrile (26), while Nd(III) and Er(III) sensitization was achieved with a fluorescein-substituted calix[4]arene (27).

Scope of the Present Work

Most of the studies described above have concerned the calix[4]arene platform and, with the exception of complexes with a biscalix[4]arene (28), bimetallic edifices have not triggered much interest. Bimetallic complexes in which lanthanide ions lie at a fixed distance are interesting for several reasons: (i) they provide good model molecules to study metal-to-metal energy transfer processes or magnetic interactions, (ii) heterobimetallic luminescent complexes combine two probes in one molecule and may be useful for imaging purposes and (iii) such assemblies are precursors for doped materials requiring the presence of metal ions at a specific distance. We have therefore concentrated most of our efforts to study bimetallic lanthanide complexes with simple calix[8]arenes and, more recently, calix[5]arenes (Figure 4).

In our work, we put a special emphasis on unraveling the relationship between the nature of the para substituent and the photophysical properties of the encapsulated lanthanide ions using the principles sketched on Figure 3. However, we also have interest in the investigation of the coordination behavior of the Ln(III) ions (1,29), especially when they are imbedded in macrocyclic receptors (30). Our approach therefore encompasses both solid state (X-ray diffraction, luminescence, magnetic susceptibility determination) and solution (NMR, UV-visible absorption and emission, mass spectrometry) studies of the structural, photophysical and, where appropriate, magnetic properties of the bimetallic edifices. One important aspect of complex formation, for instance in extraction processes, is the rate with which macrocyclic edifices form. To make up a lack of information in this field, we have undertaken a detailed investigation of the kinetics of formation of the 2:1 complex with *p-tert*-butylcalix[8]arene and determined the reaction mechanism and the composition of the reaction intermediates. We believe that the data gathered in the described studies will shed light on the nature of the Ln(III)-calixarene interaction and, therefore, make easier the design of practical devices and processes based on these systems.

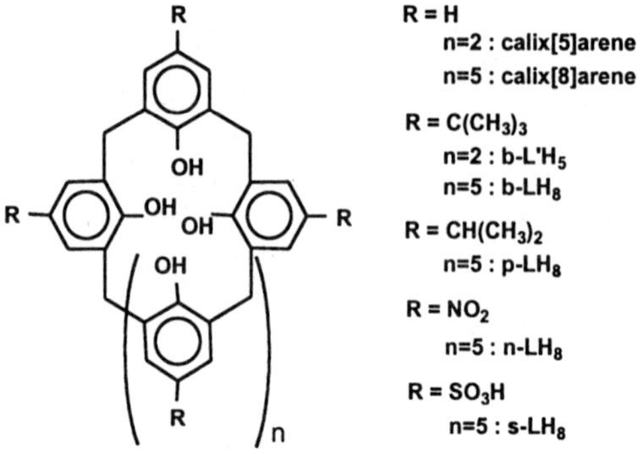

R = H
n=2 : calix[5]arene
n=5 : calix[8]arene

R = C(CH$_3$)$_3$
n=2 : b-L'H$_5$
n=5 : b-LH$_8$

R = CH(CH$_3$)$_2$
n=5 : p-LH$_8$

R = NO$_2$
n=5 : n-LH$_8$

R = SO$_3$H
n=5 : s-LH$_8$

Figure 4. Scheme and notation for the investigated calixarenes.

Bimetallic Complexes With Calix[8]arenes

p-tert-Butylcalix[8]arene, b-LH$_8$

The X-ray crystal structure determination of the homobimetallic Eu complex with b-LH$_8$ (*2*), [Eu$_2$(b-LH$_2$)(DMF)$_5$]·4DMF, shows both Eu(III) ions encompassed by the ligand in essentially identical environments with pseudo C_4 symmetry, as confirmed by luminescence measurements (*31*). The ligand adopts a double-bladed propellor shape and the metal ions are 8-coordinate, bonded to two bridging phenoxide donor atoms of the macrocyclic ligand, one bridging DMF molecule, two monodentate DMF molecules and three other O atoms from the ligand. The Eu-Eu distance is 3.69 Å.

The absorption spectrum of the 2:1 complex displays a band centered at 400 nm and assigned to the LMCT transition (ε = 720 M^{-1}·cm^{-1}), in addition to the weak f-f transitions. However, the 4f-LMCT mixing results in an unusually intense 0-0 transition, with ε = 5.1 M^{-1}·cm^{-1} (*31*) which may be used to monitor the complex formation (Figure 5, left). Since no kinetic data are available for the formation of lanthanide complexes with calixarenes, we have taken advantage of this peculiar property to investigate the mechanism of formation of such a macrocyclic assembly. Analysis of the $^5D_0 \rightarrow ^7F_0$ transition, which is unique for a given Eu(III) chemical environment, upon mixing the lanthanide salt and the ligand in the presence of an excess of triethylamine, points to a very fast formation of the 1:1 complex (half-life in the ms range) which then slowly transforms into the bimetallic 2:1 assembly with a half-life of 5,500 s in the presence of an 18-fold excess of triethylamine. It is noteworthy that this second reaction is 6 times slower in the presence of a 4-fold excess of Et$_3$N and becomes incomplete when only one equivalent of base per Eu(III) ion is added (*32*).

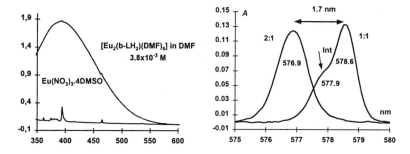

Figure 5. Absorption spectra (10 cm cells) at 298 K. Left: [Eu$_2$(b-LH$_2$)(DMF)$_5$] and [Eu(NO$_3$)$_3$(DMSO)$_4$] 3.8x10^{-3} M in DMF. Right: $^5D_0 \leftarrow {^7F_0}$ transition for the bimetallic 2:1 (9.9x10^{-3} M) and the monometallic 1:1 (2.7x10^{-3} M) complexes; the shoulder at 577.9 nm arises from an intermediate of the reaction.

This kinetic aspect is understandable since the conformation of the calixarene in the 1:1 [Eu(NO$_3$)(b-LH$_6$)(DMF)$_4$] complex is like the one adopted in the free ligand (33), while a substantial conformational change occurs upon formation of the bimetallic edifice, leading to the two-bladed propeller conformation of the macrocycle. The $^5D_0 \leftarrow {^7F_0}$ spectrum of the 1:1 complex is comprised of two components. We have investigated this transition under numerous conditions of temperature, pressure, concentration, and also upon addition of acid and base, which allowed us to propose that two reaction intermediates are involved in the 1:1 to 2:1 transformation (Figure 6), one charged species corresponding to the dissociation of the nitrate anion and a neutral species which forms upon the third deprotonation of the calixarene.

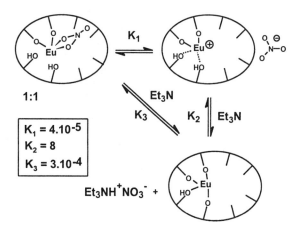

Figure 6. Proposed reaction intermediates.

A complete study of the fast reaction step by a stopped-flow method, monitoring the LMCT band and by conventional UV-visible spectrophotometry (0-0 transition) led us to propose the following mechanism for the formation of the complexes in DMF, that was further confirmed by a multi-level stochastic simulation (*34*). The main path to the formation of the bimetallic complex is reaction (5). A variable temperature thermodynamic study of the equilibria involved in the complex formation demonstrates that the energetic drive for the formation of both 1:1 and 2:1 complexes is in fact provided by triethylamine which captures the protons released in the process. Taking the presence of NEt$_3$ into account leads to LogK_3 = 4 ($\Delta_r G^0(3)$ = -23 kJ·mol^{-1}) and to LogK_5 = 5 ($\Delta_r G^0(5)$ = -29 kJ·mol^{-1}). The larger value of K_5 with respect to K_3 points to a cooperative effect induced by the change in the conformation of the macrocycle.

$$Et_3NH^+ \underset{k_2}{\overset{k_1}{\rightleftharpoons}} Et_3N + H^+ \quad (1)$$

$$b\text{-}LH_8 \underset{k_4}{\overset{k_3}{\rightleftharpoons}} b\text{-}LH_7^- + H^+ \quad (2)$$

$$b\text{-}LH_7^- + [EuNO_3]^{2+} \underset{k_6}{\overset{k_5}{\rightleftharpoons}} [Eu(b\text{-}LH_6)(NO_3)] + H^+ \quad (3)$$

$$[Eu(b\text{-}LH_6)(NO_3)] \underset{k_8}{\overset{k_7}{\rightleftharpoons}} [Eu(b\text{-}LH_5)] + H^+ + NO_3^- \quad (4)$$

$$[Eu(b\text{-}LH_6)(NO_3)] + [EuNO_3]^{2+} \underset{k_{10}}{\overset{k_9}{\rightleftharpoons}} [Eu_2(b\text{-}LH_2)] + 3H^+ + NO_3^- \quad (5)$$

$$2[Eu(b\text{-}LH_5)] \underset{k_{12}}{\overset{k_{11}}{\rightleftharpoons}} [Eu_2(b\text{-}LH_2)] + b\text{-}LH_8 \quad (6)$$

The photophysical properties of [Ln$_2$(b-LH$_2$)(DMF)$_5$] (*31*) and the energy-transfer processes taking place in these assemblies (*35*) have been reported; for Ln = Eu, the low-lying LMCT state (extending far beyond 25,000 cm^{-1}) quenches most of the metal-centered luminescence. On the other hand, the Tb(III) ion is conveniently sensitized and the detection limit of the Tb bimetallic complex in DMF is as low as 10^{-10} M. It also displays cathodoluminescence upon excitation by direct current (8.5 mA, 1.44 mm spot size) the maximum intensity of which can be estimated to 50 Cd·m^{-2} (*36*). Although the latter value does not compare favorably with commercial Tb-doped inorganic phosphors, the luminescence of which is 10-100 times more intense, this result is encouraging since the grafting of substituents with a better antenna effect seems to be within easy reach (*14*).

Magnetic Interactions

The Gd-Gd distance estimated from the crystal structure of the Eu(III) bimetallic edifice amounts to 3.66 Å in [Gd$_2$(b-LH$_2$)(DMF)$_5$], so that antiferromagnetic coupling can be expected. Indeed, close examination of the magnetic susceptibility dependence upon temperature of a sample of the bimetallic Gd complex synthesized according to reference (2) shows a slight deviation from Curie law below 20 K. Using the Lu(III) bimetallic complex for the diamagnetic correction, the fitting of the χ versus T curve with a model taking into account two very weakly antiferromagnetically coupled S = 7/2 ions yields g = 1.97 and J = -0.063 cm^{-1}. The same measurements made on the 2:2 complex between p-tert-butylcalix[5]arene and Gd(III) in which the estimated Gd-Gd distance is 3.86 Å yield g = 1.98 and J = -0.073 cm^{-1}. These values are consistent with the parameters found for [Gd(L)$_2$]·2CHCl$_3$ where L stands for tris{[(2-hydroxybenzyl)amino]ethyl}amine and in which the Gd-Gd distance is equal to 3.98 Å: g = 2.00 and J = -0.045 cm^{-1} (37).

p-Nitro-calix[8]arene, n-LH$_8$

Upon reaction of n-LH$_8$ with Ln(NO$_3$)$_3$ in DMF containing an excess of triethylamine, bimetallic complexes are isolated whose elemental analyses correspond to the formula [Ln$_2$(n-LH$_2$)(DMF)$_x$](DMF)$_y$(ROH)$_z$ (38). Solvation is somewhat difficult to control but the crystal structure of the Eu(III) complex (x = 5, y = 2, z = 1, R=H) is similar to the one found for the b-LH$_8$ edifice, particularly as far as the ligand conformation is concerned (39). Some of the Eu-O distances are different in the two complexes, leading to metal ions less centered in the cavity of the n-LH$_2$$^{6-}$ macrocyclic anion compared to b-LH$_2$$^{6-}$ and, therefore, to a longer Eu-Eu distance, 3.81 Å versus 3.69 Å. The ligand conformation can be viewed as if the calix[8]arene would consist of two calix[4]arenes in the cone conformation and placed side by side in a "transoid" configuration (Figure 7). This general geometric arrangement is maintained in solution, as demonstrated by a 2D-NMR study conducted on [Lu$_2$(L)(DMF)$_5$], with L = b-LH$_2$$^{6-}$ and n-LH$_2$$^{6-}$. For instance, 17 signals are observed for the 34 protons of the n-LH$_2$$^{6-}$ edifice implying a two-fold symmetry element. However, the complexation of the lanthanide ions does not make the structure completely rigid and the ligand undergoes a pseudo-rotation of the phenyl rings resulting in a fast racemization process in which protons a and b exchange their chemical environment (Figure 7). The kinetic parameters of the latter are : k(294 K) = 7.0 ± 0.7 and 1.1 ± 0.1 s^{-1} and $\Delta G^{\ddagger}$ = 67.1 ± 0.4 and 71.6 ± 0.4 kJ·mol^{-1} for b-LH$_2$$^{6-}$ and n-LH$_2$$^{6-}$, respectively (39).

The replacement of p-tert-butyl groups by electron-attracting nitro groups produces a bathochromic shift of the $^1\pi\pi^*$ and $^3\pi\pi^*$ states so that the ligand-to-metal energy transfer is enhanced for Eu(III) and suppressed for Tb(III), since the $^3\pi\pi^*$ state now lies under the $^5D_4$(Tb) excited state. In addition, the LMCT state of the Eu(III) complex undergoes a bathochromic shift compared to the situation for b-LH$_8$, decreasing the 4f-LMCT mixing, and resulting in a smaller molar absorption coefficient for the $^5D_0\leftarrow{}^7F_0$ transition (0.84 vs 5.10 M^{-1}·cm^{-1}). Therefore, the simple change from a tert-butyl to a nitro substituent switches the sensitizing properties of the calix[8]arene : b-LH$_8$ is a good Tb(III) sensitizer, but quenches Eu(III), while n-LH$_8$ insures a good energy transfer to Eu(III) but is unable to excite Tb(III).

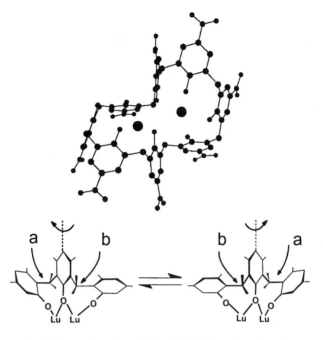

Figure 7. Ligand conformation in [Eu$_2$(n-LH$_2$)(DMF)$_5$] (top) and conformational exchange mechanism occurring in DMF solutions of [Lu$_2$(L)(DMF)$_5$], L = b-LH$_2^{6-}$, n-LH$_2^{6-}$ (bottom).

p-Sulfonato-calix[8]arene, s-LH$_8$

The ligand was synthesized according to Scharff *et al.* (*40*) who have determined its pK$_a$'s : 4.10 and 4.84 for deprotonation of the last two sulfonate groups, and 7.70, 9.10, and > 11 for the ionization of the phenol groups. In water and in DMF, deprotonation of the OH groups leads to a sizable red shift of the $\pi \rightarrow \pi^*$ transitions (about 2,000 cm^{-1}). The emission spectrum of a solution of s-LH$_8$ 5x10^{-4} M in DMF containing 0.15 M Et$_3$N presents an fluorescence band at 28,820 with a shoulder at 25,970 cm^{-1} assigned to emission from $^1\pi\pi^*$ states (lifetime : 4.5 ± 2 ns) and a triplet state emission can be evidenced at 22,730 cm^{-1} at 77 K on a solid state sample. Therefore, a good sensitization of both Eu(III) and Tb(III) is expected. The titration of a 5x10^{-3} M solution of s-LH$_8$ in DMF containing 0.15 M Et$_3$N by Eu(NO$_3$)$_3$·4DMSO was followed upon monitoring the ^{5}D$_0 \leftarrow {}^7$F$_0$ transition which occurs at 17,322 cm^{-1} with ε = 1.42 M^{-1}·cm^{-1}. The data reported in Figure 8 point to the formation of a 2:1 complex. When the ratio [Eu]$_t$/[s-LH$_8$]$_t$ is larger than 2.4, precipitation takes place, probably in view of an interaction with sulfonate groups.

Elemental analyses of several dried precipitates were not completely reproducible and indicated the formation of polymetallic species containing 3-4 Ln(III) ions. The 2:1 complex could not be isolated but both the energy of the 0-0 transition, which depends upon the nephelauxetic effects of the coordinated ligands, and its intensity are consistent with a structure similar to those evidenced for the bimetallic edifices with b-LH$_8$ and n-LH$_8$. The solutions, however only display a faint metal centered luminescence upon excitation through the ligand levels. This is due to two main effects. First, the $^3\pi\pi^*$ state of the complexes is at relatively high energy: it can be located at 24,400 cm^{-1} on the phosphorescence spectrum of a solution containing 2 equivalents of Gd(III) nitrate, that is about 6,700 cm^{-1} higher than the $^5D_0$ level, which is not optimum for an efficient energy transfer. Second, the sizable molar absorption coefficient of the $^5D_0 \leftarrow {}^7F_0$ transition points to some 4f-LMCT interaction which could explain the low efficiency of the ligand-to-metal energy transfer. The LMCT transition could not be located with certainty, but we note a weak and very broad feature in the absorption spectrum of the solution with $R = 2$ extending as low as 450 nm (maximum around 330 nm), which may be assigned to this transition.

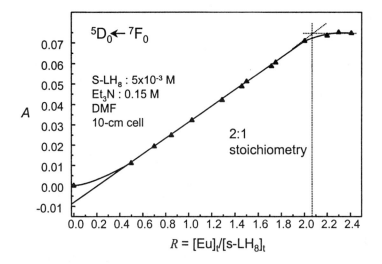

Figure 8. Titration of s-LH$_8$ by Eu(NO$_3$)$_3$·4DMSO in DMF, at 298 K.

On the other hand, the energy difference between the ligand $^3\pi\pi^*$ state and the Tb($^5D_4$) excited level is 3,900 cm^{-1} and under the same experimental conditions, a solution containing Tb(NO$_3$)$_3$ is strongly luminescent, displaying the typical metal-centered emission with a lifetime of 1.3 ± 0.1 ms at room temperature. That is, the energy transfer onto the metal ion is efficient and no back transfer occurs (41).

Complexes With *p-tert*-Butylcalix[5[arene, b-L'H$_5$

Odd-number calix[n]arenes are more difficult to synthesize than even-number calix[n]arenes (*42*), which may explain why the complexation properties of calix[5]arenes have not yet been extensively investigated, despite their appealing chemistry (*43*). A new interest was stirred for these macrocycles after the work of F. Arnaud-Neu and collaborators who demonstrated the efficiency of a crown ether derivative of calix[5]arene to selectively extract the Cs(I) cation (*44*). The inclusion of a tetrafluoroborate anion in the calix[5]arene cavity was also realized (*45*) and an improved synthesis has been newly published (*46*). Until recently, the only study on the interaction between Ln(III) ions and a calix[5]arene concerned water-soluble inclusion complexes with *p*-sulfonatocalix[5]arene. We have therefore undertaken a systematic investigation of the complexing ability of molecular receptors based on the calix[5]arene platform.

To start with, *p-tert*-butylcalix[5]arene was studied. It soon turned out that the Ln(III)/b-L'H$_5$ interaction in acetonitrile, in the presence of triethylamine, is relatively weak and leads to a mixture of 1:1 and possibly 2:1 complexes. We have therefore switched to NaH to deprotonate the calixarene and to tetrahydrofuran (THF) as solvent. Surprisingly, the addition of DMSO adducts of lanthanide nitrates to the solution of the deprotonated calixarene did not yield a 1:1 nitrato complex, but, rather, a 2:2 dimer, similar to the one isolated with the parent calix[4]arene (*47*). A single-crystal analysis of [Eu$_2$(b-L'H$_2$)$_2$(DMSO)$_4$] showed the 8-coordinated metal ion bonded to eight oxygen atoms, five from a calix[5]arene trianion, one of them bridging to the second Eu(III) ion, two from DMSO molecules and one from the second, bridging, anionic ligand (Figure 9). Interestingly, the Eu-Eu distance, 3.89 Å, is close to the distances observed in the bimetallic edifices with calix[8]arenes. The bridging nature of the calixarene trianion implies a deformation of the cone conformation of the ligand and a special feature of the structure is the coordination of one DMSO molecule to Eu(III) through the hydrophobic cavity of the calixarene, thus combining coordination to the metal ion and inclusion in the calixarene (*48*).

In THF, the Eu(III) dimer presents a LMCT transition with a maximum at 24,445 cm^{-1} (ε = 720 M^{-1}·cm^{-1}) extending to about 15,000 cm^{-1} so that emission from the $^5D_0$(Eu) level is almost completely quenched. On the other hand, a conveniently located $^3\pi\pi^*$ state (24,000 cm^{-1}) transfers energy on Tb(III) leading to sizable luminescence from the $^5D_4$ (Tb) state. However, the absolute quantum yield, 5.1% in THF, remains modest because of a back transfer process identified through the $^5D_4$ lifetime dependence on temperature : 1.12 ± 0.04 ms at 10 K, and 0.21 ± 0.01 ms at 295 K. Analysis of this dependence yields an "activation barrier" of 180 cm^{-1} pointing to a Tb-O phonon assisted back transfer. Presently, we are investigating the influence of the para substituent and of functional groups attached to the lower rim on the structural and photophysical properties of the resulting complexes.

191

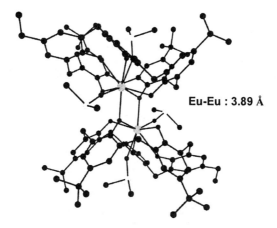

Figure 9. Molecular structure of the [Eu$_2$(b-L'H$_2$)$_2$(DMSO)$_4$] dimer.

Conclusions

Calixarene-based lanthanide bimetallic edifices offer a great variety of interesting properties. First, as has been demonstrated by J. M. Harrowfield, neutral complexes can be obtained and this feature could be of interest in medical magnetic resonance imaging. Second, lanthanide ions are immobilized at rather short distances, typically 3.60-3.90 Å, in these assemblies, which renders them amenable to weak antiferromagnetic interactions, to the study of metal-to-metal energy transfer processes, and which makes them interesting precursors for doped materials in which metal ions have to be implemented at a given distance. Third, photophysical properties can be easily tuned by modifying the substituent in the *para* position of the phenol rings. The metal-centered light absorption and emission processes, particularly in Eu^{III}-containing edifices, are excellent reporter of the local environment in which the metal ion is embedded and consequently, high-resolution analysis of these processes is an informative method to study equilibria as well as kinetic processes in solution. This could bring specific and valuable information in the investigation of extraction and separation processes involving f-elements.

Finally, given the extensive chemistry being presently developed to graft functional arms on calixarenes, both on the lower and upper rims, there is no doubt that elegant and efficient edifices with predetermined physico-chemical (photophysical or magnetic) properties will be engineered in a near future for use either in biomedical analyses, materials sciences or industrial processes.

Acknowledgements

This research is supported through grants from the Swiss National Science Foundation. We thank the Fondation Herbette (Lausanne) for the gift of spectroscopic equipment and Professor N. Re (Università di Parma, Italy) for calculating the antiferromagnetic coupling of the Gd bimetallic complexes.

References

1. Bünzli, J.-C.G. In *Rare Earths*; Saez Puche, R., Caro, P., Eds.; Editorial Complutense: Madrid, 1998; pp 223-259.
2. Furphy, B.M.; Harrowfield, J.M.; Kepert, D.L.; Skelton, B.W.; White, A.H.; Wilner, F.R. *Inorg. Chem.* **1987**, *26*, 4231.
3. Bünzli, J.-C.G.; Harrowfield, J.M. In *Calixarenes : A Versatile Class of Macrocyclic Compounds*; Vicens, J., Böhmer, V., Eds.; Kluwer Academic Publ.: Dordrecht, 1991; pp 211-231.
4. Arnaud-Neu, F.; Cremin, S.; Cunningham, D.; Harris, S.J.; McArdle, P.; McKervey, M.A.; McManus, J.P.; Schwing-Weill, M.-J.; Ziat, K. *J. Incl. Phenom. Mol. Recogn. Chem.* **1991**, *10*, 329.
5. Delmau, L.H.; Simon, N.; Schwing-Weill, M.-J.; Arnaud-Neu, F.; Dozol, J.F.; Eymard, S.; Tournois, B.; Böhmer, V.; Grüttner, C.; Musigmann, C.; Tunayar, A. *Chem. Commun.* **1998**, 1627.
6. Bünzli, J.-C.G. In *Lanthanide Probes in Life, Chemical and Earth Sciences. Theory and Practice*; Bünzli, J.-C.G., Choppin, G.R., Eds.; Elsevier Science Publ. B.V.: Amsterdam, 1989; pp 219-293.
7. Blasse, G.; Grabmaier, B.C. *Luminescent Materials*; Springer-Verlag: Berlin - Heidelberg, 1994.
8. Hemmilä, I.; Ståhlberg, T.; Mottram, P. *Bioanalytical Applications of Labelling Technologies*; Wallac Oy: Turku, 1995.
9. Weissman, S.I. *J. Chem. Phys.* **1942**, *10*, 214.
10. Sabbatini, N.; Guardigli, M.; Manet, I. In *Handbook on the Physics and Chemistry of Rare Earths*; Gschneidner, K.A.Jr., Eyring, L., Eds.; Elsevier Science Publ.: Amsterdam, 1996; Vol. 23, Ch. 154.
11. Haas, Y.; Stein, G. *Chem. Phys. Lett.* **1971**, *8*, 366.
12. Blasse, G. *Struct. Bonding* **1976**, *26*, 45.
13. Malta, O.L. *J. Luminesc.* **1997**, *71*, 229.
14. Steemers, F.J.; Verboom, W.; Reinhoudt, D.N.; Vandertol, E.B.; Verhoeven, J.W. *J. Am. Chem. Soc.* **1995**, *117*, 9408.
15. Latva, M.; Takalo, H.; Mukkala, V.M.; Matachescu, C.; Rodriguez-Ubis, J.-C.; Kankare, J. *J. Luminesc.* **1997**, *75*, 149.
16. Alexander, V. *Chem. Rev.* **1995**, *95*, 273.
17. Jones, P.L.; Amoroso, A.J.; Jeffery, J.C.; McCleverty, J.A.; Psillakis, E.; Rees; LH; Ward, M.D. *Inorg. Chem.* **1997**, *36*, 10.
18. Piguet, C.; Bünzli, J.-C.G. *Chimia* **1998**, *52*, 579.

19 Roundhill, D.M. *Prog. Inorg. Chem.* **1995**, *43*, 533.
20 Sabbatini, N.; Guardigli, M.; Mecati, A.; Balzani, V.; Ungaro, R.; Ghidini, E.; Casnati, A.; Pochini, A. *J. Chem. Soc., Chem. Commun.* **1990**, 878.
21 Hazenkamp, M.F.; Blasse, G.; Sabbatini, N.; Ungaro, R. *Inorg. Chim. Acta* **1990**, *172*, 93.
22 Sato, N.; Shinkai, S. *J. Chem. Soc., Perkin Trans.2* **1993**, 621.
23 Sato, N.; Yoshida, I.; Shinkai, S. *Chem. Lett.* **1993**, 1261.
24 Georgiev, E.M.; Clymire, J.; McPherson, G.L.; Roundhill, D.M. *Inorg. Chim. Acta* **1994**, *227*, 293.
25 Rudkevich, D.M.; Verboom, W.; Vandertol, E.B.; Van Staveren, C.J.; Kaspersen, F.M.; Verhoeven, J.W.; Reinhoudt, D.N. *J. Chem. Soc., Perkin Trans.2* **1995**, 131.
26 Ulrich, G.; Ziessel, R.; Manet, I.; Guardigli, M.; Sabbatini, N.; Fraternali, F.; Wipff, G. *Chem. Eur. J.* **1997**, *3*, 1815.
27 Wolbers, M.P.O.; Vanveggel, F.C.J.M.; Peters, F.G.A.; Vanbeelen, E.S.E.; Hofstraat, J.W.; Geurts, F.J.; Reinhoudt, D.N. *Chem. Eur. J.* **1998**, *4*, 772.
28 Wolbers, M.P.O.; Vanveggel, F.C.J.M.; Heeringa; RHM; Hofstraat, J.W.; Geurts, F.A.J.; Vanhummel, G.J.; Harkema, S.; Reinhoudt, D.N. *Liebigs Annalen* **1997**, 2587.
29 Bünzli, J.-C.G.; Milicic-Tang, A. In *Handbook on the Physics and Chemistry of Rare Earths*; Gschneidner, K.A.Jr., Eyring, L., Eds.; Elsevier Science Publishers B.V.: Amsterdam, 1995; pp 306-366.
30 Bünzli, J.-C.G. In *Handbook on the Physics and Chemistry of Rare Earths*; Gschneidner, K.A.Jr., Eyring, L., Eds.; Elsevier Science Publ.: Amsterdam, 1987; pp 321-394.
31 Bünzli, J.-C.G.; Froidevaux, P.; Harrowfield, J.M. *Inorg. Chem.* **1993**, *32*, 3306.
32 Besançon, F., PhD Thesis, University of Lausanne, 1996.
33 Harrowfield, J.M.; Ogden, M.I.; Richmond, W.R.; White, A.H. *J. Chem. Soc., Dalton Trans.* **1991**, 2153.
34 Bünzli, J.-C. G., Besançon, F., and Schaad, O., unpublished results.
35 Froidevaux, P.; Bünzli, J.-C.G. *J. Phys. Chem.* **1994**, *98*, 532.
36 Sievers, R. E. and Andersen, W., personal communication, 1998.
37 Liu, S.; Gelmini, L.; Rettig, S.J.; Thompson, R.C.; Orvig, C. *J. Am. Chem. Soc.* **1992**, *114*, 6081.
38 Bünzli, J.-C.G.; Ihringer, F. *Inorg. Chim. Acta* **1996**, *246*, 195.
39 Bünzli, J.-C.G.; Ihringer, F.; Dumy, P.; Sager, C.; Rogers, R.D. *J. Chem. Soc., Dalton Trans.* **1998**, 497.
40 Scharff, J.-P.; Mahjoubi, M.; Perrin, R. *New J. Chem.* **1991**, *15*, 883.
41 Ihringer, F., PhD Thesis, University of Lausanne, 1997.
42 Gutsche, C.D. *Calixarenes*; Royal Society of Chemistry: Cambridge, 1989.
43 Asfari, Z.; Vicens, J. *Acros Organics Acta* **1995**, *1*, 18.
44 Arnaud-Neu, F.; Arnecke, R.; Böhmer, V.; Fanni, S.; Gordon, J.L.M.; Schwing-Weill, M.-J.; Vogt, W. *J. Chem. Soc., Perkin Trans.2* **1996**, 1855.

45 Steed, J.W.; Johnson, C.P.; Juneja, R.K.; Atwood, J.L.; Burkhalter, R.S. *Supramolecular Chemistry* **1996**, *6*, 235.
46 Dumazet, I.; Ehlinger, N.; Vocanson, F.; Lecocq, S.; Lamartine, R.; Perrin, M. *J. Incl. Phenom. Mol. Recogn. Chem.* **1997**, *29*, 175.
47 Furphy, B.M.; Harrowfield, J.M.; Ogden, J.S.; Skelton, B.W.; White, A.H.; Wilner, F.R. *J. Chem. Soc., Dalton Trans.* **1989**, 2217.
48 Charbonnière, L.J.; Balsiger, C.; Schenk, K.J.; Bünzli, J.-C.G. *J. Chem. Soc., Dalton Trans.* **1998**, 505.

Chapter 15

Molecular Recognition by Azacalix[3]arenes

Philip D. Hampton, Si Wu, Panadda Chirakul, Zsolt Bencze, and Eileen N. Duesler

Department of Chemistry, University of New Mexico, Albuquerque, NM 87131

The azacalix[3]arene macrocycles **1** are ideal hosts for the binding of metal ions and alkylammonium ions due to the presence of heteroatoms in the macrocycle ring and the ease of modifying their structure through the R, R', and R" substituents. Synthetic routes to the azacalix[3]arene macrocycles will be discussed from the perspective of their ability to generate the azacalix[3]arenes **1** free from the corresponding azacalix[4]arenes **2**. Azacalix[3]arene macrocycle **1a** exhibits the ability to bind trivalent metal ions and form complexes where the macrocycle is either a neutral (H_3L) or a trianionic (L^{3-}) donor. Crystal structures of well-defined yttrium(III) [**3**: $Y(H_3L)Cl_3$] and lanthanum(III) [**4**: La(L)] complexes are reported. Molecular recognition studies have been performed on the azacalix[3]arenes **1**. In contrast with the O-unsubstituted macrocycles (**1**, R"= H) which exhibit no detectable binding of alkali or ammonium ions, the O-methylated macrocycle **1j** is observed to extract alkali metal and alkylammonium picrates.

The hexahomotriazacalix[3]arene macrocycles **1** (Figure 1), abbreviated to azacalix[3]arenes in this paper, are interesting targets for molecular recognition and ion separation studies since their structure can be modified at not only the upper-rim (R) and lower-rim (R"), but also within the macrocycle cup at the inner-rim (R') (*1- 4*). The azacalix[3]arenes have received only limited attention for their molecular recognition behavior; this is probably due to the difficulty of synthesizing the macrocycles free from the corresponding cyclic tetramers, the azacalix[4]arenes **2** (*1*).

Takemura and co-workers first reported that the azacalix[3]arenes **1b-e** could be synthesized (Eq. 1) by heating 2,6-bis(hydroxymethylphenols) **5** in the presence of a primary amine in refluxing toluene (*2-4*). The azacalix[3]arenes were reported to be isolated without contamination by the corresponding azacalix[4]arenes **2**. They also reported that azacalix[3]arene **1b** could be O-alkylated with picolyl chloride to yield macrocycle **1f** but the cone vs. partial-cone selectivity of this reaction was not discussed (*4*). Macrocycle **1b** was reported to extract UO_2^{2+} from aqueous solutions.

© 2000 American Chemical Society

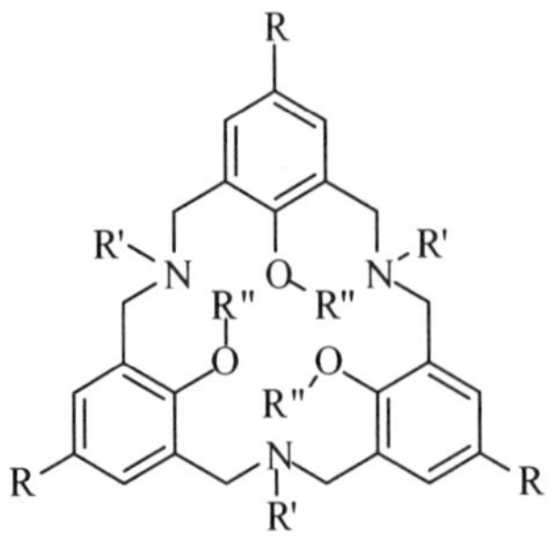

1a: R= CH_3, R'= $CH_2CO_2CH_3$, R"= H
1b: R= CH_3, R'= CH_2Ph, R"= H
1c: R= t-Bu, R'= CH_2Ph, R"= H
1d: R= CH_3, R'= (S)–$CH(CH_3)Ph$, R"= H
1e: R= CH_3, R'= CH_2(2-pyridyl), R"= H
1f: R= CH_3, R'= CH_2Ph, R"= CH_2(2-pyridyl)
1g: R= CH_3, R'= $CH_2CH(CH_3)_2$, R"= H
1h: R= CH_3, R'= $CH_2CH=CH_2$, R"= H
1i: R= CH_3, R'= H, R"= H
1j: R= CH_3, R'= CH_2Ph, R"= CH_3

Figure 1. Structures of Azacalix[3]arene Macrocycles 1a-j

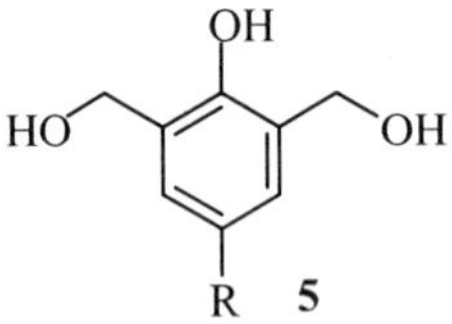

$$\text{5} \xrightarrow[\text{toluene, 135°}]{\text{R'NH}_2} \text{1b-e} \quad \text{(Eq. 1)}$$

Several years later, we developed an alternative approach to the azacalix[3]arenes **1** which is compatible with volatile amines that involves a cyclooligomerization reaction between 2,6-bis(chloromethyl)phenols **6** and primary amines at 60° in DMF (*1*). The use of glycine methyl ester as the primary amine resulted in a mixture of azacalix[3]arene **1a**, and the corresponding azacalix[4]arene **2** which could be separated by recrystallization. The crystal structure of this azacalix[3]arene exhibited a cone-shaped conformation where the three nitrogen R' groups are on the same face of the macrocycle and within the cone of the cupped macrocycle ligand. With all other primary amines that we examined, inseparable mixtures of the azacalix[3]arene **1** and azacalix[4]arene **2** macrocycles were obtained. Azacalix[3]arene **1a** exhibited no significant extraction of alkali metal or ammonium picrates. The absence of molecular recognition exhibited by these macrocycles was attributed to strong, intramolecular hydrogen-bonding which prevented both the phenolic oxygens and amines from participating in molecular recognition processes.

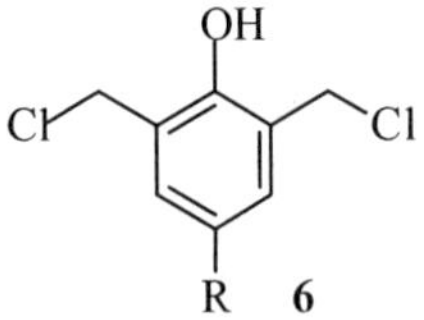

$$\text{6} \xrightarrow[\text{DMF, 60°}]{\text{R'NH}_2, \text{K}_2\text{CO}_3} \text{1 + 2} \quad \text{(Eq. 2)}$$

The difficulties associated with these syntheses led us to develop new routes to the azacalix[3]arene macrocycles **1** which would provide the macrocycles in high purity, free from contamination by azacalix[4]arenes **2**, and with the ability to easily modify the R, R', and R" substituents. In this paper, we describe several synthetic approaches to the azacalix[3]arene macrocycles **1** and studies of their molecular recognition behavior.

Results and Discussion

Synthetic Routes to the Azacalix[3]arenes

Four cyclooligomerization approaches to the azacalix[3]arene macrocycles **1** have been examined: (1) reaction of monomers **5** with primary amines (Eq. 1), (2) reaction of the monomers **6** with primary amines (Eq. 2), (3) condensation of the aminomethyl-chloromethyl monomers **7** (Eq. 3), and (4) cyclization of the aminomethyl-salicylaldehyde **8** (Eq. 4). The first route (Eq. 1), which was reported by Takemura and co-workers to yield only azacalix[3]arenes **1**, in our hands resulted in inseparable mixtures of the azacalix[3]arene **1** and azacalix[4]arene **2** macrocycles. Similar results were obtained with the second (Eq. 2) and third (Eq. 3) routes (5).

The fourth approach (Eq. 4), the cyclization of monomer **8** to form the iminocalix[3]arene **9**, was examined as a route to the *N*-unsubstituted azacalix[3]arenes **1** (R'= R"= H) which we believed would be ideal precursors to a wide range of *N*-substituted azacalix[3]arenes **1f**. Modification of the secondary amines in **1f** by Michael addition, alkylation, reductive alkylation could be used to introduce R' substituents on the nitrogen atoms. Unfortunately, this cyclooligomerization reaction yielded mostly polymeric material and low yields of macrocycles (<5%). The iminocalix[3]arene **9** was the major macrocyclic product in the reaction. Reduction with NaBH$_4$ yielded a mixture of the *N*-unsubstituted azacalix[3]arene **1** (R'= R"= H) and azacalix[4]arene **2**. Separation of the imine-linked macrocycles was not possible due to the lability of these molecules, and the azacalix[n]arenes (n= 3, 4) exhibited too similar of chromatographic properties to allow for separation of the mixture. The monomers **7** and **8** were synthesized from monoaldehydes **10** and **11** (Figure 2). Monoaldehyde **10** was prepared by protection of bis(hydroxymethyl)phenol **5** as its acetonide (2,2-dimethoxypropane, H$_2$SO$_4$, acetone), followed by oxidation (PCC, CH$_2$Cl$_2$) and deprotection (dilute HCl) (*5,6*).

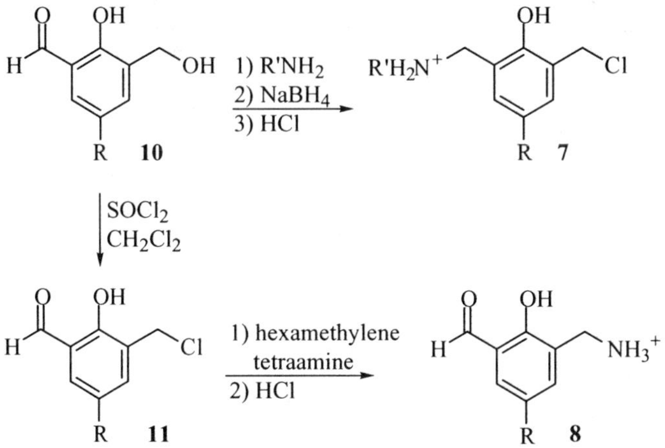

Figure 2. Synthesis of Azacalix[3]arene Precursors 7 and 8

The failure of the above cyclooligomerization approaches led us to examine a convergent approach to the selective synthesis of the azacalix[3]arene macrocycles **1** by the coupling of the triamine dimer **12** and monomer **6**, as shown in Equation 5. Under the conditions of the coupling reaction, the condensation of **12** with **6** proceeds in exceptionally high yield (~95%) and without the formation of other macrocyclic products, i.e. the azacalix[4]arenes **2**. The high yield for this cyclization step is likely due to intramolecular hydrogen bonding which favors cyclization to form azacalix[3]arenes. (*6*). The triamine dimer **12** was prepared from monoaldehyde **11** by reacting with 1.5 equiv. of a primary amine followed by reduction (NaBH$_4$) to yield **12** in 50-60% yields. Azacalix[3]arenes **1b**, **1d**, **1g**, and **1h** have been prepared by this approach. Deprotection of azacalix[3]arene **1h** to form the *N*-unsubstituted azacalix[3]arene **1i** has been accomplished under palladium catalyzed conditions. We are currently investigating the introduction of *N*-substituents onto macrocycle **1i**.

$$\text{12} + \text{6} \xrightarrow[\text{DMF, 25°}]{\text{K}_2\text{CO}_3} \text{1b,d,g,h} \quad (\text{Eq. 5})$$

Isolation of Group 3 Metal-Ion Complexes of the Azacalix[3]arenes

The azacalix[3]arene macrocycles **1** have the potential of binding trivalent metal ions as neutral (H$_3$L) and trianionic (L^{3-}) donors; complexes of both types have been observed depending on the reaction conditions. The reaction of macrocycle **1a** with a stoichiometric amount of MX$_3$ [M= Sc, Y, La; X= Cl, OSO$_2$CF$_3$ (= OTf)] resulted in the formation of M(H$_3$L)X$_3$ complexes. In the reaction of YCl$_3$ with **1a**, X-ray quality crystals of the Y(H$_3$L)Cl$_3$ complex (**3**) were obtained. The crystal structure of this complex is shown in Figure 3. Crystallographic data and selected structural parameters for complex **3** are listed in Tables I and II, respectively. The metal binds to the three phenolic oxygens and the three ring nitrogen atoms are protonated resulting in a metal complex of an overall neutral ligand (H$_3$L). A similar complex has been reported by Orvig and co-workers for lanthanide complexes of ligand **13**. (*7-9*). The average dihedral angle between the phenolic oxygens and the aryl rings (27.5°) is very close to that observed in the crystal structure of macrocycle **1a** (26.8°) which indicates that there is very little change in the conformation of the macrocycle on binding of Y to the phenolic oxygens. The ^{1}H NMR spectrum of complex **3** indicates an N-H signal and coupling between the N-H and the adjacent methylenes.

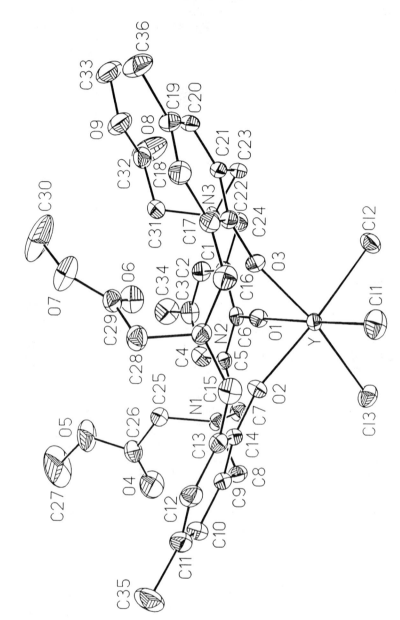

Figure 3. ORTEP drawing of Y(H$_3$L)Cl$_3$ (**3**). Ellipsoids are drawn at the 30% probability level.

The reaction of M(H₃L)X₃ complexes with a strong base (i.e. NaOCH₃) results in their deprotonation to form M(H$_n$L)$^{n+}$ complexes. In the presence of three equivalents of NaOCH₃, the M(H₃L)X₃ complexes are gradually converted into new compounds which do not exhibit an N-H signal. X-ray quality crystals of the La(L) complex **4** could be obtained in this way by the deprotonation of the La(H₃L)(OTf)₃ complex. The crystal structure of this complex is shown in Figure 4; crystallographic data and selected structural parameters are listed in Tables I and II.

Table I. X-ray Crystallographic Data for Y(H₃L) • 3 CH₃CN (3) and La(L) • 2 CH₃OH (4)

	Complex 3	Complex 4
chem. formula	$C_{42}H_{51}Cl_3N_6O_9Y$	$C_{38}H_{50}LaN_3O_{11}$
fw	979.1	863.72
cryst. system	monoclinic	orthorhombic
space group	C2/c	Pna2₁
a, Å	22.011(2)	19.716(2)
b, Å	18.278(2)	10.183(1)
c, Å	24.464(2)	20.090(2)
β, deg	104.71 (1)	90
V, Å³	9519.8(16)	4033.4(7)
Z	8	4
T, K	293(2)	293(2)
λ, Å	0.71073	0.71073
ρ_{calcd}, g cm⁻³	1.366	1.422
μ, cm⁻¹	14.50	11.19
transm.	0.8166–0.9681	0.855–0.984
$R(F_o)$	0.0828[a]	0.0505[a]
$R_W(F_o)$	0.0511[b]	0.0811[c]

[a] $R(F_o) = \Sigma\ (|F_o| - |F_c|) / \Sigma\ |F_o|$
[b] $R_w(F_o) = [\Sigma\ (w(F_o - F_c)^2) / \Sigma\ (wF_o^2)]^{1/2}$
[c] $R_w(F_o) = [\Sigma\ (w(F_o^2 - F_c^2)^2) / \Sigma\ (wF_o^2)^2]^{1/2}$

The La(L) complex is structurally similar to that previously reported for the La complex of *p-tert*-butylhexahomotrioxacalix[3]arene (**14**) (*10*). The La centers in complexes **4** and **14** are coordinated to the three phenolic oxygen atoms and the three ring heteroatoms (O, N). For complex **4**, the complex possesses a tricapped trigonal prismatic structure where the remaining coordination sites are occupied by the ester carbonyl groups. The C_3 symmetry of the complex is observed both in the solid state and in solution, since ¹H and ¹³C NMR spectra provide evidence for inequivalence due to the three-fold rotation axis in the molecule at 298 K. At higher temperatures, the complex exhibits dynamic behavior resulting in loss of the C_3 symmetry, presumably due to rapid decoordination of the ester groups. The average La-O bond lengths and La-O-C bond angles in complexes **4** (2.335 Å, 109°) and **14** (2.343 Å, 112°) are essentially identical for the two complexes. In contrast with the shallow cupping observed in the crystal structure of macrocycle **1a** (26.8° dihedral angle), the macrocycle in complex **4** is more cupped (39.4° dihedral angle). The La-N bond lengths in complex **4** are extremely long (2.981 Å) compared to a lanthanum complex of a substituted tetraazacyclododecane (*11*) (*ca.* 2.807 Å) and of triethylenetetraaminehexaacetic acid (*12*) (*ca.* 2.718 Å) indicating a weak coordination of the nitrogen atoms to the La center in complex **4**.

Table II: Selected Bond Lengths (Å), Angles (deg), and Dihedral Angles[c] (deg) for Y(H₃L)Cl₃ (3) and La(L) (4) Complexes.

Complex **3**			
Y-Cl1	2.631(2)	C6-O1-Y	109.0(3)
Y-Cl2	2.648(2)	C14-O2-Y	108.6(3)
Y-Cl3	2.654(2)	C22-O3-Y	108.5(3)
Y-O1	2.226(4)		
Y-O2	2.230(4)	ΦA[c]	30.0(5)
Y-O3	2.178(3)	ΦB[c]	25.5(5)
d[a]	1.405(5)	ΦC[c]	26.9(5)
Complex **4**			
La-O1	2.328(3)	C1-O1-La	109.0(3)
La-O2	2.349(3)	C10-O2-La	108.6(3)
La-O3	2.328(4)	C18-O3-La	108.5(3)
La-O4	2.675(4)	C29-O4-La	124.2(3)
La-O6	2.677(4)	C32-O6-La	122.6(4)
La-O8	2.662(3)	C35-O8-La	123.7(3)
La-N1	3.027(4)		
La-N2	2.937(4)	ΦA[c]	38.8(5)
La-N3	2.980(4)	ΦB[c]	39.7(5)
d[b]	1.300(5)	ΦC[c]	39.7(5)

[a] The displacement of Y out of the plane containing O1, O2, and O3.
[b] The displacement of La out of the plane containing O1, O2, and O3.
[c] Dihedral angles are defined as the angle between the (O1,O2,O3) plane and the mean planes of the aryl rings C1-C6 (ΦA), C9-C14 (ΦB), and C17-C22 (ΦC).

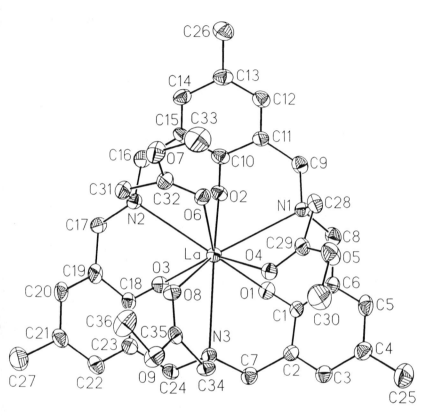

Figure 4. ORTEP drawing of La(L) (**4**). Ellipsoids are drawn at the 30% probability level.

Alkali-Metal and Alkylammonium Ion Extraction by Azacalix[3]arenes

The extraction of alkali metal and alkyl ammonium ions was examined for macrocycles **1b** and **1j**. Macrocycle **1j** was prepared by the *O*-methylation of azacalix[3]arene **1b**. ^{1}H NMR spectra of macrocycle **1j** indicated that the cone and partial-cone isomers of this macrocycle are rapidly equilibrating on the NMR timescale at room temperature. A similar dynamic behavior was reported by Shinkai and co-workers for the *O*-methylated *p-tert*-butylhexahomotrioxacalix[3]arene macrocycle (*13*). Picrate extraction studies indicated no detectable extraction of alkali metal ion (Li, Na, K, Rb, or Cs), NH_4^+, or $iPrNH_3^+$ picrates by azacalix[3]arene **1b**. Under identical conditions, the *O*-methylated azacalix[3]arene **1j** exhibited the following extraction behavior: Li^+ 10%, Na^+ 20%, K^+ 23%, Rb^+ 21%, Cs^+ 13%, NH_4^+ 13%, and $iPrNH_3^+$ 20%. These extraction studies indicate that macrocycle **1j** exhibits a greater extraction ability than does macrocycle **1b** which is likely due to the removal of intramolecular hydrogen bonding and the enhanced donor ability of the ether groups. Since **1f** is a mixture of rapidly interconverting cone and partial cone isomers, efforts are underway to prepare pure cone and partial-cone conformers of the azacalix[3]arenes through modification of the R" groups, and examine their molecular recognition chemistry.

Conclusions

A new, convergent synthetic route to the azacalix[3]arene macrocycles **1** has been developed which allows for synthesis of the macrocycles in high yield and without contamination by azacalix[4]arene macrocycles **2**. The synthesis is general and allows for the preparation of azacalix[3]arenes with a wide range of R and R' substituents. An *N*-unsubstituted azacalix[3]arene **1i** has been synthesized and the introduction of nitrogen substituents is currently being examined. The azacalix[3]arene macrocycles readily form complexes with group 3 metal-ions. With the ability to vary the structure of the azacalix[3]arene macrocycles through the R, R', and R" groups, it will be possible to examine the effect of variations in the geometric and electronic structure of the macrocycles on their metal ion binding affinity and selectivity.

Experimental Section

Synthesis of Y(H$_3$L)Cl$_3$ (2).

A solution of 14.8 mg YCl$_3$ (0.076 mmol) in 5 mL of acetonitrile as added to a solution of 52.4 mg azacalix[3]arene **1a** (*1*) (0.074 mmol) in 5 mL acetonitrile and the reaction was stirred for 30 min. Pale yellow crystals formed in the solution overnight. The crystals were collected, and dried *in vacuo* for 4 hours. Yield: 92%. ^{1}H NMR (CD$_3$CN, 298 K): δ 10.64 (br s, 3H, NH), 6.97 (s, 6H, Ar-H), 5.53 (d, J$_{AB}$ = 12.2 Hz, 6H, ring C*H*$_2$), 4.17 (d of d, J$_{AB}$ = 12.6 Hz, J$_{AX}$ = 9.4 Hz, 6H, ring C*H*$_2$), 3.52 (d, J$_{CX}$ = 3.3 Hz, 6H, NHC*H*$_2$CO$_2$), 3.47 (s, 9H, CO$_2$C*H*$_3$), 2.44 (s, 9H, *p*-C*H*$_3$). ^{13}C NMR (CD$_3$CN, 298 K): δ 167.00 (*C*=O), 163.47 (1-aryl), 133.55 (3,5-aryl), 125.51 (4-aryl), 119.19 (2,6-aryl), 58.73 (ring *C*H$_2$), 53.38 (N*C*H$_2$CO$_2$), 48.86 (CO$_2$*C*H$_3$),

20.21 (p-CH$_3$). Anal. Calcd. for C$_{36}$H$_{45}$N$_3$O$_9$Cl$_3$Y: C, 50.33; H, 5.28; N 4.89. Found: C, 49.05; H, 5.38; N, 4.89. An X-ray quality crystal was obtained by crystallization of the complex under the above conditions; X-ray analysis showed a composition of Y(H$_3$L)Cl$_3$ • 3 CH$_3$CN.

Synthesis of La(L) (3).

In an inert atmosphere glove box, a solution of 104.8 mg La(OTf)$_3$ (0.149 mmol) in 10 mL of methanol was combined with a solution of 87.1 mg azacalix[3]arene **1a** (*1*) (0.149 mmol) in 10 mL methanol and the mixture was stirred for 30 min. A solution of 0.45 mmol NaOCH$_3$ in methanol was prepared by the addition of 10.7 mg NaH (0.45 mmol) to 5 mL of methanol, and this solution was combined with the above solution. A white precipitate formed immediately on mixing of the two solutions. The suspension was heated at reflux for 5 minutes resulting in a clear solution. Yellow, triangular plate crystals formed overnight at room temperature. The crystals were collected, and dried *in vacuo* for 12 hours. Yield: 93%. ^{1}H NMR (CDCl$_3$, 318 K): δ 6.67 (s, 6H, Ar-H), 4.50 (d, J= 11.4 Hz, 6H, ring C*H*$_2$), 3.06 (br s, 6H, NC*H*$_2$CO$_2$), 2.92 (s, 9H, CO$_2$C*H*$_3$), 2.82 (d, J= 11.4 Hz, 6H, ring C*H*$_2$), 2.12 (s, 9H, p-C*H*$_3$). ^{13}C NMR (CD$_3$OD, 298 K): δ 179.00 (C=O), 160.12 (1-aryl), 133.02 (s, 3,5-aryl), 132.05 (s, 3,5-aryl), 130.22 (s, 2,6-aryl), 129.44 (s, 2,6-aryl), 125.56 (4-aryl), 63.83 (s, N-CH$_2$-aryl), 60.38 (s, N-CH$_2$-aryl), 56.68 (NCH$_2$CO$_2$), 52.76 (CO$_2$CH$_3$), 20.53 (p-CH$_3$). A suitable crystal for single crystal X-ray analysis was obtained under the above conditions; the crystal analyzed for La(L) • 2 CH$_3$OH.

X-ray Crystallographic Studies on Complexes 3 and 4.

A pale yellow, monoclinic crystal of **2** used for X-ray study (0.41 x 0.46 x 0.96 mm) was sealed in a glass capillary under nitrogen. A total of 15177 independent reflections (R$_{int}$ = 1.75%) were measured on a Siemens R3m/V diffractometer using graphite monochromated Mo-Kα radiation, with ω scan mode, 2θ$_{max}$ = 50.0°. After absorption correction (semi-empirical), the structure was solved as reported previously (*10*). A yellow, orthorhomic crystal of **4** (0.46 x 0.39 x 0.16 mm) was also sealed in a glass capillary under nitrogen. A total of 10862 independent reflections (R$_{int}$ =2.24%) were measured following similar procedures as for **3**. Crystallographic data and basic details of data collection for **3** and **4** are presented in Table I.

Synthesis of Azacalix[3]arene 1i via *O*-Methylation of 1b

A mixture of azacalix[3]arene **1b** (90 mg, 0.125 mmol) and NaH (450 mg, 18.8 mmol) in *N,N*-dimethylformamide (1 mL) and tetrahydrofuran (1 mL) was heated at 65° C. After 1 h, iodomethane (0.23 mL, 3.75 mmol) was added dropwise to the reaction mixture; after complete addition, the reaction mixture was stirred at 65° for 3 h. The reaction was cooled to room temperature and filtered. The filtrate was concentrated *in vacuo*, and the residue was extracted with dichloromethane (3 x 5 mL). The combined dichloromethane extracts were washed with water (3 x 5 mL), dried over sodium sulfate and concentrated under reduced pressure. The residue was chromatographed on silica gel with ethyl acetate as the eluent to give the product in 85% yield. ^{1}H NMR (250 MHz, CD$_3$CN) δ 7.49 - 7.24 (m, 15 H, Ph), 6.93 (s, 6H, Ar*H*), 3.72 (s, 6 H, *CH*$_2$Ph), 3.42 (s, 9H. O*CH*$_3$), 3.26 (s, 12H, ring *CH*$_2$), 2.15 (s, 9H, *p*-*CH*$_3$).

Alkali Metal and Ammonium Picrate Extraction Studies

Caution! Picric acid and its derivatives are potentially explosive materials. They should be prepared without heating and stored as solutions in water. Metal picrates (2.5 x 10^{-4} M) were prepared *in situ* by dissolving the metal hydroxide (0.010 mol) in 100 mL of 2.5 x 10^{-4} M picric acid. Deionized water was used in the preparation of all solutions. Treatment of ammonium hydroxide (0.10 mol) and isopropylamine (0.10 mol) with an aqueous solution of picric acid, as described above, afforded solutions of ammonium picrate and isopropylammonium picrate.

Two-phase solvent extraction was carried out between 5 mL of the aqueous solution of the alkali metal or ammonium picrate ([picrate] = 2.5 x 10^{-4} M) and 5 mL of chloroform containing the either **1b** or **1i** ([macrocycle] = 2.5 x 10^{-4} M). The amount of picrate extracted into the organic phase was determined based on the decrease in absorption by picrate ion in the aqueous phase. The percent extraction was the amount of picrate extracted from the aqueous phase divided by the amount originally in the aqueous phase.

References

1. Hampton, P. D.; Tong, W.; Wu, S.; Duesler, E. N. *J. Chem. Soc. Perkin Trans. 2* **1996**, 1127.

2. Takemura, H.; Yoshimura, K.; Shinmyozu, T.; Khan, I. U.; Inazu, T. *Tet. Lett.*, **1992**, *39*, 5775.

3. Khan, I. U.; Takemura, H.; Suenaga, M.; Shinmyozu, T.; Inazu, T. *J. Org. Chem.* **1993**, *58*, 3158.

4. Takemura, H.; Shinmyozu, T.; Miura, H.; Khan, I. U.; Inazu, T. *J. Incl. Phenom. Mol. Recogn. Chem.* **1994**, *19*, 193.

5. Bencze, Z.; Tong, W.; Hampton, P. D. Unpublished observations.

6. Chirakul, P.; Hampton, P. D. Manuscript in preparation to *J. Org. Chem.*

7. Liu, S.; Gelmini, L.; Rettig, S. J.; Thompson, R. C.; Orvig, C. *J. Am. Chem. Soc.* **1992**, *114*, 6081.

8. Caravan, P.; Hedlund, T.; Liu, S.; Sjöberg, S.; Orvig, C. *J. Am. Chem. Soc.* **1995**, *117*, 11230.

9. Yang, L.-W.; Liu, S.; Wong, E.; Rettig, S. J.; Orvig, C. *Inorg. Chem.* **1995**, *34*, 2164.

10. Daitch, C. E.; Hampton, P. D.; Duesler, E. N.; Alam, T. M. *J. Am. Chem. Soc.* **1996**, *118*, 7769.

11. Morrow, J. R.; Amin, S.; Lake, C. H.; Churchhill, M. R. *Inorg. Chem.* **1993**, *32*, 4566.

12. Wang, R.-Y.; Li, J.-R.; Jin, T.-Z.; Xu, G.-X.; Zhou, Z.-Y.; Zhou, X.-G. *Polyhedron* **1997**, *16*, 1361.

13. Araki, K; Inada, K.; Otsuka, H.; Shinkai, S. *Tetrahedron* **1993**, *49*, 9465.

Chapter 16

Metal Ion Complexation and Extraction Behavior of Some Acyclic Analogs of *tert*-Butyl-calix[4]arene Hydroxamate Extractants

Timothy N. Lambert[1], Matthew D. Tallant[1], Gordon D. Jarvinen[2], and Aravamudan S. Gopalan[1,3]

[1]Department of Chemistry and Biochemistry, New Mexico State University, Las Cruces, NM 88003-8001
[2]Los Alamos National Laboratory, Los Alamos, NM 87545

Recently, we reported the preparation of two new calix[4]arene based hydroxamate extractants **8** and **9** designed for the selective complexation of actinide(IV) ions and some results of their metal ion extraction studies; however, these ligand systems did not achieve the preferential extraction of the An(IV) over Fe(III). In order to understand the complexation behavior of **8**, we have prepared the related acyclic tetrahydroxamate **10**, trihydroxamate **11**, as well as the monohydroxamate **12** and examined their complexation/extraction of Th(IV) and Fe(III) cations into chloroform from aqueous nitrate solutions. Our results show that the trihydroxamate **11** is a selective extractant of Th(IV) over Fe(III) at pH 1-2 and in fact shows greater promise than the tetrahydroxamate **10**. Our results also suggest that not all four hydroxamate moieties of **10** are involved in the actinide complexation process in the pH range of study.

A variety of separation technologies and processes are currently being explored for the remediation of high level and transuranic radioactive waste (*1,2*). In connection with these efforts, there is an urgent need to develop cost effective and efficient chelating and extracting agents to selectively remove actinides, such as plutonium and americium, from waste streams as well as from contaminated soil and water (*3-7*).
To design chelators that have the requisite properties of selectivity and high binding constants for a particular metal ion one must consider its charge, size and specific coordination chemistry (*8*). Because of their larger size, actinide ions

[3]Corresponding author (E-mail: agopalan@nmsu.edu).

typically have a higher coordination number (eight or more) and a more flexible ligand geometry (dodecahedral, square antiprism). It is generally accepted that selective actinide complexation can be achieved by taking advantage of their larger coordination sphere relative to the smaller transition metals (9-12). Raymond and others have exploited the similarity in the coordination chemistry of Pu(IV) and Fe(III) to develop a number of cyclic and acyclic chelators for the binding of plutonium, some of which are potentially useful for *in vivo* biodecorporation of this metal ion (13-16). Since the Pu(IV) ion is a hard Lewis acid, the preferred ligands that have been incorporated into these synthetic chelators have been catecholates, hydroxypyridinonates and hydroxamates (17,18).

A number of reports have appeared on the properties and applications of calix[n]arenes, a unique class of molecules (19-27). Calixarenes that are immobilized in the cone conformation present an ideal platform for the introduction of various ligand groups onto the same face of the molecule and hence for the construction of selective extractants for actinides (28,29). The use of calixarenes for the complexation of a variety of metal ions including some transition and f-block elements has been reported (30-32).

Shinkai and coworkers have developed some calixarene based uranophiles (having carboxylate groups appended to the lower rim) which exhibit remarkable selectivity for the linear uranyl (UO_2^{2+}) ion, whose coordination geometry (planar, penta or hexacoordinate) is quite different from that of the An(IV) cations (33-35). They have also synthesized similar p-*tert*-butylcalix[n]arene derivatives with hydroxamate groups and examined their ability to extract uranium and other transition metal ions from aqueous solution (36,37).

Some of the calixarene based extractants for the tetra- and trivalent actinides that have been reported are shown in Figure 1. The calixarene derivatives **1** having groups analogous to CMPO (octyl(phenyl)N,N-diisobutylcarbamoyl methylphosphine oxide) appended to the upper rim have been found to be better extractants for actinides than CMPO itself (38). More recently, it has been shown that calix[4]arenes of this type show selectivity for the trivalent light lanthanides and trivalent actinides in their extraction from highly saline (4 M $NaNO_3$) or acidic media (3 M HNO_3) into chloroform (39). Some resorcin[4]arene cavitands of the type **2** have also been examined for the selective extraction of Eu(III) from acidic/saline media into dichloromethane (40-42). Another novel class of calixarene derivatives with phosphine oxide groups **3** attached to the lower rim has been synthesized and shown to have high efficiency in the extraction of Th(IV) and Pu(IV) from simulated nuclear waste (43). The synthesis of a lower rim functionalized CMPO derivative **4**, as well as a tetraiminocarboxylate calix[4]arene derivative **5**, with some preliminary actinide extraction studies has recently been published (44,45). In addition, the synthesis of some novel 3-hydroxy-2-pyridinone (3,2-HOPO) derivatives of 4-*tert*-butylcalix[4]arenes, **6** and **7**, has recenty been disclosed from our laboratory (46). These 3,2-HOPO derivatives show considerable selectivity for the preferential extraction of the Th(IV) ion over Fe(III) under acidic conditions.

The overall goal of our research program is to develop organic chelators capable of the specific binding/removal of actinides, such as Pu(IV), from process waste streams in the presence of more abundant and competing metal ions such as Fe(III), Al(III), alkali and alkaline earth metal ions (47,48). Recently, we reported the preparation of two new calix[4]arene based hydroxamate extractants **8** and **9**,

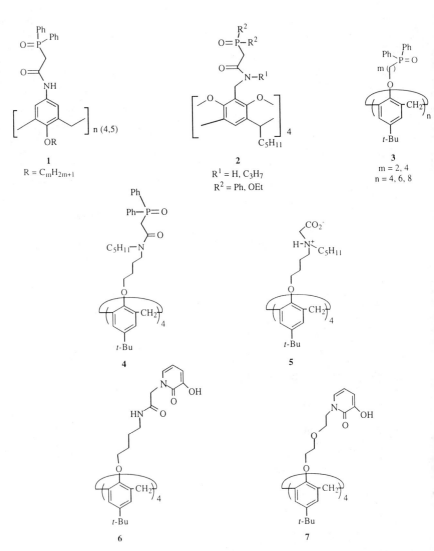

Figure 1. Calixarene based extractants reported for actinide (III or IV) ions

Figure 2, designed for the selective complexation of An(IV) ions and some results of their metal ion extraction studies (49). The primary hydroxamate **8** was the more efficient extractant of Th(IV) at pH 2 removing greater than 90% of this metal ion from the aqueous phase in comparison to approximately 20% for the secondary hydroxamate **9** under identical conditions. However, in competitive studies, hydroxamates **8** and **9** were found to be selective for the extraction of Fe(III) (83 and 98%) over Th(IV) (24 and 3%) at pH 2 from an aqueous 0.10 M NaCl solution.

In order to ascertain the importance of the calix[4]arene backbone in selective metal ion binding, and to develop a more systematic understanding of the actinide chelation/extraction properties of this class of ligands, we have now prepared the related acyclic tetrahydroxamate **10**, trihydroxamate **11**, as well as the monohydroxamate **12**, Figure 2. Their complexation/extraction behavior with Th(IV), Fe(III), Eu(III) and Cu(II) has been examined using liquid-liquid extraction and spectrophotometric studies.

RESULTS AND DISCUSSION

Syntheses

The syntheses of ligands **10-12** were achieved in good yields via synthetic routes analogous to that previously reported for calixarene **8** (49). The synthetic route for the preparation of the acyclic tetrahydroxamate **10** is given in Scheme 1 (47% overall yield).

Single Metal Ion Extraction Studies

The ability of chelators **8** and **10-12** to extract Th(IV), Fe(III), Eu(III) and Cu(II) ions from aqueous $NaNO_3$ (0.10 M) at pH 1 and 2 into chloroform has been examined and the results are presented in Table I. The protocol for the metal ion extraction experiments followed procedures described earlier (46). Thorium(IV) and Eu(III) were chosen for these studies, as they are surrogates for Pu(IV) and Am(III) present in radioactive waste streams. A solution of 0.10 M $NaNO_3$/1% HNO_3 was adjusted to pH 1 or 2 with concentrated aqueous NaOH, followed by addition of the desired concentration of the metal ion(s) of interest. Equal volumes (4 mL:4 mL) of the aqueous solution containing the metal ion(s) at the specified pH and chloroform containing excess of the ligand (four to sixteen-fold excess, depending on the ligand) were contacted for 2 h at ambient temperature with gentle shaking. In order to have a valid comparison, the molar concentrations of the ligands were adjusted so as to provide the same equivalents (4 mM) of the hydroxamate binding units in each case. The layers were separated carefully by centrifugation and the concentrations of the metal ions in the aqueous layers were determined by ICP analysis. The percent metal ion extracted by the ligands could then be determined. In the case of calix[4]arene tetrahydroxamate **8**, a significant amount of precipitation occurred in these extraction experiments (much more at pH 1 compared to pH 2). This precipitation led to results with poor reproducibility and obviously prevents the reliable comparison of calixarene

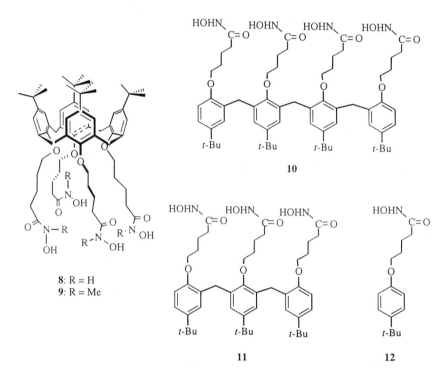

Figure 2. Calix[4]arene tetrahydroxamates and acyclic analogs

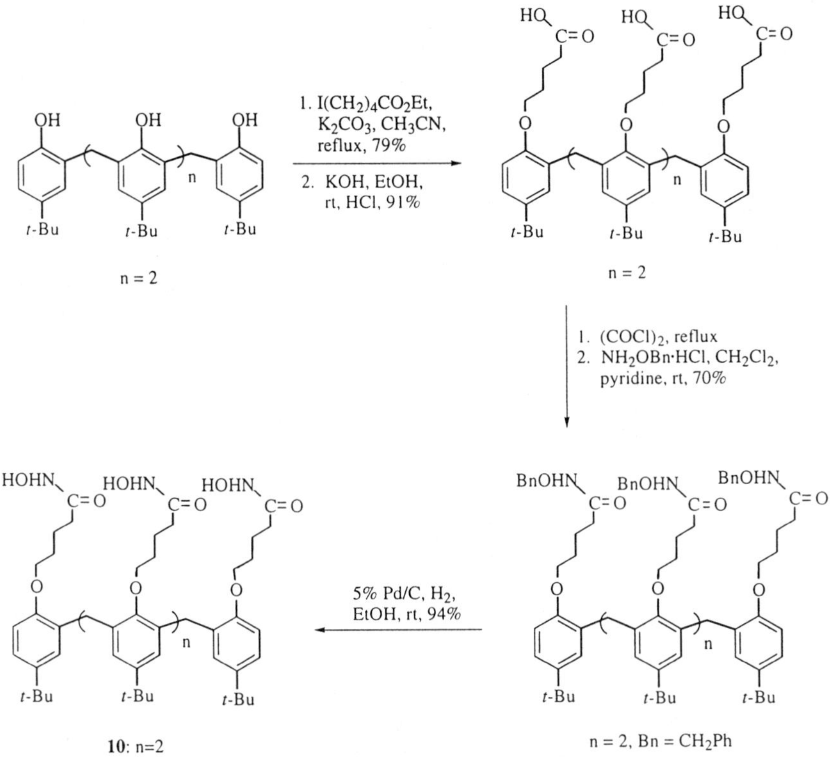

*Scheme 1. Synthetic route to tetrahydroxamate extractant **10***

8 to extractants **10-12** at pH 1, and to a certain extent at pH 2. Hence, the discussion that follows focuses to a large extent on the acyclic analogs **10** and **11**, and the monohydroxamate **12**.

As seen in Table I, at pH 1 the acyclic tetrahydroxamate **10** and trihydroxamate **11** are very effective for the extraction of Th(IV) with 99% removal. The monohydroxamate **12** only extracts 13% under these conditions. At pH 2, all the hydroxamate ligands are highly efficient in the extraction of Th(IV) into chloroform with greater than 99% of the available metal being removed.

The extraction capabilities of these ligands for Fe(III) are quite different than that found for Th(IV) and show a strong pH dependence, Table I. Surprisingly, in the presence of excess ligand concentration, the monohydroxamate **12** is the best extractant for Fe(III) with 64% removal at pH 1. In contrast, the tetrahydroxamate **10** and trihydroxamate **11** were less efficient (30 and 33% respectively). At pH 2, the efficiency for the removal of Fe(III) for all ligands increases. For example, the tetrahydroxamate **10** and trihydroxamate **11** are able to extract 83% and 85% respectively while the monohydroxamate **12** still remains the most effective extractant (92%).

The ability of these ligands to extract Eu(III) and Cu(II) into chloroform under similar conditions (0.10 M NaNO$_3$, pH 2) was also examined. None of the ligands extracted Eu(III) or Cu(II) to any appreciable amount under these conditions. In fact, the extraction of Eu(III) was less than 2% while the removal of the cupric ion was less than 3% in all cases.

Table I. Th(IV) and Fe(III) Single Metal Extraction Studies - % Extracted by Hydroxamate Ligands[a]

Initial pH	Th(IV)				Fe(III)			
	8[d]	10	11	12	8[d]	10	11	12
1[b]	-	99	99	13	-	30	33	64
2[c]	>99	>99	>99	>99	85	83	85	92

[a] into chloroform from 0.10 M NaNO$_3$, [8] = 1.00 mM, [10] = 1.00 mM, [11] =1.33 mM, [12] = 4.00 mM;
[b] [Th^{4+}] = 0.23 mM, [Fe^{3+}] = 0.23 mM; [c] [Th^{4+}] = 0.22 mM, [Fe^{3+}] = 0.18 mM. [d] precipitation occurred with this ligand at pH 1. Precipitation was less obvious but still present at pH 2.

Competitive Extraction Studies - Th(IV) vs. Fe(III)

Clearly the selective extraction of Th(IV) or Fe(III) in the presence of trivalent lanthanides or divalent copper can be achieved using any of the hydroxamate ligands **10-12** at pH 2 or lower; however, it was more pertinent to our goals to ascertain whether the selective extraction of Th(IV) could be achieved in the presence of Fe(III) under these conditions. These results prompted us to conduct competitive extraction studies with these two metal ions in order to more accurately evaluate the potential of these new ligands to serve as actinide selective extractants. At pH 1, when an aqueous mixture of equimolar quantities (0.22 mM) of Th(IV) and Fe(III) was contacted with a slight molar excess of the ligand of interest (all balanced to 1 mM total hydroxamate) in chloroform, both the tetrahydroxamate **10** and trihydroxamate **11** were found to selectively extract Th(IV) (53 and 75% respectively) over Fe(III) (18 and 28% respectively), Table II. The extraction efficiency of the monohydroxamate **12** drops significantly under these conditions but it appears to extract slightly more Fe(III). Again, significant precipitation of the calix[4]arene tetrahydroxamate **8** occurred at this pH preventing its reliable comparison to the other ligands.

Table II. Competitive Extraction Studies at pH 1 - Th(IV) vs. Fe(III) - % Extracted by Hydroxamate Ligands[a]

Ligand	%E$_{Th}$	%E$_{Fe}$	D$_{Th}$	D$_{Fe}$	S$_{Th/Fe}$
10	53	18	1.1	0.22	5.0
11	75	28	3.0	0.39	7.7
12	16	20	0.18	0.25	0.72

[a] $D = (\Sigma[M^{n+}]_o/\Sigma[M^{n+}]_w)$, $S_{a/b} = D_a/D_b$, %E = [D/(D+1)] x100%, into chloroform from 0.10 M NaNO$_3$; [Th^{4+}]=0.22 mM, [Fe^{3+}]=0.22 mM, [**10**] = 0.25 mM, [**11**] = 0.33 mM, [**12**] = 1.00 mM

Competitive metal ion extraction studies conducted at pH 2 gave similar results, Table III. The tetrahydroxamate **10** and trihydroxamate **11** were still found to selectively extract Th(IV) (56 and 83% respectively) over Fe(III) (40 and 44% respectively). At pH 2, one can see that the monohydroxamate **12** indeed has the opposite selectivity with 60% of the Fe(III) being removed in contrast to 31% for the removal of Th(IV). At pH 2, the calixarene **8** also appears to be selective for Fe(III) extraction, consistent with earlier results (49), although the numbers must be regarded with some caution due to the observed precipitation.

It was surprising to find that the trihydroxamate **11** was an efficient and more selective extractant for Th(IV) over Fe(III) than the corresponding tetrahydroxamate **10** ($S_{Th/Fe}$ = 5.9 vs. $S_{Th/Fe}$ = 1.9, respectively at pH 2). This suggested that Th(IV) was being extracted as a trihydroxamate species by both these ligands, with the trihydroxamate **11** forming a more stable actinide complex. To investigate this further, a competitive study between these two metal ions with equal molar concentrations (0.25 mM) of **10** and **11** was performed, Table IV. In this experiment, the concentration of hydroxamate groups available for complexation was higher for tetrahydroxamate **10** than trihydroxamate **11** (1.00 mM and 0.75 mM respectively). The extraction behavior of **10** and **11** was quite similar under these conditions once again suggesting that the fourth hydroxamate arm of **10** provided no advantage in enhancement of the actinide extraction process.

Table III. Competitive Extraction Studies at pH 2 - Th(IV) vs. Fe(III) - % Extracted by Hydroxamate Ligands[a]

Ligand	%E Th	%E Fe	D Th	D Fe	S Th/Fe
8[b]	37	51	0.58	1.0	0.58
10	56	40	1.3	0.67	1.9
11	83	44	4.7	0.80	5.9
12	31	60	0.45	1.5	0.30

[a] $D = (\Sigma[M^{n+}]_o / \Sigma[M^{n+}]_w)$, $S_{a/b} = D_a/D_b$, $\%E = [D/(D+1)] \times 100\%$, into chloroform from 0.10 M $NaNO_3$; $[Th^{4+}]=0.17$ mM, $[Fe^{3+}]=0.18$ mM, **[8]** = 0.25 mM, **[10]** = 0.25 mM, **[11]** = 0.33 mM, **[12]** = 1.00 mM; [b] some precipitation observed

Table IV. Competitive Extraction Studies at pH 2 - Th(IV) vs. Fe(III) (Equimolar Extractant Concentration) - % Extracted by Hydroxamate Ligands[a]

Ligand	%E Th	%E Fe	D Th	D Fe	S Th/Fe
10	54	48	1.2	0.93	1.3
11	57	47	1.3	0.88	1.5

[a] $D = (\Sigma[M^{n+}]_o / \Sigma[M^{n+}]_w)$, $S_{a/b} = D_a/D_b$, $\%E = [D/(D+1)] \times 100\%$, into chloroform from 0.10 M $NaNO_3$; $[Th^{4+}]=0.25$ mM, $[Fe^{3+}]=0.24$ mM, **[10]** = 0.25 mM, **[11]** = 0.25 mM

One would have predicted **10** to be a better extractant for Th(IV) than **11**. The tetrahydroxamate **10** was expected to form a strong neutral 1:1 ML tetrahydroxamato complex with Th(IV) in the extraction process by loss of four protons, one from each of the ligand moieties. If extractant **10** only uses three of its hydroxamate groups in the coordination of the actinide(IV) ion to give a ML species, either dissociation of a non participating hydroxamate group leading to a neutral complex or co-extraction of an anion, such as nitrate, must occur in order to maintain electrical neutrality. The 1:1 Th-**11** complex, on the other hand, cannot be neutral and an anion must be transported into the organic layer to ensure charge neutrality, making this a less favorable process. One would also predict that both **10** and **11** would have similar effectiveness for the extraction of Fe(III) and the results are consistent with this expectation.

Ligand Molar Variation Studies

Ligand molar variation studies were performed to understand the surprising extraction preference of the trihydroxamate **11** for the Th(IV) cation, Figure 3. The intent was to determine the stoichiometry of the extracted species for Th(IV) and Fe(III) and to establish the extent of hydroxamate coordination/participation in the ligand-metal complex for extractant **11**. At pH 2, traditional breakthrough curves, Figures 3a and 3b, for the trihydroxamate **11** with Fe(III) and Th(IV) both gave saturation break-points of approximately 1.4 - 1.5 suggesting the formation of a neutral $(M_2L_3)_n$ extracted species for both metals. A log-log plot analysis of the trihydroxamate **11** with Fe(III), Figure 3c, also showed that the extracted species appeared to be a $(M_2L_3)_n$ species (m = 1.5, r^2 = 0.992). On the other hand, the corresponding plot for Th(IV), Figure 3d, does not show a linear relationship suggesting that at least two complexes of different stoichiometries are involved in the extraction process.

If **11** extracts Fe(III) at pH 2 as a neutral trishydroxamato $(M_2L_3)_n$ species, this would indicate that only two of the hydroxamic acids per ligand molecule are complexing with the cations while the third remains undissociated. Such a preference may be due to geometric constraints of the ligand or due to the high pK_a's (8-10) of the hydroxamic acids(*50*). For the complexation of Th(IV) it appears that on average, three hydroxamic acid groups in each ligand molecule are involved in the binding of the cations.

Spectroscopic Analyses

As Fe^{3+}-hydroxamate complexes are colored and well characterized in the literature (*51*) the nature of the extracted Fe^{3+}-ligand complexes from these studies were investigated using UV-VIS spectroscopy, Figure 4 (*52*). For each of the ligands studied, the predominant complex in chloroform at pH 1 and 2 that was observed from the Fe(III) extraction studies (Table I) appeared to be an iron-trishydroxamato species with λ_{max} ~430nm, Figure 4a. UV-VIS spectra of the chloroform layers from the ligand variation study of **11** with Fe(III) also support the formation of a trishydroxamato complex, even at extremely low ligand concentrations, Figure 4b.

Based on literature precedents, we propose this species to be a neutral ferric-trishydroxamato complex. The preference to extract neutral complexes has been noted in the extraction of the uranyl ion by calix[n]arenes (*36*). In general, such behavior is anticipated for the extraction of a metal ion into a hydrophobic layer by lipophillic acidic chelating agents (*53*). One would also expect Th(IV) to be extracted as a neutral complex with these ligands. Unfortunately, Th(IV)-hydroxamato complexes are not colored and therefore cannot be examined using UV-VIS spectroscopy.

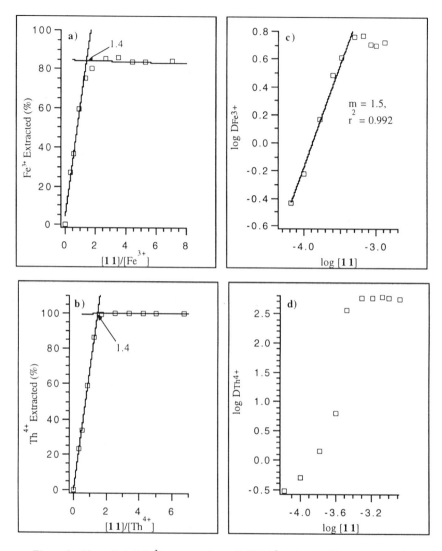

Figure 3. Plots of a) %Fe^{3+} extracted vs. [11]/[Fe^{3+}] b) %Th^{4+} vs. [11]/[Th^{4+}] c) log $D_{Fe^{3+}}$ vs. log [11] d) log $D_{Th^{4+}}$ vs. log [11]; [Fe^{3+}] = 0.19 mM, [Th^{4+}] = 0.20 mM, pH 2, 0.10 M $NaNO_3$; where m = slope.

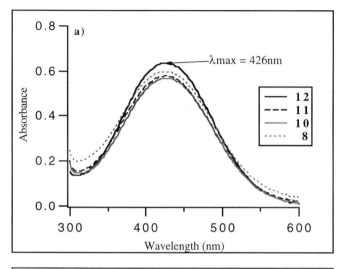
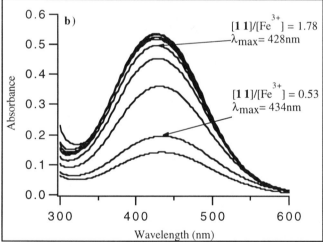

Figure 4. UV-VIS spectra in CHCl$_3$ of the extracted species of a) Fe(III) complexes at pH 2 (Table I) and b) 11-Fe(III) complexes from ligand molar variation study

CONCLUSIONS

In contrast to the *tert*-butylcalix[4]arene tetrahydroxamate **8**, the acyclic tetrahydroxamate **10** and trihydroxamate **11** preferentially extracted Th(IV) over Fe(III) in the pH range of 1-2. This suggests that the calix[4]arene platform may not be vital to achieve selective complexation of the larger actinide cation in these systems. Furthermore, the trihydroxamate **11** was surprisingly a more efficient extractant than tetrahydroxamate **10** for the actinide ion. This indicates that all four hydroxamate groups of **10** do not bind to the same actinide ion to impart a greater extraction efficiency with this ligand. It is important to point out that in contrast to **10** and **11**, the monohydroxamate **12** extracted Fe(III) preferentially in competitive studies and was the most efficient extractant of Fe(III) at pH 1 and 2. Given this fact, it is not clear why the monomer is a much less efficient extractant of Th(IV) at pH 1. For the extraction of Fe(III) with all of these ligand systems, spectroscopic studies clearly indicate that the iron-trishydroxamato species (λ_{max} ~ 430nm) is dominant in the organic layer under a wide range of conditions.

Given that the pK_a's of hydroxamic acids are usually in the range of 8-10, it may be difficult to ensure full ligand participation in actinide binding with tetrahydroxamate systems at this pH range. Also, one cannot ignore the possibility of geometric constraints that inhibit the formation of a 1:1 complex and hence lead to formation of aggregates, such as $(M_2L_3)_n$, to achieve efficiency in the extraction process. Of course, this makes it difficult to predict the metal ion selectivity in the extraction processes involving these hydroxamate ligand systems. Further studies are necessary if we hope to understand the complexation and stoichiometry of extraction of these fascinating metal-ligand systems.

ACKNOWLEDGMENTS

Portions of this work were supported by the Waste-Management Education and Research Consortium of New Mexico. Dr. Hollie Jacobs (NMSU) is thanked for helpful discussions. Dr. Gary Rayson (NMSU), Mr. Patrick Williams (NMSU) and the NMSU-SWAT lab are thanked for their assistance in metal ion analyses.

LITERATURE CITED

1. Proceedings of the First Hanford Separation Science Workshop, PNL-SA-21775, Pacific Northwest Laboratory, Richland, WA, July 23-25, 1991.
2. U. S. Department of Energy Office of Environmental Management Technology Development, Efficient Separations and Processing Crosscutting Program-Technology Summary, DOE/EM-0249, June 1995.
3. Gopalan, A.; Huber, V.; Jacobs, H. In *Waste Management from Risk to Remediation;* Bhada, R., Ed.; ECM Press: Albuquerque, NM, 1994, pp 227-246.
4. Cecille, L.; Casarci, M.; Pietrelli, L. *New Separation Chemistry Techniques for Radioactive Waste and Other Specific Applications;* Elsevier: London, 1991.

5. Horwitz, E.; Schulz, W. In *Solvent Extraction and Ion Exchange in the Nuclear Fuel Cycle;* Logsdail, D. H.; Mills, A. L., Eds.; Ellis Horwood Limited: Chichester, 1985, pp 137-144.
6. *Separations of f Elements;* Nash, K. L.; Choppin, G. R., Eds.; Plenum Press: New York, 1995.
7. Nash, K. L.; Choppin, G. R. *Sep. Sci. Tech.* **1997**, *32*, 255.
8. Martell, A. E.; Hancock, R. D. *Metal Complexes in Aqueous Solutions;* Plenum Press: New York, 1996.
9. Raymond, K. N.; Garrett, T. M. *Pure and Appl. Chem.* **1988**, *60*, 1807.
10. Raymond, K. N.; Smith, W. L. *Structure and Bonding (Berlin)* **1981**, *43*, 159.
11. Martell, A. E.; Hancock, R. D.; Motekaitis, R. J. *Coord. Chem. Rev.* **1994**, *133*, 39.
12. Hancock, R. D.; Martell, A. E. *Chem. Rev.* **1989**, *89*, 1875.
13. Uhlir, L. C.; Durbin, P. W.; Jeung, N.; Raymond, K. N. *J. Med. Chem.* **1993**, *36*, 504.
14. Xu, J.; Kullgren, B.; Durbin, P. W.; Raymond, K. N. *J. Med. Chem.* **1995**, *38*, 2606.
15. Stradling, G.N. *Radiat. Prot. Dosim.* **1994**, *53*, 297.
16. Stradling, G. N. *Journal of Alloys and Compounds* **1998**, *271-273*, 72.
17. For a review on hydroxamate siderophores see Miller, M. J. *Chem. Rev.* **1989**, *89*, 1563.
18. For a recent review on plutonium selective ligands see O'Boyle, N. C.; Nicholson, G. P.; Piper, T. J.; Taylor, D. M.; Williams, D. R.; Williams, G. *Appl. Radiat. Isot.* **1997**, *48*, 183.
19. Gutsche, C. D. *Calixarenes;* Royal Society of Chemistry: Cambridge, 1989.
20. *Calixarenes: A Versatile Class of Molecules;* Vicens, J.; Böhmer, V., Eds.; Kluwer: Dordrecht, 1991.
21. Gutsche, C. D. *Aldrichimica Acta,* **1995**, *28*, 3.
22. Böhmer, V. *Angew. Chem. Int. Ed. Engl.* **1995**, *34*, 713.
23. Shinkai, S. *Tetrahedron* **1993**, *49*, 8933.
24. Takeshita, M.; Shinkai, S. *Bull. Chem. Soc. Jpn.* **1995**, *68*, 1088.
25. Ikeda, A.; Shinkai, S. *Chem. Rev.* **1997**, *97*, 1713.
26. Roundhill, D. M. *Prog. Inorg. Chem.* **1995**, *43*, 533.
27. Casnati, A. *Gazz. Chim. Ital.* **1997**, *127*, 637.
28. Schwing-Weill, M-J.; Arnaud-Neu, F. *Gazz. Chim. Ital.* **1997**, *127*, 687.
29. Archimbaud, M.; Henhe-Napoli, M. H.; Lilienbaum, D.; Desloges, M.; Montagne, C. *Radiat. Prot. Dosim.* **1994**, *53*, 327.
30. Arnaud-Neu, F. *Chem. Soc. Rev.* **1994**, 235.
31. Nagasaki, T.; Shinkai, S. *Bull. Chem. Soc. Jpn.* **1992**, *65*, 471.
32. Georgiev, E. M.; Clymire, J.; McPherson, G. L.; Roundhill, D. M. *Inorg. Chim. Acta* **1994**, *227*, 293.
33. Shinkai, S.; Koreishi, H.; Ueda, K.; Arimura, T.; Manabe, O. *J. Am. Chem. Soc.* **1987**, *109*, 6371.
34. Shinkai, S.; Shiramama, Y.; Satoh, H.; Manabe, O.; Arimura, T.; Fujimoto, K.; Matsuda, T. *J. Chem. Soc. Perkin Trans. 2* **1989**, 1167.
35. Also see Dinse, C.; Baglan, N.; Cossonnet, C.; Le Du, J. F.; Asfari, Z.; Vicens, J. *Journal of Alloys and Compounds* **1998**, *271-273*, 778.
36. Nagasaki, T.; Shinkai, S. *J. Chem. Soc. Perkin Trans. 2* **1991**, 1063.

37. Araki, K; Hashimoto, N.; Otsuka, H.; Nagasaki, T.; Shinkai, S. *Chem. Lett.* **1993**, 829.
38. Arnaud-Neu, F.; Böhmer, V.; Dozol, J. -F.; Grüttner, C.; Jakobi, R. A.; Kraft, D.; Mauprivez, O.; Rouquette, H.; Schwing-Weill, M. -J.; Simon, N.; Vogt, W. *J. Chem. Soc. Perkin Trans. 2* **1996**, 1175.
39. Delmau, L. H.; Simon, N.; Schwing-Weill, M. -J.; Arnaud-Neu, F.; Dozol, J. -F.; Eymard, S.; Tournois, B.; Böhmer, V.; Grüttner, C.; Musigmann, C; Tunayar, A. *J. Chem. Soc. Chem. Commun.* **1998**, 1627.
40. Boerrigter, H.; Verboom, W.; de Jong, F.; Reinhoudt, D. N. *Radiochim. Acta* **1998**, *81*, 39.
41. Boerrigter, H.; Verboom, W.; Reinhoudt, D. N. *J. Org. Chem.* **1997**, *62*, 7148.
42. Boerrigter, H.; Verboom, W.; Reinhoudt, D. N. *Liebigs Ann./Recueil* **1997**, 2247.
43. Malone, J. F.; Marrs, D. J.; McKervey, M. A.; O'Hagan, P.; Thompson, N.; Walker, A.; Arnaud-Neu, F.; Mauprivez, O.; Schwing-Weill, M.-J.; Dozol, J.- F.; Rouquette, H.; Simon, N. *J. Chem. Soc. Chem. Commun.* **1995**, 2151.
44. Lambert, T. N.; Jarvinen, G. D.; Gopalan, A. S. *Tetrahedron Lett.* **1999**, *40*, 1613.
45. For lower rim CMPO's also see Barboso, S.; Carrera, A. G.; Matthews, S. E.; Arnaud-Neu, F.; Böhmer, V.; Dozol, J.-F.; Rouquette, H.; Schwing-Weill, M.-J. *J. Chem. Soc. Perkin Trans.* 2, **1999**, 719.
46. Lambert, T. N.; Dasaradhi, L.; Huber, V. J.; Gopalan, A. S. *J. Org. Chem.* **1999**, *16*, 6097.
47. Gopalan, A.; Huber, V.; Koshti, N.; Jacobs, H.; Zincircioglu, O.; Jarvinen, G.; Smith, P. In *Separations of f Elements;* Nash, K. L.; Choppin, G. R., Eds.; Plenum Press: New York, 1995, pp 77-98.
48. Gopalan, A.; Jacobs, H. K.; Stark, P. C.; Koshti, N. M.; Smith, B. F.; Jarvinen, G. D.; Robison, T. W. *International Journal of Environmentally Conscious Design and Manufacturing* **1995**, *4 (3-4)*, 19.
49. Dasaradhi, L.; Stark, P. C.; Huber, V. J.; Smith, P. H.; Jarvinen, G. D.; Gopalan, A. S. *J. Chem. Soc. Perkin Trans. 2* **1997**, 1187.
50. Martell, A. E.; Smith, R. M.; Motekaitis, R. J. *NIST Critical Stability Constants of Metal Complexes Database*, 1993.
51. Albrecht-Gary, A. -M.; Crumbliss, A. L. "Coordination Chemistry of Siderophores: Thermodynamics and Kinetics of Iron Chelation and Release", *Metal Ions in Biological Systems*, **1998**, *35*, 239.
52. The chloroform layers were carefully isolated and their spectra obtained against a blank of chloroform. No attempt was made to dry the layers before obtaining their spectra as good phase separation was achieved after centrifugation.
53. Nash, K. L. In *Metal-Ion Separation and Preconcentration;* Bond, A. H.; Dietz, M. L.; Rogers, R. D., Eds.; ACS Symposium Series 716; American Chemical Society: Washington, DC., 1999, pp. 52-78.

Chapter 17

Calixarenes as Ligands in Environmentally-Benign Liquid–Liquid Extraction Media

Aqueous Biphasic Systems and Room Temperature Ionic Liquids

Ann E. Visser, Richard P. Swatloski, Deborah H. Hartman, Jonathan G. Huddleston, and Robin D. Rogers[1]

Department of Chemistry and Center for Green Manufacturing, The University of Alabama, Tuscaloosa, AL 35487

Calixarene partitioning in Aqueous Biphasic Systems (ABS) and Room Temperature Ionic Liquids (RTIL) has been studied to establish fundamental correlations between the nature of the solute and partitioning behavior in these novel systems that represent alternative liquid/liquid separation technologies. Sulfonated, water-soluble calix[6] and calix[4]arenes quantitatively partition to the polymer-rich phase of an ABS, but remain in the aqueous phase when contacted with a RTIL. In contrast, the distribution studies of unsubstituted calixarenes indicate their affinity for the RTIL. Substitution of hydrophilic or functional groups on the upper rim of the calixarenes tailors the solubility of the macrocycle necessary for adapting traditional complexants for use in novel solvent systems.

Separation processes are commonplace throughout science, from nuclear waste remediation to organic synthesis, as a means to ultimately segregate components of a mixture. To accommodate each process, conditions for separation are tailored to make them as efficient as possible, often employing one or more separation steps for complicated mixtures. Thus, solvent extraction (SX) and the associated formation of two-phase systems for partitioning and sequestering solutes in an organic solvent from an aqueous phase are of major importance in separations science (1,2). Solvent extraction has the advantages of rapid kinetics, high selectivity, and high throughput. This technology is regularly used on a large scale in industry.

[1]Corresponding author.

Green Chemistry and Alternative Separations Technologies

The wide-spread usage of traditional SX systems has come to a crossroads as the emphasis for sustainable technology, or "Green Chemistry," takes into consideration the overall environmental impact of both the process and waste streams generated in a variety of industrial processes (*3-6*). In particular, organic solvents have come under intense scrutiny as many are classified Volatile Organic Compounds (VOCs) and are associated with increasing regulations regarding their usage, disposal, and human health awareness. No doubt that environmental regulations for these chemicals will be a major factor for industry, as predicted in *Vision 2020: 1998 Separations Roadmap*, a recent report from the American Institute of Chemical Engineers and the Department of Energy (*3*). The report cited specific areas of emphasis, including separation science and the development of novel solvents, that will be crucial towards the continued growth of the chemical industry and compliance with societal standards and demands. Excerpts from another report (*7*) reference a publication by the U. S. National Research Council in which concerns under "High Priority Research Needs and Opportunities" include separation systems that are more "selective and efficient, to improve selectivity among solutes in separations."

We are investigating Aqueous Biphasic Systems (ABS) (*5,6,8-16*) and Room Temperature Ionic Liquids (RTIL) (*5,17,18*) as alternatives to VOCs in liquid/liquid separations. ABS are composed of sufficient concentrations of both a water-soluble polymer and certain salts such that a two-phase system forms concurrently with the salting-out of the polymer. Polyethylene glycol (PEG)-based ABS have demonstrated their utility in such areas as nuclear waste processing (*12,19,20*), small organic molecule partitioning (*9,21,22*), and metal ion separations (*6,10,12,23-25*). Extensive work has been carried out in PEG-ABS as that polymer is non-toxic, commercially available, and inexpensive. Since both phases are aqueous (each layer is over 80% water on a molar basis), non-traditional metal complexants can be used that are not viable for use in organic solvent liquid/liquid extraction systems.

We have also recently established the use of RTIL in liquid/liquid separations (*5,17,18*). RTIL are liquids composed entirely of ions, the most widely studied comprised of organic cations (e.g., alkyl-methyl-imidazolium ion or alkyl-pyridinium) and hydrophobic anions (PF_6^-). These RTIL exhibit solvent-like properties and can function as diluents for a wide range of materials (*18,26-28*). Certain RTIL are stable to air and moisture at room temperature, are immiscible with water, and, unlike traditional organic solvents, have no vapor pressure.

It is the chemistry of the anion that governs the majority of the properties of these liquids and their true utility is realized when the synthesis is varied to include different anions; PF_6^- salts are water immiscible and air stable (*17,18,29*), BF_4^- are not stable in air but under certain compositions are water immiscible (*29*), and superacidic, albeit air and water sensitive, systems may be present with $AlCl_4^-$ as the anion (*27*). Liquid/liquid extraction from aqueous systems can be carried out with ionic liquids, since the chemistry of PF_6^- renders RTIL formed from it capable of forming a two-phase system with aqueous media (*10,17,27,30*).

Alternative Separation Technologies with Calixarene Extractants

Selective separations can be induced or enhanced via the introduction of extractants. In particular, metal ion extraction in SX has relied heavily on the development of extractants that provide both selective metal ion complexation and increased preference of the metal for the extracting phase. For example, the chemistry of macrocyclic ligands, such as calixarenes, has focused on their size-selective ion complexation and transport properties (*31-36*). These basket-like molecules offer unique chemistry as the outside of the molecule is lipophilic while the core of the molecule provides a hydrophilic region. It is this behavior that renders them ideal for complexing metal ions and subsequent transport to a hydrophobic environment. Calixarenes also provide a rigid platform for hanging extracting groups at the upper rim (*32*) or the lower rim (*33,37*). The lower rim is hydrophilic while the upper rim is either hydrophilic or hydrophobic, depending on the substitution. Figure 1 shows the location and orientation of the sulfonic acid groups on the water soluble *p*-sulfonatocalix[4]arene (*38*).

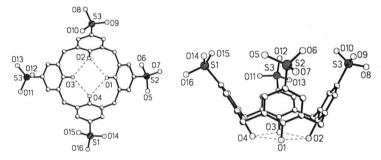

Figure 1. Crystal structure of p-sulfonatocalix[4]arene. Coordinates from (38).

It is the match between the solute size and the cavity size that allows only certain molecules or ions to enter and possibly bind, highlighting the selective nature exhibited by this class of molecules. For example, certain crown ether-functionalized calixarenes show an enhanced selectivity for cesium in the presence of other alkali cations (*39*). Thus, a considerable amount of calixarene science has focused on understanding the solid state and aqueous phase chemistry in order to capitalize on their potential.

The structure of a calixarene with upper and lower functional groups provides a platform to further tailor the characteristics of the molecule to match the hydrophobicity of the extracting phase. In general, when using an extractant, optimal results will be obtained when the extractant quantitatively partitions to the extracting phase regardless of system pH. In traditional organic solvent-based liquid/liquid separations, the complexants are modified through the addition of long alkyl chains that increase the hydrophobicity of the molecule and, hence, its retention in the organic phase. Calixarenes have shown their utility as host molecules for alkali and alkaline-earth metals (*40*), transition metals (*38,40*), and lanthanides (*34*).

The new Green chemistry paradigm applied to separations will require not only the development of VOC alternatives that are inherently less polluting and more environmentally-friendly, but also the adaptation of traditional SX extractants to these systems. This will entail modifying the extractants and understanding their behavior in these novel systems to sustain efficient separations.

Experimental

Polyethylene glycol (average molecular weight 2000) and all salts and dyes were obtained from Aldrich. Calix[4] and calix[6]arenes were obtained from Acros Organics (Fisher) and the 4-sulfonic acid, *tert*-butyl, and unsubstituted derivatives of each calixarene were used without further purification. $^{59}FeCl_3$ and $^{152}EuCl_3$ were obtained from Amersham. All solutions were made with deionized water polished to 18.3•MΩ·cm using a Barnstead Nanopure filtration system. RTIL were prepared in our laboratory following an established synthetic procedure (*17*) and subsequently stored in plastic bottles and equilibrated with water.

Standard curves were constructed for the soluble calixarenes in both phases of each RTIL/aqueous system and each ABS/salt system. Equal volumes of both phases were contacted, vortexed, and centrifuged. The phases were then separated and a known amount of calixarene was added to each phase and the absorbance at 283 nm was measured. No spectral interferences were observed for the RTIL phase or the PEG-rich phase. There was a linear relationship between the absorbance and concentration, indicating these systems were within the limit of Beer's Law.

Calixarene solutions were made in both the PEG and RTIL phase from which equal volumes of 40% PEG-2000 or appropriate RTIL solutions were taken and contacted with aqueous phases. The systems were vortexed and centrifuged twice for two minutes each and then the phases were separated. The absorbance of the solutions were measured in 1 mm length quartz cuvets from which the blank measurement was automatically subtracted. Concentrations of the calixarene in each phase were calculated using the equation for the standard curve. All metal ion partitioning experiments were carried out as above using radiotracers to monitor ion distribution. The distribution ratios were obtained by sampling 100 μL aliquots from both phases for gamma ray emission analysis with a Packard Cobra II Auto-Gamma counting system. Experiments were conducted in duplicate and the results agree to within 5%. The spectrophotometrically and radiometrically determined distribution ratios were calculated as:

$$D = \frac{\text{Concentration or Activity in the RTIL or PEG - rich phase}}{\text{Concentration or Activity in the aqueous or salt - rich phase}}$$

Calixarene Partitioning in Aqueous Biphasic Systems

In traditional SX, unsubstituted or *t*-butyl-calixarenes have a high affinity for the organic solvent (*31,35*). Thus, these molecules must be modified for optimal behavior in an entirely aqueous two-phase system. In ABS, the unique problem arises that the metal ion extractant must prefer the PEG-rich phase, not the salt-rich phase, even though both phases are aqueous. The partitioning of dyes has been studied in detail in ABS (*11,41*) and particular success for PEG-rich phase affinity was found for aromatic dyes that contained a sulfonic acid group that dramatically increased the water solubility and the preference for the PEG-rich phase. Figure 2 shows quantitative partitioning for several sulfonated azo dyes.

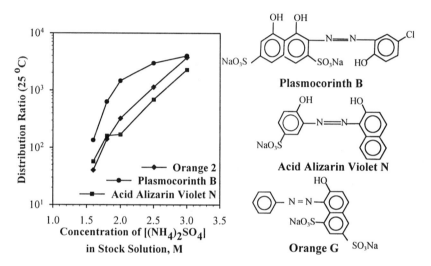

Figure 2. Partitioning and structures of three sulfonated dyes in PEG-2000 ABS prepared by mixing equal volumes of 40% PEG-2000 and $(NH_4)_2SO_4$ stock solutions of increasing concentrations.

For these complexing dyes, the azo group coordinates metal ions, as shown by the Cr^{3+} complex of acid alizarin violet N in Figure 3. Metal ions are coordinated to both nitrogens and the presence of an oxygen in the position *ortho* to the azo groups appears to play an important role in the successful complexation of the metal ions. Complexation reduces the hydration sphere around the metal ion and facilitates transport into an organic solvent. Thus, sulfonated, water-soluble azo dyes can be used as metal extractants due to their preference for the PEG-rich phase and their metal complexing ability.

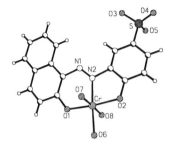
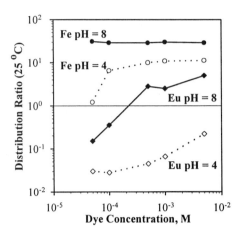

Figure 3. Structure of the complex in Cr(AAV)(OH$_2$)$_3$·4H$_2$O. Formation of a 2:1 complex dehydrates the metal ion.

Figure 4. D Values for Fe^{3+} and Eu^{3+} in 40% PEG-2000/3.5 M (NH$_4$)$_2$SO$_4$ ABS with increasing AAV concentration.

Acid alizarin violet N (AAV) enhances the partitioning of several metal ions to the PEG-rich phase, as shown for Fe(III) and Eu(III) in Figure 4. In contrast to the distribution ratios observed in the absence of dye (0.059 for Eu^{3+} and 0.066 for Fe^{3+}), there is a significant enhancement in the presence of AAV. The partitioning results can be correlated to the complexation constants for each metal ion with the molecule. At pH 4 and 8, the sulfonic acid groups are ionized (log K_{PCV} = 0.8 and log K_{AAV} = 0.88 (42)). Analysis of the results indicates that a 2:1 dye:metal complex forms for all cases except Fe^{3+} at pH 8 where lower dye concentrations need to be studied before such a determination can be made.

The unsubstituted and *tert*-butyl derivatives of both calix[4] and calix[6]arene are hydrophobic and indeed these ligands do not dissolve in either the salt-rich or PEG-rich phase of PEG-2000/(NH$_4$)$_2$SO$_4$ ABS, precluding their study in these systems. The upper-rim sulfonated-calixarene derivatives, however, are water soluble and we studied their partitioning behavior in an ABS prepared by mixing equal aliquots of 40% (w/w) PEG-2000 and 3.5 M (NH$_4$)$_2$SO$_4$.

In the pH range 2–10, the affinity of sulfonated-calix[4] and calix[6]arene for the PEG-rich phase is apparently independent of hydroxyl group ionization, although all sulfonic acid groups are deprotonated. A speciation diagram (Figure 5) for *p*-sulfonatocalix[4]arene indicates the first H is lost in acidic conditions, corresponding to a pKa$_1$ value less than 1 (35).

Both sulfonated calix[4] and calix[6]arene display a high affinity for the polymer-rich phase (Figure 6). Partitioning of a series of sulfonated indigo dyes (41) has shown that the D values decrease as the number of sulfonic acid groups increases.

Here, increasing both the size of the molecule and the number of sulfonic acid groups from four to six may increase the hydrophilicity and result in slightly lower D values.

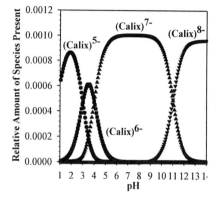

Figure 5. Speciation diagram for p-sulfonatocalix[4]arene as a function of pH. Data from (35).

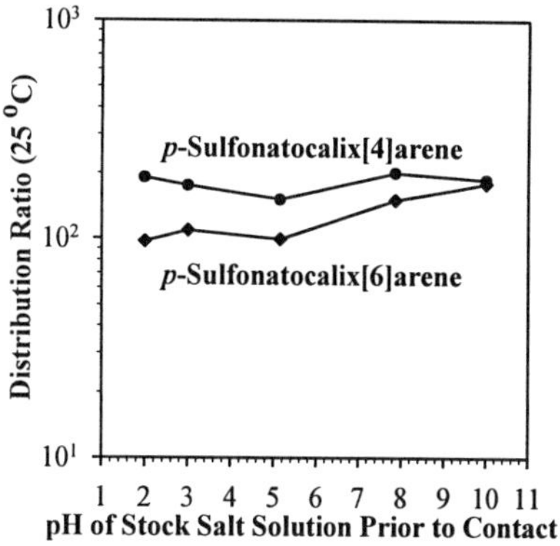

Figure 6. Distribution ratios for p-sulfonatocalix[4]- and calix[6]arene in 40% PEG-2000 with either 3.5 M $(NH_4)_2SO_4$ or 4 M NaOH adjusted to the appropriate pH.

Previous work has determined a correlation between a solute's Gibbs free energy of hydration (ΔG_{hyd}) and phase preference (16). Available thermodynamic data for the sulfonated calixarenes indicates ΔG_{hyd} values of –21.3 and –15.3 kJ/mol for calix[4] and calix[6]arene, respectively (43), and supports their preference for the PEG-rich phase, but not the order of partitioning.

Calixarene Partitioning in Room Temperature Ionic Liquids

We have recently demonstrated the use of 1-alkyl-3-methylimidazolium hexafluorophosphate ([Rmim][PF_6]) RTIL (Figure 7) for the removal of aromatic solutes such as benzene and its derivatives from aqueous solutions (5,17,18). The resulting distribution ratios for the aromatic organic molecules in the ionic liquid/aqueous system were correlated to their 1-octanol/water partition coefficients (log P) and show that a hydrophobic environment is present in this particular ionic liquid. These results indicate very different requirements for enhanced solute partitioning in liquid/liquid separations with RTIL in comparison to an ABS.

The partitioning of ionizable aromatic acids has been studied in these systems and the observed phase preference depends upon the degree of ionization (17). The ionic liquid will tolerate the addition of neutral aromatic species, but if the aqueous phase pH is adjusted sufficiently to ionize the solute, the ions prefer the aqueous phase (17). This observation opens the area of pH-dependent partitioning for the development of RTIL technology with metal ion extractants.

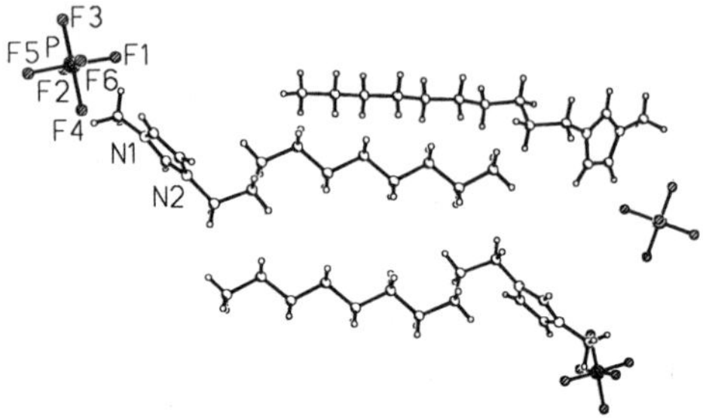

Figure 7. Structure of the low melting 1-decyl-3-methylimidazolium hexafluorophosphate.

Distribution ratios for metal ions in [bmim][PF_6]/aqueous systems indicate that despite the ionic nature of RTIL, metal ions remain in the aqueous phase due to their hydration. Thus, as with traditional SX, an extractant is necessary to enhance their affinity for the ionic liquid phase (6,30).

The extractants 1-(2-pyridylazo)-napthol (PAN) and 1-(2-thiazolylazo)-napthol (TAN) have been used to show that traditional metal ion extraction is possible in RTIL. These ligands remain in the ionic liquid phase from pH 1–13 and their pH dependant complexation of metal ions results in extraction to the RTIL phase at high pH and to the aqueous phase at low pH (Figure 8). Typical metal ion D values in the [bmim][PF_6]/H_2O system without an extractant are less than 0.01 for those metals shown in Figure 8.

The hydrophobic nature of [bmim][PF_6] and other alkyl derivatives provides a suitable environment for solubilizing both the unsubstituted calix[4] and calix[6]arenes, however, the *tert*-butyl calixarenes were not soluble in this RTIL.

Figures 9a and 9b show the distribution of the sulfonated and unsubstituted calixarenes, respectively, in [bmim][PF_6]/water as a function of aqueous phase pH. As expected, the sulfonated calixarenes have little, if any, affinity for the ionic liquid although the presence of a small amount of water in the octyl ionic liquid ([omim][PF_6]) could explain a measurable, albeit small, D value in those systems. Figure 9b indicates that the unsubstituted calixarenes have a high affinity for the ionic liquid phase over the entire range of pH values studied. Thus, the unsubstituted calixarenes are potential extractants for use in RTIL systems and future studies will explore this possibility.

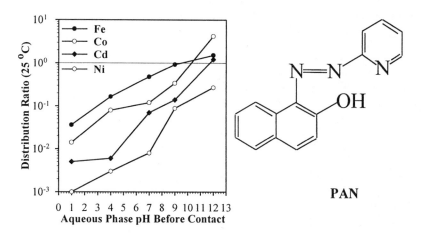

Figure 8. Distribution ratios for metal ions between [bmim][PF$_6$] with 0.1 mM PAN and water as a function of pH.

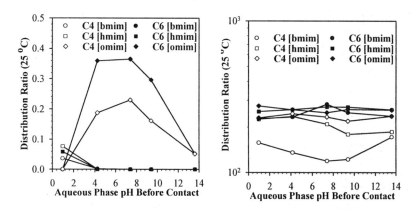

Figure 9. Distribution of sulfonated (a) and unsubstituted (b) calixarenes in 1-alkyl-3-methylimidazolium hexafluorophosphate ([Rmim][PF$_6$], R = butyl, hexyl, octyl) RTIL.

Conclusions

ABS and RTIL have utility in alternative separation technologies and offer similar physical properties, but drastically different solvating properties from traditional liquid/liquid separations employing VOCs as the extracting phase. The novelty stems from the differences in the behavior of ABS and RTIL as solvents; namely their ability to form a biphasic system and remove a variety of solutes from aqueous systems without the incorporation of VOCs.

New separation systems require the understanding of solute behavior as it applies to extractants or other organic molecules and metal ions. ABS require less structured aqueous extractant molecules that prefer a hydrophobic environment while RTIL exhibit similar solubilizing properties as many less polar organic solvents, without the deleterious properties associated with VOCs.

While the unique environments of ABS and RTIL allow for the implementation of new solutes as extractants, we have demonstrated that simple modifications of known metal ion extractants can extend their utility to other nontraditional liquid/liquid separation systems. These 'Green' separations systems should provide fertile ground for future development in support of sustainable industrial technologies.

Acknowledgements

Support of this work by the U.S. National Science Foundation (Grant No. CTS-9522159) for the ABS studies and by the Division of Chemical Sciences, Office of Basic Energy Sciences, Office of Energy Research, U.S. Department of Energy (Grant No. DE-FG02-96ER14673) for the RTIL studies is gratefully acknowledged.

References

1. Barakat, N.; Burgard, M.; Asfari, Z.; Vicens, J.; Montavon, G.; Duplâtre, G. Solvent Extraction of Alkaline-Earth Ions by Dicarboxylated Calix[4]arenes. *Polyhedron* **1998**, *17*, 3649.
2. Musikas, C.; Schulz, W. W. Solvent Extraction in Nuclear Science and Technology. In *Principles and Practices of Solvent Extraction*; Rydberg, J.; Musikas, C.; Choppin, G. R., Eds.; Marcel Dekker: New York, 1992; pp 413-447.
3. Adler, S.; Beaver, E.; Bryan, P.; Rogers, J. E. L.; Robinson, S.; Russomanno, C. *Vision 2020: 1998 Separations Roadmap*; Center for Waste Reduction Technologies of the American Institute of Chemical Engineers, AIChE: New York, 1998.

4. Anastas, P. T.; Warner, J. C. *Green Chemistry: Theory and Practice*; Oxford University Press: Oxford, 1998.
5. Spear, S. K.; Visser, A. E.; Willauer, H. D.; Swatloski, R. P.; Griffin, S. T.; Huddleston, J. G.; Rogers, R. D. Green Separation Science & Technology: Replacement of Volatile Organic Compounds in Industrial Scale Liquid/Liquid or Chromatographic Separations. In *Green Chemistry and Engineering*; ACS Symposium Series; American Chemical Society: Washington, DC, 1999; In Press.
6. Visser, A. E.; Griffin, S. T.; Ingenito, C. A.; Hartman, D. H.; Huddleston, J. G.; Rogers, R. D. Aqueous Biphasic Systems as a Novel Environmentally-Benign Separations Technology for Metal Ion Removal. In *Metal Separation Technologies Beyond 2000: Integrating Novel Chemistry with Processing*; Liddell, K. C.; Chaiko, D. J., Eds.; The Minerals, Metals & Materials Society: Warrendale, PA, 1999; pp 119-130.
7. *Separation and Purification: Critical Needs and Opportunities* (Report to the National Research Council by the Committee on Separations Science and Technology, Chairman C. J. King); National Academy; Washington, DC, 1987.
8. *Partitioning in Aqueous Two-Phase Systems; Theory, Methods, Uses, and Applications to Biotechnology*; Walter, H.; Brooks, D. E.; Fisher, D., Eds.; Academic Press: Orlando, FL, 1985.
9. Willauer, H. D.; Huddleston, J. G.; Griffin, S. T.; Rogers, R. D. Partitioning of Aromatic Molecules in Aqueous Biphasic Systems. *Sep. Sci. Technol.* **1999**, 1069.
10. Rogers, R. D.; Bond, A. H.; Bauer, C. B. Metal Ion Separations in Polyethylene Glycol-Based Aqueous Biphasic Systems. *Sep. Sci. Technol.* **1993**, *28*, 1091.
11. Rogers, R. D.; Bond, A. H.; Bauer, C. B.; Griffin, S. T.; Zhang, J. Polyethylene Glycol-Based Aqueous Biphasic Systems for Extraction and Recovery of Dyes and Metal/Dye Complexes. Shallcross, D. G.; Paimin, R.; Prvcic, L. M., Eds.; In *Value Adding Through Solvent Extraction Proceedings of ISEC '96*; The University of Melbourne: Parkville, Victoria, Australia, 1996; Vol. 2, pp 1537-1542.
12. Rogers, R. D.; Zhang, J. New Technologies for Metal Ion Separations: Polyethylene Glycol Based-Aqueous Biphasic Systems and Aqueous Biphasic Extraction Chromatography. In *Ion Exchange and Solvent Extraction*; Marinsky, J. A.; Marcus, Y., Eds.; Marcel Dekker: New York, 1997; Vol. 13, pp 141-193.
13. Rogers, R. D.; Bauer, C. B.; Bond, A. H. Novel Polyethylene Glycol-Based Aqueous Biphasic Systems for the Extraction of Strontium and Cesium. *Sep. Sci. Technol.* **1995**, *30*, 1203.
14. Huddleston, J. G.; Griffin, S. T.; Zhang, J.; Willauer, H. D.; Rogers, R. D. Metal Ion Separations in Aqueous Biphasic Systems and with ABEC™ Resins. In *Aqueous Two-Phase Systems*; Kaul, R., Ed.; In *Methods in Biotechnology*; Walker, J. M., Ed.; Humana Press, Totowa, NJ, 1999; In Press.

15. Rogers, R. D.; Bond, A. H.; Bauer, C. B.; Zhang, J.; Jezl, M. L.; Roden, D. M.; Rein, S. D.; Chomko, R. R. Metal Ion Separations in Polyethylene Glycol-Based Aqueous Biphasic Systems. In *Aqueous Biphasic Separations: Biomolecules to Metal Ions*; Rogers, R. D.; Eiteman, M. A., Eds.; Plenum Press: New York, 1995; pp 1-20.
16. Rogers, R. D.; Bond, A. H.; Bauer, C. B.; Zhang, J.; Griffin, S. T. Metal Ion Separations in Polyethylene Glycol-Based Aqueous Biphasic Systems: Correlation of Partitioning Behavior with Available Thermodynamic Data. *J. Chromatogr., B* **1996**, *680*, 221.
17. Huddleston, J. G.; Willauer, H. D.; Swatloski, R. P.; Visser, A. E.; Rogers, R. D. Room Temperature Ionic Liquids as Novel Media for 'Clean' Liquid-Liquid Extraction. *Chem. Commun.* **1998**, 1765.
18. Rogers, R. D.; Visser, A. E.; Swatloski, R. P.; Hartman, D. H. Metal Ion Separations in Room Temperature Ionic Liquids: Potential Replacements for Volatile Organic Diluents. In *Metal Separation Technologies Beyond 2000: Integrating Novel Chemistry with Processing*; Liddell, K. C.; Chaiko, D. J., Eds.; The Minerals, Metals & Materials Society: Warrendale, PA, 1999; pp 139-147.
19. Rogers, R. D.; Zhang, J.; Bond, A. H.; Bauer, C. B.; Jezl, M. L.; Roden, D. M. Selective and Quantitative Partitioning of Pertechnetate in Polyethylene-Glycol Based Aqueous Biphasic Systems. *Solv. Extr. Ion Exch.* **1995**, *13*, 665.
20. Rogers, R. D.; Zhang, J.; Griffin, S. T. The Effects of Halide Anions on the Partitioning Behavior of Pertechnetate in Polyethylene Glycol-Based Aqueous Biphasic Systems. *Sep. Sci. Technol.* **1997**, *32*, 699.
21. Huddleston, J. G.; Willauer, H. D.; Herrington, J. F.; Carruth, A. D.; Griffin, S. T.; Rogers, R. D. Extraction of Organic Molecules Utilizing Aqueous Biphasic Systems and the Physicochemical Properties of the Phases. *J. Chromatogr., A* **1999**, Submitted.
22. Rogers, R. D.; Willauer, H. D.; Griffin, S. T.; Huddleston, J. G. Partitioning of Small Organic Molecules in Aqueous Biphasic Systems. *J. Chromatogr., B* **1998**, *711*, 255.
23. Huddleston, J. G.; Griffin, S. T.; Zhang, J.; Willauer, H. D.; Rogers, R. D. Metal Ion Separations in Aqueous Biphasic Systems and Using Aqueous Biphasic Extraction Chromatography. In *Metal Ion Separation and Preconcentration, Progress, and Opportunities*; Dietz, M. L.; Bond, A. H.; Rogers, R. D., Eds.; ACS Symposium Series 716, American Chemical Society: Washington, DC, 1999; pp 79-100.
24. Rogers, R. D.; Bauer, C. B. Water Soluble Calixarenes as Possible Metal Ion Extractants in Polyethylene Glycol-Based Aqueous Biphasic Systems. *J. Radioanal. Nucl. Chem.* **1996**, *208*, 153.
25. Rogers, R. D.; Griffin, S. T. Partitioning of Mercury in Aqueous Biphasic Systems and on ABEC™ Resins. *J. Chromatogr., B* **1998**, *711*, 277.
26. Freemantle, M. Designer Solvents, Ionic Liquids May Boost Clean Technology Development. *Chem. Eng. News* 1998, 76 (March 30), 32.
27. Seddon, K. R. Ionic Liquids for Clean Technology. *J. Chem. Tech. Biotechnol.* **1997**, *68*, 351.

28. Wilkes, J. S.; Levisky, J. A.; Wilson, R. A.; Hussey, C. L. Dialkylimidazolium Chloroaluminate Melts: A New Class of Room-Temperature Ionic Liquids for Electrochemistry, Spectroscopy, and Synthesis. *Inorg. Chem.* **1982**, *21*, 1263.
29. Suarez, P. A. Z.; Einloft, S.; Dullius, J. E. L.; de Souza, R. F.; Dupont, J. Synthesis and Physical-Chemical Properties of Ionic Liquids Based on 1-*n*-butyl-3-methylimidazolium Cation. *J. Chim. Phys.* **1998**, *95*, 1626.
30. Dai, S.; Ju, Y. H.; Barnes, C. E. Solvent Extraction of Strontium Nitrate by a Crown Ether Using Room-Temperature Ionic Liquids. *J. Chem. Soc. Dalton Trans.* **1999**, 1201.
31. *Calixarenes: A Versatile Class of Macrocyclic Compounds*; Vicens, J.; Böhmer, V., Eds.; In *Topics in Inclusion Science*; Davies, J. E. D., Ed.; Kluwer Academic Publishers: Dordrecht, 1991; Vol. 3.
32. Arnaud-Neu, F.; Barboso, S.; Charbonniere, L. J.; Schwing-Weill, M. J. Complexation of Lanthanides(III) and Thorium(IV) by Phosphorylated Calixarenes. In *Abstracts of Papers for the 217th American Chemical Society National Meeting*, Anaheim, CA; American Chemical Society: Washington, DC, 1999; I&EC 244.
33. Böhmer, V. CMPO-Substituted Calixarenes. In *Abstracts of Papers for the 217th American Chemical Society National Meeting*, Anaheim, CA; American Chemical Society: Washington, DC, 1999; I&EC 158.
34. Bünzli, J.-C. G.; Ihringer, F.; Dumy, P.; Sager, C.; Rogers, R. D. Structural and Dynamic Properties of Calixarene Bimetallic Complexes: Solution *versus* Solid-State Structure of Dinuclear Complexes of Eu^{III} and Lu^{III} with Substituted Calix[8]arenes. *J. Chem. Soc., Dalton Trans.* **1998**, 497.
35. Gutsche, C. D. *Calixarenes*; In *Monographs in Supramolecular Chemistry*; Stoddart, J. F., Ed.; Royal Society of Chemistry: Cambridge, 1989; No. 1.
36. Bünzli, J.-C. G.; Harrowfield, J. MacB. Lanthanide Ions and Calixarenes. In *Calixarenes: A Versatile Class of Macrocyclic Compounds*; In *Topics in Inclusion Science*; Davies, J. E. D., Ed.; Kluwer Academic Publishers: Dordrecht, 1991; Vol. 3, pp 211-231.
37. Talanova, G. G.; Talanov, V. S.; Hwang, H.-S.; Bartsch, R. A. Separations of Soft Heavy Metal Cations by Lower Rim Functionalized Calix[4]arenes. In *Abstracts of Papers for the 217th American Chemical Society National Meeting*, Anaheim, CA; American Chemical Society: Washington, DC, 1999; I&EC 156.
38. Yordanov, A. T.; Gansow, O. A.; Brechbiel, M. W.; Rogers, L. M.; Rogers, R. D. The Preparation and X-ray Crystallographic Characterization of Lead(II) Calix[4]arenesulfonate Complex. *Polyhedron* **1999**, In Press.
39. Sachleben, R. A.; Bryan, J. C.; Engle, N. L.; Franconville, B.; Haverlock, T. J.; Hay, B. P.; Urvoas, A.; Moyer, B. A. Calix[4]arene Crown-6 Ethers: Recent Developments in Enhanced Cesium-Selective Extractants. In *Abstracts of Papers for the 217th American Chemical Society National Meeting*, Anaheim, CA; American Chemical Society: Washington, DC, 1999; I&EC 173.
40. Izatt, S. R.; Hawkins, R. T.; Christensen, J. J.; Izatt, R. M. Cation Transport from Multiple Alkali Cation Mixtures Using a Liquid Membrane System Containing a Series of Calixarene Carriers. *J. Am. Chem. Soc.* **1985**, *107*, 63.

41. Huddleston, J. G.; Willauer, H. D.; Boaz, K.; Rogers, R. D. Separation and Recovery of Food Coloring Dyes Using Aqueous Biphasic Extraction Chromatographic Resins. *J. Chromatogr., B* **1998**, *711*, 237.
42. *Stability Constants of Metal Ion Complexes*; The Chemical Society: London, 1971; Supp. No. 1.
43. Shinkai, S.; Araki, K.; Manabe, O. NMR Determination of Association Constants for Calixarene Complexes. Evidence for the Formation of a 1:2 Complex with Calix[8]arene. *J. Am. Chem. Soc.* **1988**, *110*, 7214.

CALIXARENE–ANION COMPLEXATION

Chapter 18

Calix[4]pyrrole-Functionalized Silica Gels: Novel Supports for the HPLC-Based Separation of Anions

Jonathan L. Sessler[1], John W. Genge[1], Philip A. Gale[2], and Vladimír Král[3]

[1]Department of Chemistry and Biochemistry,
The University of Texas, Austin, TX 78712
[2]Department of Chemistry, University of Southampton,
Southampton SO17 1BJ, United Kingdom
[3]Department of Analytical Chemistry, Institute of Chemical Technology,
16628 Prague 6, Technická 5, Czech Republic

The coupling of calix[4]pyrroles to silyl capped aminopropyl silica gel produces functionalized silica gels that may be used to effect the HPLC-based separation of various anionic substrates including medium length oligonucleotides under neutral, isochratic conditions.

Calix[4]arenes (e.g., **1**), a class of molecules whose structure was first proposed by Zinke (*1*), are cyclic tetramers composed of substituted phenols linked by sp^3-hybridized *meso*-type carbons. Increasingly in recent years and especially since Gutsche's seminal synthetic studies (*2*), an inspirational amount of time and energy has been dedicated to studying these bowl-shaped molecules (*3*). The X-ray crystal structure of a *p-tert*-butylcalix[4]arene-toluene inclusion complex was elucidated by Andretti, Ungaro, and Pochini in 1979 (*4*) and this provided the first direct proof of the proposed "vase-like" structure. Among their other important contributions, these same workers also studied the role of calix[4]arenes in the area of liquid and gas-solid chromatography (*5*), a theme that continues to be developed with great success as evidenced *inter alia* by the use of calixarenes to separate fullerenes (*6, 7*), remove radioactive metal ions from solution

(8), and by the publication of the present ACS Volume. Our own efforts, by contrast, have focused on the use of a class of heterocalixarene analogues termed calix[4]pyrroles (9).

1

2

The calix[4]pyrroles are formally related to calix[4]arenes. Originally termed octaalkylporphyrinogens, they are tetrapyrrolic macrocycles linked in the α-position via sp^3 hybridized *meso*-type carbons. Like the calixarenes, this class of macrocycle, represented by the prototypical octamethyl system **2**, is venerable in the extreme. Indeed, they were first reported by Baeyer over 100 years ago (10). Since Baeyer's day, many octaalkylporphyrinogens have been synthesized (11-13). Nonetheless, as a general rule this class of compounds has not been extensively explored. Indeed, apart from Floriani's elegant studies, involving the use of octaalkylporphyrinogens to stabilize unusual organometallic structures (14) and our own more recent efforts in the area (9, 15-21), the generalized substrate binding properties of these systems appear to have been largely ignored.

One of the reasons the calix[4]pyrroles may have remained poorly explored for so long is that they resemble porphyrinogens. Porphyrinogens containing hydrogens at the *meso*-carbons are prone to oxidation and are generally converted easily into the corresponding porphyrin. Fully *meso*-substituted porphyrinogens (calix[4]pyrroles) on the other hand are not susceptible to such oxidation processes (i.e., they cannot form the corresponding porphyrins). As such, they resemble calix[4]arenes and are, we propose, much more accurately described as being calix[4]pyrroles (9) than porphyrinogens. This renaming, which has precedent within the annals of heterocyclic chemistry (22), led us to consider that the calix[4]pyrroles, like their calixarene "cousins," might act as highly effective supramolecular receptor systems. However, possessing as they might "bowl shaped" NH hydrogen bonding donor functionality, the calix[4]pyrroles were expected to act as receptors for anionic and/or neutral substrates (9), rather than for the electron deficient materials that are so typically bound by calix[n]arenes (3).

Recently, we succeeded in showing that the above hypotheses do in fact have a basis in chemical reality. Specifically, we showed that calix[4]pyrroles, while possessing so-called 1,3-alternate conformations in the absence of bound substrates, adopt cone conformations readily and bind phosphates, halides, carboxylates and other anions both in solution and in the solid state (c.f., e.g., Figure 1) (*15, 17, 19*). We also found that the calix[4]pyrroles bind neutral substrates, albeit weakly, *via* hydrogen bonds (*16*) and it was these findings, taken in concert, that led us to consider that the calixpyrroles might have an important role to play in separation science (*21*).

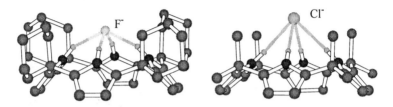

Figure 1. X-ray structures of a tetraspirocyclohexylcalix-[4]pyrrole·fluoride anion complex and the chloride anion complex of 2. This figure was generated from data published in reference 15.

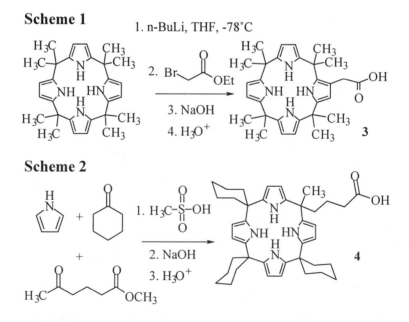

Schemes 1 and 2. Synthesis of "β-hook" and "meso-hook" calixpyrroles 3 and 4.

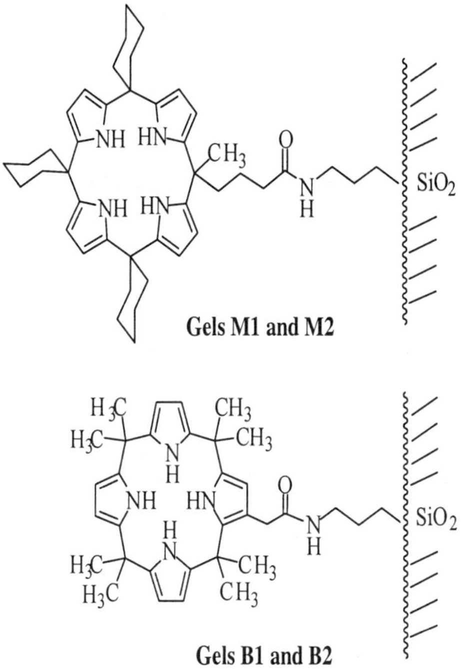

Figure 2. Schematic representations of calix[4]pyrrole functionalized silica gels, **Gels M1, M2, B1,** and **B2**, that are used as HPLC solid supports in this study. **Gels M2** and **B2** are identical to **Gels M1** and **B1** in all respects except that the loading levels are about a factor of 10 lower. See text for details.

Strengthening the above impression were previous studies of sapphyrin-based HPLC supports. The sapphyrins are a prototypic class of so-called expanded porphyrins that act as very effective anion binding agents (23) and which can be functionalized easily so as to generate modified silica gels (24, 25). While these latter gels, when packed in columns and used as reverse phase HPLC solid supports, can indeed be used to separate certain anions, the observation of broad peaks and long retention times, coupled with a need for high buffer concentrations (greater than 1.0 M in the case of short (3- to 9-mer) oligonucleotide mixtures), led us to consider that these systems were less than ideal. We intuited that it was the overly large binding constants, and correspondingly slower rates of substrate disassociation, that were responsible for the less than stellar performance. In any event, the fact that over 30 synthetic steps were required to generate the requisite sapphyrin-modified gels, provided a powerful incentive to try strategies based on calix[4]pyrroles. In this case, it appeared as if only five steps (or fewer!) would be required to prepare the functionalized silica gels. Further, because the neutral calixpyrroles were expected to be far less efficacious anion binding agents than the sapphyrins (species that remain monoprotonated at pH 7), it was expected that better resolution would be obtained in the case of oligonucleotide separations. As described below, these expectations were in fact fully realized.

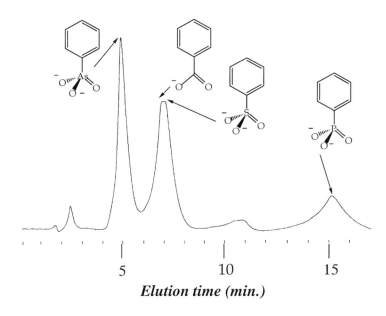

*Figure 3. HPLC-based separation of various phenyl substituted anions on calixpyrrole modified silica gel column, prepared from **Gel M1**. Anions were eluted as 1 mM aqueous solutions under the following conditions: Flow rate 0.3 ml/min.; mobile phase, aqueous phosphate buffer (50 mM) at pH = 7.0; column temperature 25° C; UV detection at 254 nm.*

Synthesis of Calix[4]pyrrole-Functionalized Silica Gels

Based on our work with sapphyrin functionalized silica gels, we reasoned that monofunctionalized calixpyrrole carboxylic acids would constitute the key precursors needed to generate calixpyrrole-modified supports. As it transpired, two facile entries into such systems were quickly found (c.f. Schemes 1 and 2); this produced the key "β-hook" and "*meso*-hook" calix[4]pyrrole carboxylic acid precursors, **3** and **4**, respectively (*17, 18*). Subsequent activation, coupling to trimethylsilyl silanol protected aminopropyl silica gel (*26*), and follow-up capping with acetyl chloride then produced the amidocalix[4]pyrrole functionalized supports **Gel M1** and **Gel M2** (*meso*-hook) and **Gel B1** and **Gel B2** (β-hook), systems that are portrayed schematically in Figure 2 (*21*). The difference between **Gel M1** and **Gel M2** (and between **Gel B1** and **Gel B2**) lies only in the extent of calix[4]pyrrole coverage. Whereas **Gels M2** and **B2** bear calix[4]pyrroles on only 10% (approximately) of the possible aminopropyl sites, **Gels M1** and **B1** were designed to lie much closer to saturation in terms of their loading levels. Commercial packing (by Alltech Associates, Deerfield, IL) then gave reverse phase HPLC columns that could be used to test readily the anion separating utility of these gels.

TABLE 1. Retention Times Recorded on HPLC Columns Derived from Gel M1 and Gel B1.

	Elution Times (min.)	
Anion	**Silica Gel M1**	**Silica Gel B1**
Chloride[b]	15.2 (± 0.1)[a]	17.9 (± 0.1)[a]
Dihydrogen phosphate[b]	20.1 (± 0.1)[a]	22.0 (± 0.1)[a]
Hydrogen sulfate[b]	16.2 (± 0.2)[a]	16.2 (± 0.1)[a]
Fluoride[b]	16.4 (± 0.2)[a]	16.9 (± 0.1)[a]
Phenyl arsenate[c]	4.9 (± 0.1)	n. d.[d]
Benzoate[c]	6.9 (± 0.1)	n. d.[d]
Benzene sulfonate[c]	7.0 (± 0.1)	n. d.[d]
Phenyl phosphate[c]	15.1 (± 0.1)	n. d.[d]

NOTES: [a]Times given for individual elution of tetrabutylammonium anions. [b]Anions were eluted as 1 mM CH_3CN solutions of their tetrabutylammonium salts under the following conditions: Mobile phase, CH_3CN; flow rate, 0.40 ml/min.; Detection, Conductivity; Column Temperature, 25° C. [c]Anions were eluted as 1 mM aqueous solutions under the following conditions: Mobile phase 50 mM phosphate buffer, pH = 7.0, Flow rate 0.3 ml/min., Column temperature 25° C, Detection UV = 254 nm. [d] n. d. = not determined.
SOURCE: The data in this table were originally published in ref. 21.

Separation of Simple Anions

Once columns based on **Gels M1** and **B1** were in hand, initial studies focused on assessing whether selective retentions of simple anions such as fluoride, chloride, bromide, hydrogen sulfate, and dihydrogen phosphate could be achieved (*21*). As illustrated by the data collected in Table 1, selective retention times were indeed observed for these anions. Further, a range of slightly more complex, phenyl substituted anions, including specifically phenyl arsenate, benzene sulfonate, benzoate, and phenyl phosphate were seen to be separated readily on columns containing **Gel M1** (c.f. Figure 3). These selectivities, thought to reflect differences in basic anion binding affinities (K_a's) rather than less obvious structural, hydrophobic, or steric effects, were considered as auguring well for the use of calix[4]pyrrole-functionalized silica gels as anion separating media.

5 R = CONHC$_4$H$_9$ **6**

To test more directly the validity of the above binding-based rationalizations, considered predicative to the success of the basic calix[4]pyrrole-centered approach to anion separation, the control calix[4]pyrrole monomers **5** and **6** were prepared (*21*). Their solution phase chloride, dihydrogen phosphate and hydrogen sulfate binding properties were then studied in dichloromethane-d_2 (for reasons of solubility) with the resulting stability constants and binding stoichiometries (1:1 in all instances) being determined using standard (*27*) ^{1}H NMR titration methods (*21*). In these critical control studies, the use of amide substituted systems (e.g., **5** and **6**), rather than simple, unsubstituted materials such as **2**, was considered essential. This is because the electronic nature of the calixpyrrole skeleton is known to influence the anion binding properties considerably (*9, 15, 18*). Further, the amide NH groups may play a role in mediating the anion coordination process (*28*).

TABLE 2. Anion Binding Affinities of Control Compounds 5 and 6

	Stability Constants (M^{-1})[a]	
Anion	Compound 5	Compound 6
Chloride	415 (± 45)	405 (± 10)
Dihydrogen phosphate	62 (± 6)	80 (± 15)[b]
Hydrogen sulfate	< 10	< 10

NOTES: [a]Determined in dichloromethane-d_2 using the appropriate tetrabutylammonium salts as the anion source. In determining these stability constants, the possible effects of ion pairing (if any) were ignored. [b]Estimated value.
SOURCE: The data in this table were originally published in ref. 21.

As can be deduced from the summary of results presented in Tables 1 and 2, basic binding affinities do indeed provide a good, first-order means of understanding the separation-based properties of calix[4]pyrrole-functionalized reverse phase HPLC supports. For instance, the fact that phenyl phosphate is retained on columns derived from **Gel M1** and **Gel B1** longer than benzene sulfonate (and is thus readily separated from it) is reflected in the observation that dihydrogen phosphate anion is bound better to model compounds **5** and **6** than hydrogen sulfate anion (*21*). Likewise, the fact that similar retention times are observed when either **Gel M1** or **Gel B1** is used to separate, e.g., $H_2PO_4^-$ could reflect the fact that nearly identical K_a's for this substrate are recorded for both **5** and **6** in dichloromethane-d_2 solution (Table 2). Indeed, one of the noteworthy aspects of this work is that these two receptors, and the modified silica gels derived therefrom, are characterized more by their similarities than their differences. We interpret this in terms of the recognition process being dominated by NH--anion hydrogen bonding interactions rather than more complex steric or electronic effects, even though these latter obviously play a small moderating role.

Separation of N-Carbobenzyloxy Protected Amino Acids

Once the above control studies were complete, attention turned toward the problem of separating more complex anionic substrates. As a first step in this direction, a series of N-carbobenzyloxy (Cbz) protected amino acids were analyzed. As illustrated in Figure 4, **Gel B1** works remarkably well for this particular anion separation problem. Not surprisingly in light of the similarities in support performance noted above, columns derived from **Gel M1** were also found to work for this purpose (*21*). In both cases, a careful look at the obviously anion-dependent elution order leads to the inference that, at least for substrates of like charge (i.e., mono- or dianions), ancillary steric, π-π stacking, or hydrophobic effects may be playing a critical, albeit secondary role in regulating the observed separation process.

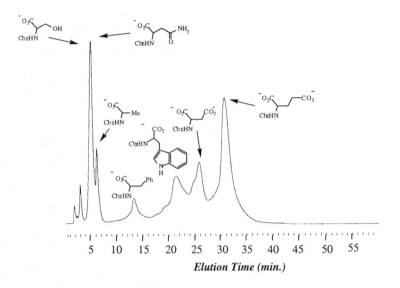

*Figure 4. HPLC-based separation of N-carbobenzyloxy (Cbz) protected amino acids on calixpyrrole modified silica gel column, prepared from **Gel B1**. Flow rate 0.3 ml/min.; mobile phase, 3:1 (v/v) aqueous acetate buffer (30 mM) at pH 7.0 / acetonitrile (isochratic); column temperature 25º C; UV detection at 254 nm.*

Separation of Nucleotides

The next series of complex anionic substrates considered is that defined by the matched adenosine nucleotides 5'-adenosine monophosphate (AMP), 5'-adenosine diphosphate (ADP), and 5'-adenosine triphosphate (ATP). To our gratification, we found that good, baseline separations of this all-important set of biological substrates could be achieved under simple isochratic elution conditions using either **Gels M1** or **B1** (c.f., e.g., Figure 5; see also ref. (*21*) wherein nearly identical results obtained with **Gel M1** is presented). In the case of both gels, which again were found to work in nearly identical fashion, the elution order was found to differ considerably from other neutral solid supports that have been used for anion separation, including specifically those used for standard reverse phase liquid chromatography (29). In particular, it is found that the more highly charged nucleotide is retained longer on **Gels M1** and **B1** without the use of ion-pairing agents (21). Presumably, this reflects the fact that the higher charge densities of the di- and triphosphate nucleotides permit more favorable interactions with the calix[4]pyrrole present on the column. Thus, it is predicted, as indeed is found by experiment, that the greater the number of phosphate groups present on the adenosine base, and the higher the corresponding anionic charge on the nucleic acid substrate system as a whole, the longer the retention times.

While presumably the actual, and perhaps relative, retention times could be varied by changing the pH of the eluting buffer, these studies, and indeed all others involving analytes of biological interest (see below), were carried out at pH 7, a pH likely to be useful in various targeted end use applications. In spite of this decision, columns derived from **Gels M1** and **B1** were tested and found to be stable at 5 < pH < 12.

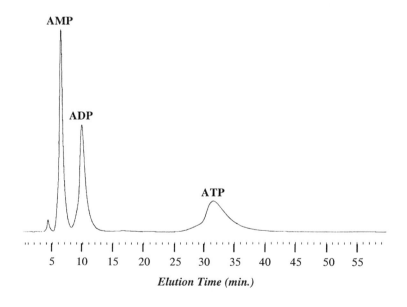

*Figure 5. HPLC-based separation of AMP, ADP and ATP on the calixpyrrole column derived from **Gel B1**. Flow rate 0.3 ml/min.; mobile phase aqueous sodium phosphate buffer (105 mM) at pH = 7.0 (isochratic); column temperature 25° C; UV detection at 262 nm.*

Separation of Oligonucleotides

An interesting off-shoot of the above mechanistic postulate is that it leads to the clear prediction that calix[4]pyrrole-functionalized silica gels could function as efficient supports for the separation of medium-length oligonucleotides. Such species play important roles both in the practice of molecular biology and medicine (e.g., anti-sense technologies) and thus define anionic separation targets of special interest (*30*). Unfortunately, it quickly became apparent from initial tests that neither **Gel M1** nor **Gel B1** would function as particularly effective HPLC supports for medium-length oligonucleotides. On the other hand **Gels M2** and **Gel B2** did prove efficacious (see Figure 6 for a representative chromatogram) (*9*). In particular, these gels allowed for the efficient separation of the individual oligonucleotide components within a mixture of unprotected oligodeoxythymidylate fragments containing between 12 and 18 nucleotide subunits. As expected, species containing a higher number of phosphate groups displayed the greater retention times, a result that was confirmed in studies involving a similar mixture of unprotected oligodeoxyadensoninates (*21*).

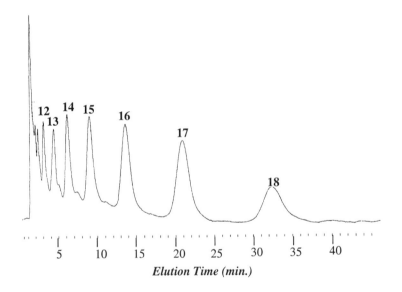

*Figure 6. Separation of dT_{12-18} on a calixpyrrole modified silica gel column derived from **Gel B2**. Flow rate 0.25 ml/min.; mobile phase 3:2 (v/v) CH_3CN / aqueous sodium chloride (50 mM)-sodium phosphate (40 mM) at pH 7.0 (isochratic); column temperature 25° C; UV detection at 265 nm.*

Presumably the lower loading levels of **Gels M2** and **B2** permit the oligonucleotide species to elute through the silica gel column faster than through **Gels M1** and **B1** while still allowing for sufficient anion recognition interactions to enable good separation. This may be due to the fact that multiple analyte-support interactions are reduced, if not precluded. In any event, the critical finding was that appropriately designed calixpyrrole-based columns can indeed be used to effect the separation of oligonucleotides under conditions where a range of other supports, including those based on sapphyrin, do not (*21, 24*).

Separations of Oligonucleotides of Similar Length

One further and especially intriguing finding to come out of the studies of **Gels M2** and **B2** was the observation that reverse phase HPLC columns derived from these supports may be used to effect the separation of ostensibly similar oligomers. For instance, as illustrated in Figure 7, the three nucleotide hexamers TCTAGA, GCATGC and CCCGGG, (C = cytidine, G = guanidine, A = adenosine, and T = thymidine) are readily separated from one another on **Gel M2** (*21, 24*).

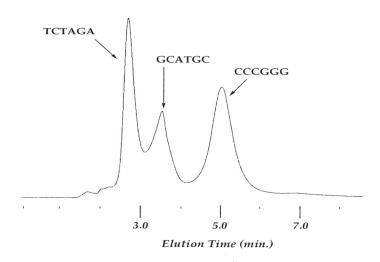

*Figure 7. HPLC-based separation of oligonucleotide hexamers, TCTAGA, GCATGC, and CCCGGG, on a modified silica gel column derived from **Gel M2**. Flow rate 0.4 ml/min.; mobile phase 1:1 CH_3CN / 50 mM sodium phosphate buffer at pH 7.0 (isochratic); column temperature 25° C; UV detection at 265 nm. Reproduced with permission from reference 21. Copyright 1998 Wiley-VCH.*

Presumably, this unexpected, but quite desirable, finding reflects the fact that various, same-length nucleotide oligomers are not identical as far as their ability to subtend solution volume or engage in ancillary hydrogen bonding interactions with the calix[4]pyrrole modified support. While such interpretations remain subject to debate, the key point is that the results displayed in Figure 7 support the contention that calixpyrrole-based solid supports could provide a new non-electrophoretic method for small anion and oligonucleotide separation. On this basis we predict that the present calix[4]pyrrole-based anion recognition, analysis, and separation approach is one that prove complementary to more classic reverse-phase (*31*), ion-pair (*32*), size exclusion (*33*), or ion exchange based strategies (*34*). Naturally, we are continuing to explore this particular possibility.

Separations of Polycarboxylates

In order to understand better the above results and to assess further the extent to which calixpyrrole-based columns could be used to separate other classes of polyanions, efforts were made to separate the anionic forms of benzoic acid, isophthalic acid, and 1,3,5-benzene tricarboxylic acid using our calix[4]pyrrole-derived solid supports. As shown in Figure 8, these functionally similar but differentially charged (and shaped) substrates could indeed be separated readily on **Gel B1** at pH 7.0 under standard reverse phase HPLC conditions (see caption to Figure 8 for exact experimental details). In this instance, the efficiency of separation, and order of elution, resembles that seen with mixtures of AMP, ADP and ATP. We consider this observation as being consistent with charge playing a dominant role in establishing the key substrate-to-support interactions and, with it, the order of anion elution.

As proved true in the case of the oligonucleotides, in those instance where the net charge in a series of substrates was held constant, it was expected that ancillary interactions, involving steric effects, relative hydrophobicity and a putative ability to form "extra" hydrogen bonds would serve to establish some degree of selectivity. In the case of columns based on **Gel B1** this expectation was realized. Indeed, as illustrated by Figure 9, the anionic form of phthalic acid is readily separated from isophthalic acid and teraphthalic acid on **Gel B1** at pH 6.9. While not yet tested by experiment, it is expected that columns based on **Gel M1** would also work to separate this set of substrates, just as they would those presented in Figure 8.

The increased retention time observed for phthalate anion as opposed to isophthalate and teraphthalate is probably due to the *ortho*-configuration of the two carboxylate groups of phthalic acid. Such a configuration is expected to permit an increased interaction between the two anionic carboxylate subunits with the hydrogen bond donors of a single, silica gel-bound calixpyrrole moiety. In any event, this result, like that involving the hexanucleotides detailed above, serves to underscore the potential utility of calixpyrrole-based columns in analytical or preparative applications that require the separation of anionic substrates bearing similar charge.

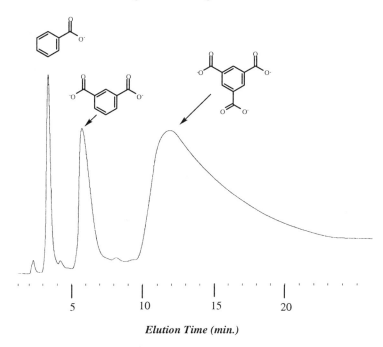

*Figure 8. Chromatogram showing the separation of related carboxylate anions on a modified silica gel column derived from **Gel B1**. Flow rate 0.3 ml/min.; mobile phase, 76:24 acetonitrile / water at pH = 7.0 (isochratic); column temperature 25° C; UV detection at 254 nm.*

Conclusion

In this review, the authors have tried to show some of the range and power that calixpyrrole modified silica gels possess with respect to high performance liquid chromatography. Not only are calix[4]pyrrole modified silica gels capable of discriminating between simple anionic species, they are also able to separate more complex polyanionic and biologically relevant nucleotides and oligonucleotides. The ability to prepare the required solid supports in high yield and with a minimum of effort leads us to suggest that the present calix[4]pyrrole-based approach to anion separation could emerge as being one of widespread utility. Bolstering this optimistic assessment is the ease with which the basic calixpyrrole skeleton can be modified. This leads us to predict that other supports, including, for instance, ones containing chiral, nonracemic functionality, might find use in the separation of other anionic substrates of analytical or biological interest.

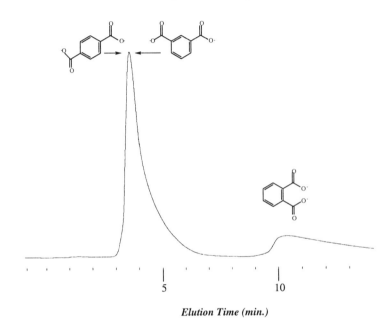

*Figure 9. Chromatogram showing the HPLC-based separation of isomeric dicarboxylate anions on a column derived from **Gel B1**. Flow rate 0.25 ml/min.; mobile phase, 76:24 acetonitrile / aqueous sodium acetate buffer (30 mM) at pH = 6.9 (isochratic); column temperature 25° C; UV detection at 254 nm.*

Acknowledgments

This work was supported by the Texas Advanced Research and Technology Programs (grants no. ATP 3658-280 and ARP 3658-102 to J.L.S.), the Czech Ministry of Education (grant no. VS 97-135 to V.K.), the National Institutes of Health (grants GM 58907 and TW 00682 to J.L.S.), and a Royal Society University Research Fellowship (to P.A.G.).

References

1. Zinke, A.; Zigeuner, G.; Hissinger, K.; Hoffman, G. *Monatsh.* **1948**, *79*, 438-439.
2. Gutsche, C. D.; Muthukrishnan, R. *J. Org. Chem.* **1978**, *43*, 4905-4906.
3. Gutsche, C. D. *Calixarenes*; Monographs in Supramolecular Chemistry; Ed. Stoddart, J.F.; The Royal Society of Chemistry: Cambridge 1989.
4. Andretti, G. D.; Ungaro, R.; Pochini, A. *J. Chem. Soc., Chem. Commun.* **1979**, 1005-1007.
5. Mangia, A.; Pochini, A.; Ungaro, R.; Andretti, G. D. *Anal. Lett.* **1983**, *16*, 1027-1036.
6. Atwood, J. L.; Koutsantonis, G. A.; Raston, C. L. *Nature,* **1994**, *368*, 229-231.
7. Suzuki, T.; Nakashima, K.; Shinkai, S. *Chem. Lett.* **1994**, 699-702.
8. Beer, P.D.; Drew, M.G.B.; Hesek, D.; Kan, M.; Nicholson, G.; Schmitt, P.; Williams, G. *J. Chem. Soc. Dalton Trans.* **1998**, 2783-2785, and references cited therein.
9. Gale, P. A.; Sessler, J. L.; Král, V. *Chem. Commun.* **1998**, 1-6.
10. Baeyer, A. *Ber. Dtsch. Chem. Ges.* **1886**, *19*, 2184-2185.
11. Chelintzev, V. V.; Tronov, B. V. *J. Russ. Phys. Chem. Soc.* **1916**, *48*, 105. See also *Chem. Abstr.* **1917**, *11*, 1418.
12. Rothemund, P.; Gage, C. L. *J. Am. Chem. Soc.* **1955**, 77, 3340-3342.
13. Brown, W. H.; Hutchinson, B. J.; MacKinnon, M. H. *Can. J. Chem.* **1971**, *49*, 4017-4022.
14. Floriani, C. *Chem. Commun.* **1996**, 1257-1263.
15. Gale, P. A.; Sessler, J. L.; Král, V.; Lynch, V. *J. Am. Chem. Soc.* **1996**, *118*, 5140-5141.
16. Allen, W. E.; Gale, P. A.; Brown, C. T.; Lynch, V. M.; Sessler, J. L. *J. Am. Chem. Soc.* **1996**, *118*, 12471-12472.
17. Sessler, J. L.; Andrievsky, A.; Gale, P. A.; Lynch, V. *Angew. Chem. Int. Ed. Engl.* **1996**, *35*, 2782-2785.
18. Gale, P. A.; Sessler, J. L.; Allen, W. E.; Tvermoes, N. A.; Lynch, V. *Chem. Commun.* **1997**, 665-666.
19. Gale, P. A.; Genge, J. W.; Král, V.; McKervey, M. A.; Sessler, J. L.; Walker, A. *Tetrahedron Lett.* **1997**, *38*, 8443-8444.
20. Sessler, J. L.; Gebauer, A.; Gale, P. A. *Gaz. Chim. Ital.* **1997**, *127*, 723-726.
21. Sessler, J. L; Gale, P. A.; Genge, J. W. *Chem. Eur. J.* **1998**, *4*, 1095-1099.
22. Musua, R. M.; Whiting, A. *J. Chem. Soc., Perkins Trans. I* **1994**, 2881-2888.

23. Král, V.; Furuta, H.; Shreder, K.; Lynch, V.; Sessler, J. L. *J. Am. Chem. Soc.* **1996**, *118*, 1595-1607, and references therein.
24. Iverson, B. L.; Thomas, R. E.; Král, V.; Sessler, J. L. *J. Am. Chem. Soc.* **1994**, *116*, 2663-2664.
25. Sessler, J. L.; Král, V.; Genge, J. W.; Thomas, R. L.; Iverson, B. L. *Anal. Chem.* **1998**, *70*, 2516-2522.
26. Tartar, A.; Gesquiere, J.-C. *J. Org. Chem.* **1979**, *44*, 5000-5002.
27. Hynes, M. J. *J. Chem. Soc., Dalton Trans.* **1993**, 311-312.
28. Beer, P.D. *Chem. Commun.* **1996**, 689-696.
29. Ramage, R.; Wahl, F. O. *Tetrahedron Lett.* **1993**, *34*, 7133-7136.
30. Gerwitz, A. M.;Sein, C. A.; Glazeer, P. M. *Proc. Natl. Acad. Sci., USA* **1996**, *93*, 3161-3163, and references therein.
31. Moriyama, H.; Kato, Y. *J. Chromatogr.* **1988**, *445*, 225-233.
32. Crowther, J. B.; Jones, R.; Hartwick, R. A. *J. Chromatogr.* **1981**, *217*, 479-490.
33. Molko, D.; Derbyshire, R.; Guy, A.; Roget, A.; Teoule, R.; Boucherle, A. *J. Chromatogr.* **1981**, *206*, 493-500.
34. Pearson, J. D.; Regnier, F. E. *J. Chromatogr.* **1983**, *255*, 137-149.

Chapter 19

Lower Rim Amide- and Amine-Substituted Calix[4]arenes as Phase Transfer Extractants for Oxyions between an Aqueous and an Organic Phase

H. Fred Koch and D. Max Roundhill

Department of Chemistry and Biochemistry,
Texas Tech University, Lubbock, TX 79409-1061

Calix[4]arene amides and amines can act as phase transfer extractants for Cr(VI), Mo(VI), W(VI), Re(VII), Se(VI) and U(VI) oxyions from water into an organic layer. Measurement of the remaining metal content in the aqueous layer by ICP-AES gives the percentage extraction from both acidic and neutral solutions. Calix[4]arene amines are better extractants than are amides, and the amines are usually better extractants from acidic than neutral solutions.

The selective extraction of metal cations and anions from aqueous solution into an organic phase is an important goal, especially if the particular ions are toxic and present in the environment in significant quantities (1, 2). For cations, the metal can be directly bound to a ligating group on the host via coordinate bonds, but for anions or uncharged molecules there may be no strong interaction between the metal center and host. Although numerous molecules act as hosts and complexants for cations, fewer act as hosts for anions (3-6). Recently, however, chemically modified calixarenes have been synthesized that are hosts for simple anions (7). From an environmental viewpoint, a series of anions for which selective hosts would be useful are the oxyanions. We present data here for a group of calix[4]arene amides and amines that act as hosts for a series of oxides, oxycations, and oxyanions.

Our chosen oxy species are those of Cr(VI), Mo(VI), W(VI), Re(VII), Se(VI), and U(VI). The Cr(VI) anions are important because of both their high toxicity (8-13) and their presence in soils and waters (14). Chromium(VI) is a carcinogen in humans and animals, and is also mutagenic and genotoxic. By contrast, water soluble Cr(III) compounds are not considered to be carcinogenic, possibly because they do not cross plasma membranes (15,16). Another tetrahedral oxyanion of environmental concern is selenate, both because of its high toxicity and because of its being a fission product of nuclear reactions (17). Perrhenate is chosen for investigation because it is a model anion for pertechnetate, another fission product. The oxycation of U(VI) is another common nuclear material that needs to be removed from soils and waters (18).

Our approach to designing oxyion phase transfer extractants is to seek compounds such as polyamines that can hydrogen bond to the guest molecule (19,20). Since metal cations are extracted by lower rim modified calix[4]arenes (21), we have therefore targeted calix[4]arenes with polyamine functionalities on their lower rims as extractants for anionic Cr(VI) (22,23). We can, however, use both calix[4]arene amines and amides as oxyion extractants. Amides are also included because they are more stable and hydrophobic than are amines, and they have both amide nitrogen and carbonyl oxygen functionalities available for hydrogen bonding with the oxyion (24).

An important aspect of developing chelate extractants for metals is that they are kinetically rapid phase transfer agents. This feature is especially important if the complexant is to be eventually bound to a polymeric support, because attachment of the chelate to such a support usually results in a decreased metal binding rate. Our extractions have therefore been carried out for only short time durations in order to gain information about the relative phase transfer rates for the complexant/oxyanion pairs. Longer contact times, however, do not cause significant changes in the results obtained.

Results and Discussion

In designing complexants for oxyanion guests, structural features can be incorporated into the host that may lead to selective binding. For hosts to be effective, their structural features must be compatible with those of the oxyanion. Although there have been recent reports of chemically modified calixarenes acting as host molecules for simple anions (25), prior to our communication reporting the use of a calix[4]arene diammonium salt, none had been used as extractants for the dianions CrO_4^{2-} and $Cr_2O_7^{2-}$ (23). Because the periphery of a calixarene can be made structurally compatible with oxyanions, and because calixarenes can act as phase transfer agents, we have targeted extractants based on calix[4]arenes.

Synthetic Strategies

We have prepared a series of calix[4]arene amides and amines in order to ascertain whether there is any particular selectivity caused by the extractant being a dication, and whether the more hydrophobic N,N-diethyl substituent on nitrogen results in the complexes having a higher extractability into chloroform from water. Although the use of amines as extractants for anions is based on precedent (23), the reason for targeting amides is the availability of both ketonic oxygens and tertiary nitrogens for hydrogen bonding with oxyanions. Furthermore, amides can undergo reversible protonation, and they also have a higher oxidative stability than do amines. The synthetic route to these amides involves treating the calixarene with either 2-bromoacetamide or 2-chloro-N,N-diethylacetamide in the presence of a base such as potassium carbonate. This reaction can involve either complete or partial substitution of the hydroxylic groups on the lower calixarene rim. Complete substitution with N,N-diethylacetamide groups occurs when the reaction is carried out in the presence of sodium iodide; in its absence monosubstitution is found. These compounds are isolated as colorless solids, and characterized by a combination of ^1H and ^{13}C NMR and IR spectroscopy.

Two routes have been used to prepare the amines, both of which involve a borane reduction step. One route involves reduction of the precursor calixarene amide, and the other uses a cyanomethoxy derivative as precursor, as shown in the scheme below (*22,26*). These compounds have been characterized by ^1H NMR and IR spectroscopy (*27*).

Extraction of Oxyanions and Oxycations

This study of the ability of these lower rim substituted calixarenes to effect the phase transfer of oxyanions has been carried out in order to determine whether extraction is observed from an aqueous to an organic layer under conditions where equal volumes of the two phases are shaken together for 1 minute. The resultant aqueous layer is then separated, and the remaining metal concentration in that layer analyzed by ICP-AES.

Scheme

These measurements are not necessarily carried out under equilibrium conditions, and in making comparisons we cannot ignore that kinetic selectivity may be controlling. Nevertheless, where the system has sufficient stability, we have kept the aqueous and chloroform layers in contact for up to 1 week, and find no further transfer between the phases, indicating that equilibrium has been rapidly established. In no case did we see any significant amount of the metal being transferred into the chloroform layer in the absence of the calix[4]arene.

The relative effectiveness of these amides as extractants for the transfer of a series of oxyanions from aqueous acidic and neutral solutions into chloroform is presented in the bar graphs shown in Figures 1 and 2 respectively, and for a set of corresponding amines in Figures 3 and 4. The amine functionalities will be protonated in acidic solution, and the amide functionalities partially protonated, but each will be unprotonated in neutral solution. These data show several features. One is that there is a somewhat higher level of extraction at the lower pH, which correlates with protonated cationic forms being the better extractants. Another is that calix[4]arene amines are better extractants than are the corresponding amides, which may be related to the relative degrees of protonation in acidic solution. Yet another is that the amide with four 2-aminoethoxy substituents on the lower rim is a particularly effective extractant from neutral solutions across the range of oxyanions tested, which may be a consequence of the presence of multiple amine sites for hydrogen bonding. An unanticipated result is that the more hydrophobic N,N-diethyl substituted derivatives are not consistently better extractants than are the unsubstituted derivatives, which may indicate that NH hydrogen bonding sites are more important than is the higher hydrophobicity.

An important aspect of interpreting these data is the solution stoichiometry of the oxyanions in both acidic and neutral solutions. In several cases, different molecular species are found under these two sets of solution conditions. In particular, some species have different overall charges in acidic and neutral aqueous solutions. For W(VI) and Mo(VI) the uncharged trioxide is extracted from acidic solution by calix[4]arene amines at least as effectively as is the dianion from neutral solution. Indeed, very good extractants for Mo(VI), W(VI), Se(VI) are available in this group of calix[4]arenes. Apparently, in rationalizing these data, a combination of charge compatibility and hydrogen bonding between the oxyion and the calix[4]arene is important in determining phase transfer extractabilities.

In developing extractants it is important to target specific solvent systems as well as individual complexants. Although in earlier work with calix[4]arenes we have used chloroform as the organic phase (*27-36*), we were aware that it would be advantageous to use a less toxic organic fluid

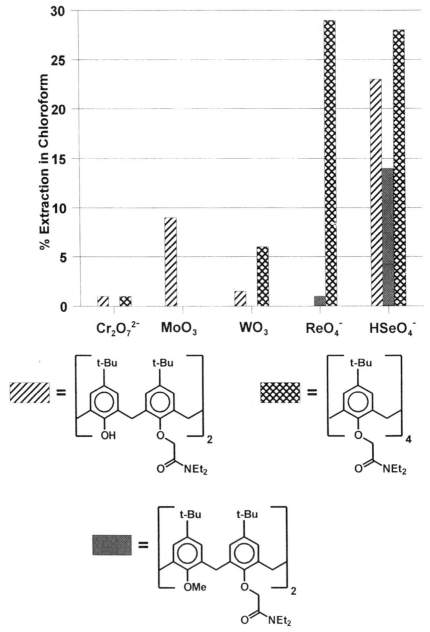

Fig. 1. Extraction by Amides (1 mM) from Aqueous Acid (pH 0.85) into Chloroform

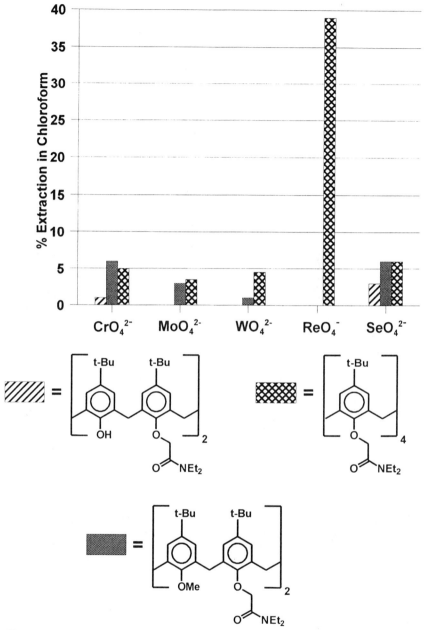

Fig. 2. Extraction by Amides (1 mM) from Water (pH 7) into Chloroform

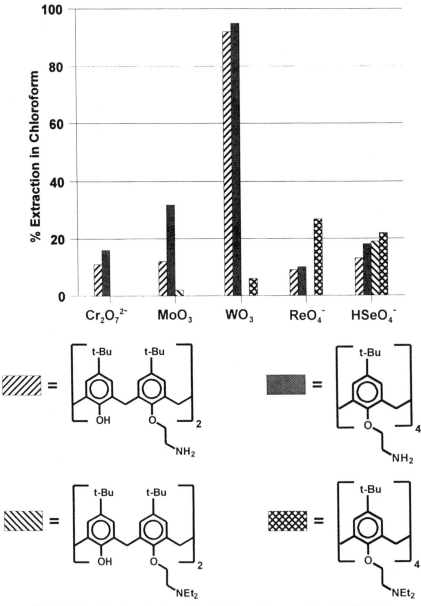

Fig.3. Extraction by Amines (1 mM) from Aqueous Acid (pH 0.85) into Chloroform

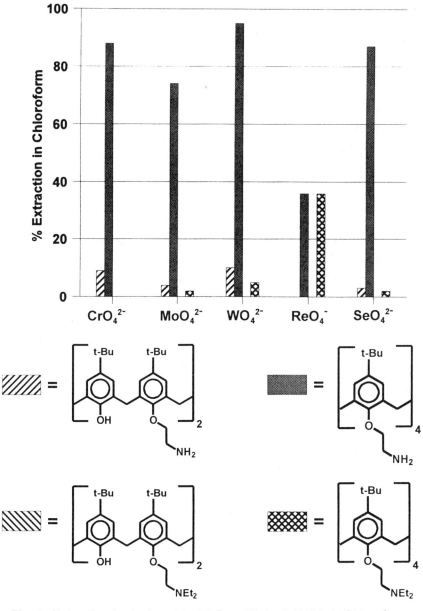

Fig. 4. Extraction by Amines (1 mM) from Water (pH 7) into Chloroform

such as an alkane. We have therefore prepared the geometrically isomeric 1,2- and 1,3-calix[4]arene amides having appended n-butyl substituents on the nitrogen in order that they can be more compatible with such an alkane phase (24). These three calix[4]arene amides have each been synthesized using the same general procedures (22). The precursor N,N-dibutyl-2-bromoacetamide is synthesized by stirring a mixture of bromoacetic acid and N,N-dibutylamine with 1,3-dicyclohexylcarbodiimide. The synthetic route to the calix[4]arene amides then involves reacting 5,11,17,23-*tert*-butylcalix[4]arene with N,N-dibutyl-2-bromoacetamide in the presence of either potassium carbonate or sodium hydride as base. The compounds are separated by column chromatography. In addition to solvent selectivity, the effect of both the geometric and conformational properties of the complexant needs also to be considered. Since the 1,2-isomer has been obtained as a separable mixture of geometric isomers, the availability of such a pair of isomers affords us the opportunity to compare their relative phase extraction properties.

Extractions are carried out with toluene or isooctane as the organic phase, along with the calix[4]arene, and the acidified aqueous solutions of the oxyions. Equal volumes of the organic and aqueous phases are then shaken for 1 minute. The resultant aqueous layer is then separated, and the remaining metal concentration in that layer analyzed by ICP-AES. These data are shown in the bar graphs in Figures 5 and 6. From these data it is apparent that the oxyions can be extracted into toluene or isooctane from an aqueous acidic solution in the presence of these hydrophobic calix[4]arene amides. The data also show differences between both these two solvents and the different isomers. From these data it appears that both UO_2^{2+} and molybdenum trioxide have a slight preference for isooctane over toluene, with the reverse trend being observed for the Cr(VI) and Se(VI) oxyanions. The differences between the isomeric calix[4]arenes are less apparent, but it does appear that there may be a preference for one isomer being a better extractant for transferring these oxyions into isooctane. Interestingly, this particular isomer is the one that has the amides in a potentially chelating geometry, with a hydrophobic *tert*-butyl group projecting into this lower rim binding cavity.

Acknowledgments

We thank the U.S. Army Research Office, the Welch foundation, the U.S. Department of Energy through the Pacific Northwest Laboratory, and NATO, for support of this research.

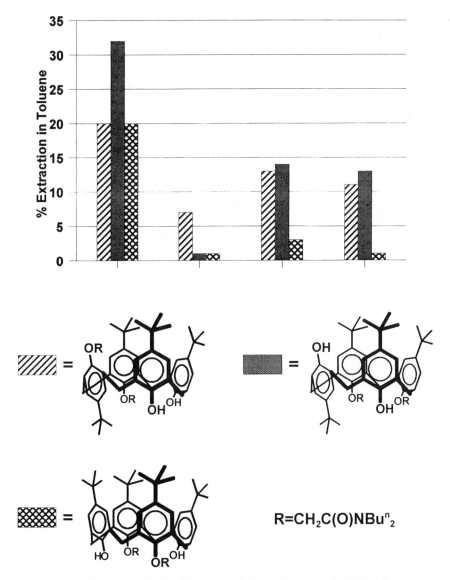

Fig. 5. Extraction by Amides (1 mM) from Aqueous Acid (pH 0.85) into Toluene

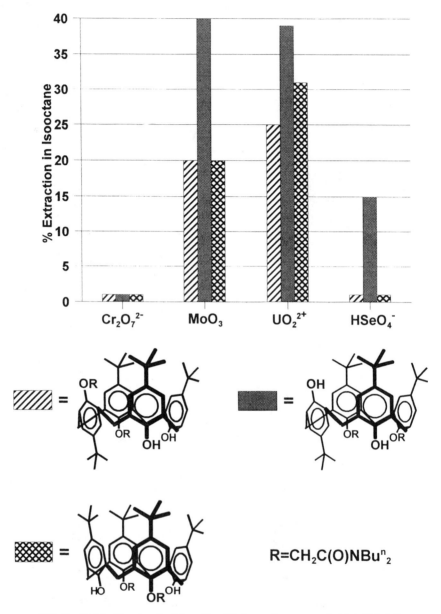

Fig. 6. Extraction by Amides (1 mM) from Aqueous Acid (pH 0.85) into Isooctane

References

1. *Heavy Metals*, Salomons, W.; Förstner, U.; Mader, P. Eds., Springer, New York, 1995.
2. Yordanov, A. T.; Roundhill, D. M. *Coord. Chem. Rev.* **1998**, *170*, 93.
3. Busch, D. H. *Chem. Rev.* **1993**, *93*, 847.
4. *The Chemistry of Macrocyclic Ligand Complexes*, Lindoy, L. F., Cambridge Univ. Press, Cambridge, U.K., 1989.
5. *Crown Ethers and Cryptands*, Gokel, G., Monographs in Supramolecular Chemistry, Fraser Stoddart, J. ed., Royal Society of Chemistry, 1991.
6. van Veggel, F. C. J. M.; Verboom W.; Reinhoudt, D. N. *Chem. Rev.* **1994**, *94*, 281.
7. Visser, H. C.; Rudkevich, D. M.; Verboom, W.; de Jong, F.; Reinhoudt, D. N. *J. Am. Chem. Soc.* **1994**, *116*, 11554.
8. *Chromium: Metabolism and Toxicity*, Burrows, D., CRC Press, Boca Raton, FL., 1983.
9. Waterhouse, J. A. H. *Br. J. Cancer* **1975**, *32*, 262.
10. Bonatti, S.; Meini M.; Abbondandolo, A. *Mutat. Res.* **1976**, *39*, 147.
11. Bianchi, V.; Zantedeschi, A.; Montaldi, A.; Majone, J. *Toxicol. Lett.* **1984**, *8*, 279.
12. De Flora, S.; Wetterhahn, K. E. *Life Chem. Rep.* **1989**, 7, 169.
13. Stearns, D. M.; Kennedy, L. J.; Courtney, K. D.; Giangrande, P. H.; Phieffer, L. S.; Wetterhahn, K. E. *Biochemistry* **1995**, *34*, 910.
14. Wittbrodt, P. R.; Palmer, C. D. *Environ. Sci. Technol.* **1995**, 29, 255.
15. Snow, E. *Pharmacol. Ther.* **1992**, *53*, 31.
16. Zhitkovich, A.; Voitkun, V.; Costa, M. *Biochemistry* **1996**, *35*, 7275.
17. Jukes, T. *Nature (London)* **1985**, *316*, 673.
18. Bulman, R. A. *Coord. Chem. Rev.* **1980**, *31*, 221.
19. Dietrich, B. *Pure & Appl. Chem.* **1993**, *65*, 1457.
20. Bradshaw, J. S.; Krakowiak, K. E.; Izatt, R. M. *Tetrahedron* **1992**, *48*, 4475.
21. Yordanov, A. T.; Mague, J. T.; Roundhill, D. M. *Inorg. Chem.* **1995**, *34*, 5084.
22. Roundhill, D. M.; Georgiev, E.; Yordanov, A. *J. Incl. Phenom. Mol. Recogn. in Chem.* **1994**, *19*, 101.
23. Wolf, N.; Georgiev, E. M.; Roundhill, D. M. *Polyhedron* **1997**, *16*, 1581.
24. Falana, O. M.; Koch, H. F.; Lumetta, G.; Hay, B.; Roundhill, D. M. *JCS, Chem. Comm.* **1998**, 503.
25. Beer, P. D. *JCS, Chem. Comm.* **1996**, 689.
26. Collins, E. M.; McKervey, M. A.; Madigan, E.; Moran, M. B.; Owens,

M.; Ferguson G.; Harris, S. J. *J. Chem. Soc., Perkin Trans 1.* **1991**, 3137.
27. Wolf, N.; Georgiev, E. M.; Yordanov, A. T.; Whittlesey, B. R.; Koch, H. F.; Roundhill, D. M. *Polyhedron* **1999**, *18*, 885.
28. Yordanov, A. T.; Wolf, N. J.; Georgiev, E. M.; Koch, H. F.; Falana, O. M.; Roundhill, D. M. *Comments on Inorg. Chem.* **1999**, *20*, 163.
29. Yordanov, A. T.; Mague, J. T.; Roundhill, D. M. *Inorg. Chim. Acta.* **1995**, *240*, 441.
30. Yordanov, A. T.; Roundhill, D. M. *New J. Chem.* **1996**, *20*, 447.
31. Yordanov, A. T.; Roundhill, D. M.; Mague, J. T. *Inorg. Chim. Acta.* **1996**, *250*, 295.
32. Yordanov, A. T.; Roundhill, D. M. *Inorg. Chim. Acta.* **1998**, *270*, 216.
33. Yordanov, A. T.; Whittlesey, B. R.; Roundhill, D. M. *Supramol. Chem.* **1998**, *9*, 13.
34. Yordanov, A. T.; Roundhill, D. M. *Inorg. Chim. Acta.* **1997**, *264*, 309.
35. Yordanov, A. T.; Falana, O. M.; Koch, H. F.; Roundhill, D. M. *Inorg. Chem.* **1997**, *36*, 6468.
36. Yordanov, A. T.; Whittlesey, B. R.; Roundhill, D. M. *Inorg. Chem.* **1998**, *37*, 3526.

CALIXARENE COMPLEXATION OF NEUTRAL MOLECULES

Chapter 20

Deep Cavities and Capsules

Dmitry M. Rudkevich and Julius Rebek, Jr.

The Skaggs Institute for Chemical Biology
and The Department of Chemistry, The Scripps Research Institute,
10550 North Torrey Pines Road, La Jolla, CA 92037

Synthesis and host-guest properties of deep open-ended cavitands and dimeric self-assembled capsules, based on calix[4]arene and resorcinarene platforms, are overviewed.

Cavitands are synthetic *open-ended* structures with enforced (rigid) cavities, molecular vessels capable of binding complementary organic compounds and ions.[1] Accordingly, the ideal cavitands are the host-molecules, *(a)* in which physical properties of the complex differ from those of the host and the guest taken separately, *(b)* where chemical reactions with the entrapped guest should be possible, *(c)* which allow selective entrance and release of a certain guest, but forbid passage to other (e.g. cell properties), and *(d)* which are synthetically accessible. Besides cyclodextrins, or perhaps in addition, calixarenes and resorcinarenes (Figure 1) are traditional cavity-forming modules, and they possess all necessary properties for the cavitand construction. They are three-dimensional and curved, large, with extended surface for interactions, fairly rigid, and readily functionalizable. Calixarenes, resorcinarenes and the analogs are widely used as platforms for molecular recognition, where complementary binding sites can be easily and selectively introduced. However, they essentially lack a sizeable intrinsic cavity; a rough estimation for calix[4]arene or resorcinarene gives a depth less than 4 Å.

Host-guest properties of cavitands are strongly dependent on the size and shape of their cavities. In this review, we concentrate on our most recent efforts to construct the next generations of these molecules - *deep cavitands* and *capsules*, that possess a really sizeable cavity (Figure 1). We also discuss the newly uncovered, unique effects of a deep cavity on complexation and physical-chemical properties of the encapsulated guest.

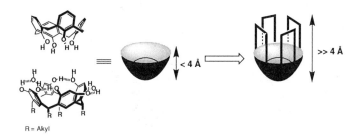

Figure 1. Calix[4]arene and resorcinarene as modules for construction of deep cavitands.

Synthesis of Deep Cavities

Within the last few years, we have synthetically attempted to deepen the cavity, rigidify the structure, and functionalize it for further applications. Thus, direct arylation of the aromatic rings in calixarenes and resorcinarene-based cavitands proved to be a promising route towards rigid and large hydrophobic cavities.[2] "Deep cavity" cavitands **1** and calix[4]arenes **2**, possessing potentially functionalizable and/or hydrogen-bonding *para*-substituents (e.g. C(O)NH-Alk, NO_2, NH_2, NH-C(O)-Alk, NH-C(O)-NH-Alk) were readily prepared from the corresponding tetrabromo derivatives and tributyltin- or boronic acid aromatics via Stille and Suzuki couplings, followed by subsequent functionalization (Figure 2). Calixarenes **2** are flexible; the "perfect" cone (C_{4v}) easily interconverts with the "pinched" C_{2v} conformer. This problem was eliminated for rigid (C_{4v}) bowl-shaped cavitands **1** based on resorcinarenes.

Other constructs **3,4** are closely related to Cram's cavitands, but possess properly positioned hydrogen bonding sites on their upper rims. The synthesis followed that published for the first generation cavitands with the appropriate changes (Figure 3, Figure 4).[3,4]

Specifically, C-undecylcalix[4]resorcinarene was coupled to 5,6-dichloropyrazine-2,3-dicarboxylic acid imide and 1,2-difluoro-4,5-dinitrobenzene in DMF in the presence of Et_3N with the formation of **3** and **5** in 48 and 80% yields, respectively. Octanitro compound **5** was subsequently reduced and acylated with the formation of octaamide **4**.

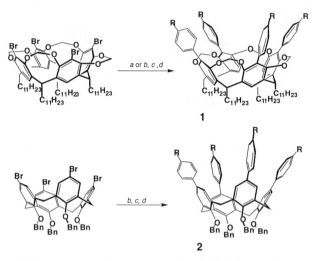

Figure 2. Synthetic route to deep cavitands 1 and calix[4]arenes 2: a. $Pd(PPh_3)_4$, p-Bu_3Sn-C_6H_4-$C(O)NH$-$(CH_2)_nCH_3$, toluene, 110°C, 1-3 d, 10-12%. b. $Pd(PPh_3)_4$, p-NO_2-C_6H_4-$B(OH)_2$, Na_2CO_3, toluene, 110°C, 1-3 d, 65-75%. c. $SnCl_2\cdot 2H_2O$, EtOH, 10 h, 80-90% for calix[4]arene, or Raney/Ni, H_2, 45 °C, toluene, 16 h, 90% for cavitand. d. Alk-COCl, EtOAc-H_2O, 1:1, K_2CO_3, r.t., 2 h, >90%, or Alk-$N=C=O$, CH_2Cl_2, r.t. 2-3 h, >90%.

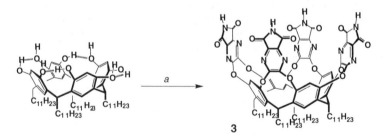

Figure 3. Synthesis of cavitand 3: a. 5,6-dichloropyrazine-2,3-dicarboxylic acid imide, Et_3N, DMF, 70°C, 6 h, 48%.

We also developed a novel synthetic strategy towards deep cavitands.[5] First, bridging of the resorcinarene hydroxyls was achieved with a simpler wall in a high yield, and then the resulting module was used to extend the walls. Cavitand 5 possessing eight NO_2 groups was employed as a module. Subsequent reduction with Raney/Ni in toluene or $SnCl_2 \cdot 2H_2O$ in EtOH/HCl followed by condensation with diketones formed fused pyrazine walls on top of the platform and gave deeper cavitands **6,7** in 13 and 49% yield, respectively (Figure 5). Cavitands **6,7** are relatively rigid and exist as C_{4v} structures at > 320 K. Octaester **6** was easily converted to the corresponding octaamide **8** and octaacid **9**. Of particular interest is cavitand **7**, which is ca 14 Å deep! According to molecular modeling, it is able to accommodate up to four benzene or toluene molecules within the internal cavity. To our knowledge, molecule **7** represents the deepest open-ended cavities synthesized to date.

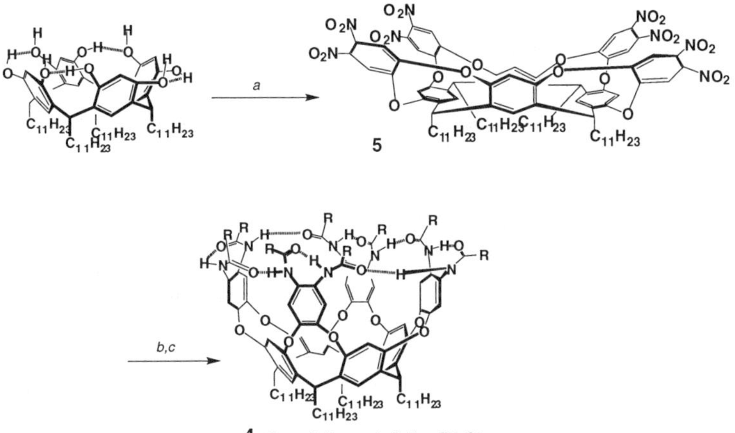

*Figure 4. Synthesis of cavitands **4**: a. 1,2-difluoro-4,5-dinitrobenzene, Et$_3$N, DMF, 70°C, 16 h, 80%. b. SnCl$_2$·2H$_2$O, aq HCl, EtOH, 4-6 h, 60-65% or Raney/Ni, H$_2$, 45 °C, toluene, 12 h, 93%. c. RCOCl, EtOAc-H$_2$O, 1:1, K$_2$CO$_3$, r.t., 2 h, 50-55%.*

Self-Folding Cavitands

In resorcinarenes, the eight hydroxy groups form a seam of intramolecular hydrogen bonds, and these interactions rigidify the structure by stabilizing the C_{4v} vase conformation.

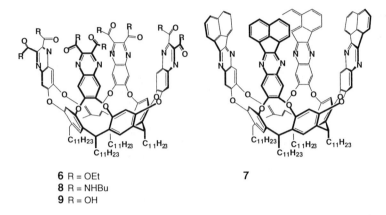

6 R = OEt
8 R = NHBu
9 R = OH

Figure 5. Preparation of deeper cavitands from 5: a. $SnCl_2·2H_2O$, *aq HCl, EtOH, 4-6 h, 60-65% or Raney/Ni, H_2, 45 °C, toluene, 12 h, 93%.* b. *diethyl-2,3-dioxosuccinate, benzene, r.t., 16 h, 13% for 6, or acenaphthoquinone, HOAc-THF, 1:25 (vol), reflux, 6 h, 49% for 7.* c. *6, n-BuNH$_2$, EtOH, reflux, 19 h, 49% for 8.* d. *6, 1N aq LiOH, THF, reflux, 1 h, quantitative yield for 9.*

When the OH hydrogens are substituted, the flexibility of the resorcinarene skeleton increases, and other conformations become preferred. To control the size and shape of the cavities, their folding and unfolding behavior and their binding properties, reversible noncovalent forces have been employed, such as *intramolecular* hydrogen bonding and solvent effects. By analogy with the similar processes in proteins, we call these molecules self-folding cavitands.[4,6]

In cavitands **4**, the vicinal secondary amides at the upper rim of the molecule form intramolecular intraannular hydrogen bonds through a seven-membered ring and, in addition, bridge all four adjacent rings. The result is a self-folded deepened vase (Figure 6). The amide groups deepen the cavity to dimensions of ca 8 x 10 Å. Although the size and shape of the cavity in **4** very much resembles those of known cavitands with quinoxaline walls, it is formed, in contrast, under thermodynamic control.

Exchange between complexed and free guest species is slow on the NMR timescale. The circle of hydrogen bonds is responsible for these features. Upon complexation with adamantanes and lactams, the ^{1}H NMR spectra exhibit two sets of signals - both for the cavitands and the corresponding caviplexes. The complexed guest species are clearly observed *upfield* of 0 ppm, a feature characteristic of inclusion in a shielded environment and similar to the shifts observed in covalently bound carceplexes. This indicates substantial energetic barriers for guest exchange *in* and *out* of the vases **4**. The *slow* rate is unprecedent for open-cavity receptors. At higher temperatures (≥60°C), the signals for complexed and free species become broad, and the guest signals eventually disappear into the baseline. It is proposed that the solvent replacement by guest takes place through an unfolding of the cavity (Figure 6).

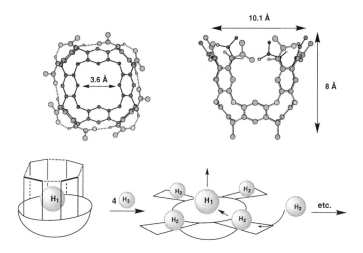

Figure 6. Intramolecular hydrogen bonding in vicinal diamides causes a self-folding in cavitands **4**. Top: two views of the energy-minimized (Amber*, MacroModel 5.5)structure **4**; the long alkyl chains and CH hydrogen are omitted for clarity. Bottom: proposed mechanism of guest/solvent exchange in **4** via folding-unfolding.

1-[N-(1-Adamantyl)]adamantanecarboxamide, taken in ca 50-fold excess, gave two *diastereomeric* 1:1 complexes with **4** in p-xylene-d_{10} solution; two sets of characteristically upfield adamantane signals can be seen (Figure 7). The adamantane guest can spin about the long axis of the cavity but is too large to tumble within it. The resulting diastereomerism has precedent in carceplexes, but was also unknown in open-ended cavities.

It was unexpected to find that an open-ended vessel could so effectively desolvate and shield guest species from the environment of the bulk solution. Although the guest exchange process is quite slow on the NMR time-scale ($k = 2\pm1$ s^{-1}), it is still faster than that observed for the completely closed hydrogen-bonded calixarene-based capsule ($k = 0.47\pm0.1$ s^{-1}) (Böhmer et al.), or for covalently bound hemicarcerands ($k \leq 0.0093$ s^{-1}) (Cram et al.).

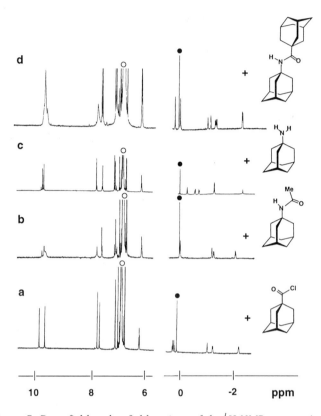

Figure 7. Downfield and upfield regions of the ^1H NMR spectra (p-xylene-d_{10}, 600 MHz, 295 K) of cavitand 4 (R = n-C_7H_{15}) during complexation with 1-substituted adamantanes. a. *With 1-adamantanecarbonyl chloride.* b. *With N-(1-adamantyl)acetamide; residual signals of guest-free cavitand 4 are also present due to weak complexation.* c. *With 1-adamantanamine. The furthest upfield signal (ca –2 ppm) is due to the NH_2 resonance; guest-free cavitand is also present due to weak complexation.* d. *With 1-[N-(1-adamantyl)]adamantane-carboxamide; two different complexes are present. The solvent signals (with corresponding satellites) and the internal standard singlet are marked "o" and "•", respectively. The host and guest concentrations are 0.5 and ≥25 mM, respectively. For the complexed 1-substituted adamantanes, all four sets of the skeleton protons can be clearly seen. Due to the effects of the nearby aromatic ring-currents, the chemical shifts of the guest signals are directly related to their position inside the cavity. The functional group at the adamantane 1-position is generally not shifted upfield in the NMR spectra, indicating that the adamantane skeleton is oriented toward the bottom of the cavity and the functional group toward the top.*

Self-Assembly of Capsules

Self-complementary molecules dimerize through solvophobic or hydrogen bonding interactions in apolar media. With appropriate curvature, such self-assembling systems generate cavities and result in encapsulation of smaller guest molecules. Inside these, reactive intermediates can be stabilized, new forms of stereoisomerism are observed and even bimolecular reactions can take place. Larger self-assembled cavities of nanoscale dimensions are still rare.

Cyclic tetraimide **3** dimerizes through hydrogen bonding (Figure 8).[3] The shape of monomers **3** is vase-like and the dimerization takes place in the rim-to-rim manner to give the large cylindrical capsules. They feature dimensions of ca 10 x 18 Å and are stabilized by a seam of bifurcated hydrogen bonds.

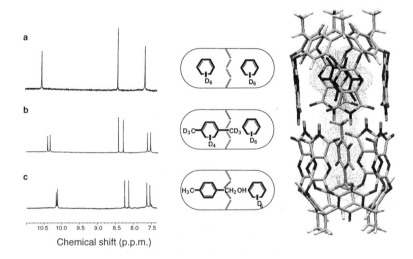

*Figure 8. Two cavitands **3** self-assemble through hydrogen bonding into a cylindrical capsule **3•3**. Two representations of the capsule are shown. Right: the energy-minimized (MacroModel 5.5, Amber* force field) structure **3•3** with benzene and p-xylene inside; the long alkyl chains and CH hydrogen atoms are omitted for viewing clarity. Center: the cartoon representation used elsewhere in this review. Left: portions of the ^1H NMR spectra (600 MHz, 295 K) of capsule **3•3** (total concentration of **3** ca 1 mM): a) in benzene-d_6; b) in benzene-d_6 - p-xylene-d_{10}, 1:1; c) complex (**3**•benzene-d_6•p-methylbenzyl alcohol•**3**) in mesitylene-d_{12}. The CH_3 singlet of the complexed p-methylbenzyl alcohol is seen at -2.77 ppm. The spectra reprinted with permission from <u>Nature</u> **1998**, 394, 764-766. Copyright (1998) Macmillan Magazines Limited.*

When both benzene and *p*-xylene were added in a 1:1 ratio to the mesitylene-d_{12} solution of **3**, an unsymmetrically filled capsule was observed exclusively; a comfortable occupancy is reached with one of each guest in the capsule. The complex is unsymmetrical because the two guests cannot squeeze past each other to exchange positions in the capsule - at least not on the NMR timescale (Figure 8). Likewise, benzene paired with *p*-trifluoromethyltoluene, *p*-chlorotoluene, 2,5-lutidine, and *p*-methylbenzyl alcohol to give new species with one of each guest inside. Even in a competition experiment involving three different solvent guests, in which toluene, benzene and *p*-xylene were added in a 2:1:1 ratio to the mesitylene-d_{12} solution of **3**, the capsule was filled preferentially (ca 90%) with benzene and *p*-xylene. Only ca 10% of the capsule with two toluenes inside was observed. In another experiment, benzene and *p*-xylene in mesitylene-d_{12} replaced encapsulated toluenes within a few minutes at room temperature.

That "molecule-within-molecule" complexes, held together only by hydrogen bonds, can show such selectivity was unexpected. Matching the overall length of two guests with the dimensions of the cavity appears to be the driving force.[3,7,8]

Capsule **3•3** also exhibits complexation of some elongated guest-molecules (Figure 9) and even smaller hydrogen bonded aggregates: 2-pyridone/2-hydroxypyridine dimer and benzoic acid dimer.[7] Finally, diastereomeric "complexes within a complex", using chiral guests, were observed. Two different species are observed in the presence of the racemic *trans*-1,2-cyclohexanediol while only one appears if only the single enantiomer is available. In either case, integration indicates there are two guests inside each capsule, but the intensity of the signals for the enantiopure vs the racemic guests indicates more of the latter- the capsule prefers to be filled with a guest and its mirror image instead of two identical molecules.[7]

"Deep cavity" tetraamides **1** (R = $NHC(O)(CH_2)_6CH_3$ and $NHC(O)(CH_2)_7CH_3$) dimerize via intermolecular hydrogen bonding (K_D = 1700±250 M^{-1}, 295 K) in toluene in an unusual manner.[2] The dimers possess egg-shaped cavities of dimensions 12 x 19 Å; the estimated volume of which is ~440 $Å^3$ (Figure 10). At the same time, one of the four amide alkyl chains from each cavitand is encapsulated in the dimeric capsule, with the terminal CH_3 groups situated in the deepest part of the cavity. Those CH_3 groups therefore are seen upfield (up to −1.8 ppm) in the ^1H NMR spectra in toluene-d_8. Due to this self-inclusion, the solvent/guest molecules inside the cavity are replaced, which is apparently the more energetically preferred scenario. The residual internal volume thus becomes 190 $Å^3$. This means that ca 57% of the cavity is already well filled by self-inclusion, and there is not enough room to accommodate any other molecules.

Self-inclusion of this sort has never been observed in the much smaller, calix[4]arene dimers. Interestingly, the short chain *n*-propionylamide derivative **1** (R = $NHC(O)CH_2CH_3$) and the long chain palmitoylamide derivative **1** (R = $NHC(O)(CH_2)_{14}CH_3$) both exist exclusively in the monomeric state in toluene-d_8.

279

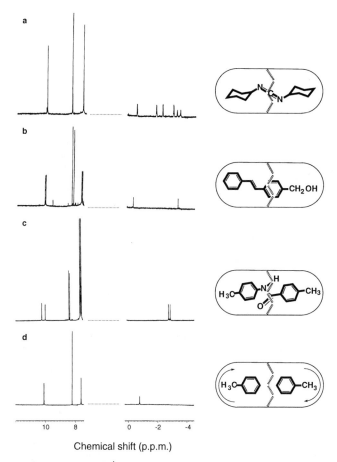

Chemical shift (p.p.m.)

Figure 9. Portions of the ^1H NMR spectra (600 MHz, 295 K) of complexes (3•3•Guest) in mesitylene-d_{12} (concentration of 3 ca 1 mM, guest concentration is ca >100 mM): a) (3•3•dicyclohexyl carbodiimide). Most of the well-resolved cyclohexyl signals for the encapsulated guest are situated between 0 and -4 ppm; b) (3•3•trans-4-stilbene methanol). The CH_2OH fragment of the encapsulated guest-molecule is situated at -0.24 ppm (d, 2 H, J = 3.5 Hz, CH_2) and -3.33 ppm (t, 1 H, J = 3.5 Hz, OH). Upon shaking with D_2O, the imide NH singlet of 3 and the OH of the encapsulated guest exchanged within a few hours; c) 3•3•p-[N-(p-tolyl)]toluamide. The CH_3 groups of the encapsulated guest are found at -2.70 and -2.81 ppm; d) (3•3•2 x toluene). The CH_3 singlet of the two complexed toluenes at -0.77 ppm broadened at temperatures below 273 K. The observed set is, in fact, the averaged one, indicating a fast - on the NMR time-scale - tumbling of the encapsulated toluene inside the capsule. Reprinted with permission from Nature 1998, 394, 764-766.Copyright (1998) Macmillan Magazines Limited.

Even though the same pattern of intermolecular hydrogen bonds are possible in their respective dimers, neither side chain is appropriate for self-inclusion. A delicate balance exists between the enthalpic and entropic contributions to this process.

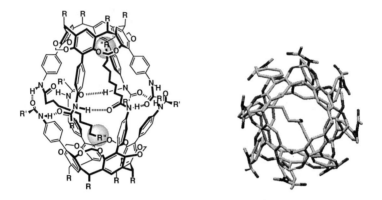

Figure 10. Dimerization of cavitand 1 (R = $C_{11}H_{23}$, R' = $NHC(O)(CH_2)_6CH_3$ or $NHC(O)(CH_2)_7CH_3$ via hydrogen bonding and self-inclusion. Right: top view of the energy-minimized (MacroModel 5.5, Amber* force field) structure 1•1; most of the long alkyl chains and CH hydrogen atoms are omitted for viewing clarity. Left: proposed structure of the dimer.

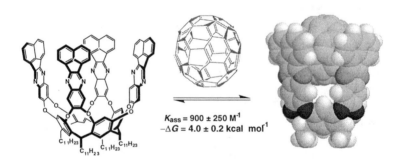

Figure 11. Cavitand 7 complexes fullerene in toluene. Right-side view: the energy-minimized CPK models of C_{60}•7 complex. The long alkyl "feet" are omitted for clarity

Deeper Cavitands

Deeper cavitands **6-9** of nanometric dimensions are now available for use as open ended molecular containers.[5] The cavities reported here feature wide openings with guest/solvent exchange fast on the NMR time scale. The uptake and release of guests involves the folding and unfolding of the host walls, motions that are influenced by solvent size and polarity. Modeling suggested that **7** can accommodate C_{60} (Figure 11) even though the round C_{60} is not well complemented by the square hole. The electron-rich walls of **7** and their considerable contact with incumbent C_{60} encouraged experiments, and strong complexation of C_{60} by **7** was indeed detected by UVvis spectroscopy in toluene solution. Specifically, addition of **7** to a toluene solution of C_{60} (1.0-2.0 x 10^{-5} M) led to a increase in the absorption of the band at 430 nm, characteristic of complexation (Figure 11). Treatment of the titration data with the Benesi-Hildebrand equation, gave a value for the association constant, K_a = 900±250 M^{-1} at 293 ± 1 K (ΔG = -4.0 ± 0.2 kcal/mol); the binding isotherm provides a good fit to a 1:1 stoichiometry. Control experiments with a noncyclic model compound and the known "shallow" cavitand showed only very weak interactions (K_a < 10 M^{-1}). Unexpectedly, the UVvis spectra showed no apparent binding of the slightly larger C_{70} in **7**. Accordingly, this example of selective complexation of fullerenes by resorcinarene-based compounds is one of the first reported.

Conclusions

Deep cavities represent novel unique species of molecular containers, distinct not only from the covalently sealed carcerands, but also from other open-cavity macrocycles. Their encapsulation behavior is derived from the considerable three-dimensional size and/or the elongated shape. They are open-ended, but guest inclusion may be slow on the NMR time scale. In some spectacular cases, the significant upfield ^{1}H NMR shifts of the complexed molecules offers the possibility to examine the structural details of the complexes "from inside", i.e. to determine the orientation of the guest and its interaction with the receptor walls inside the cavity. The uptake and release of guests may, in some cases, involve the folding and unfolding of the host, achieved by either varying solvent polarity and/or temperature. As the constant flow of the substrate into and the product out of the cavity can be achieved, there is a possibility of using these containers as reaction vessels. Catalysis may also be possible. The introduction of additional (catalytically) useful sites appears synthetically likely. Another application is in analysis and sensing. Extended lipophilic surfaces should strongly attract organic guests in aqueous solution. Both covalent bonds and noncovalent interactions can be utilized to control size and shape of the cavities. Dimers have also been prepared by means of hydrogen bonding. The pairwise selection of two different molecules by self-assembled deep capsules points directly on their use as selective reaction chambers for bimolecular processes, and suggest that orientations of the reacting partners can be controlled.

Acknowledgements: The work described in this paper is the result of the efforts of our coworkers Drs. S. Ma, G. Hilmersson, T. Heinz and F. C. Tucci. We also thank Dr. B. O'Leary of Massachusetts Institute of Technology for the computer graphics of capsule 3•3. We are grateful to the Skaggs Research Foundation and the National Institutes of Health for support.

References

1. Cram; D. J.; Cram, J. M. *Container Molecules and their Guests*; Royal Society of Chemistry: Cambridge, **1994**.
2. Ma, S.; Rudkevich, D. M.; Rebek, Jr., J. *J. Am. Chem. Soc.* **1998**, *120*, 4977-4981.
3. Heinz, T.; Rudkevich, D. M.; Rebek, Jr., J. *Nature* **1998**, *394*, 764-766.
4. Rudkevich, D. M.; Hilmersson, G.; Rebek, Jr., J. *J. Am. Chem. Soc.* **1998**, *120*, 12216-12225.
5. Tucci, F. C.; Rudkevich, D. M.; Rebek, Jr., J. *J. Org. Chem.* **1999**, *64*, 4555-4559.
6. Rudkevich, D. M.; Hilmersson, G.; Rebek, Jr., J. *J. Am. Chem. Soc.* **1997**, *119*, 9911-9912.
7. Heinz, T.; Rudkevich, D. M.; Rebek, Jr., J. *Angew. Chem., Int. Ed. Engl.* **1999**, *38*, 1136-1139.
8. Tucci, F. C.; Rudkevich, D. M.; Rebek, Jr., J. *J. Am. Chem. Soc.* **1999**, *121*, 4928-4929.

Chapter 21

Calixarene Metalloreceptors: Demonstration of Size and Shape Selectivity inside a Calixarene Cavity

Stephen J. Loeb and Beth R. Cameron

School of Physical Science, Chemistry and Biochemistry,
University of Windsor, Windsor, Ontario N9B 3P4, Canada

Calix[4]arenes substituted with amido groups on the upper rim, can be used a starting materials for the synthesis of calixarene ligands. These ligands can be ortho-palladated to form calixarene metalloreceptors capable of coordinating a substrate inside the bowl-shaped cavity of the calixarene. The principles of size and shape selectivity can be demonstrated employing substituted aromatic amines as model substrates.

Metalloreceptors are complexes of macrocyclic or multidentate ligands which contain preorganized, non-covalent binding sites for secondary interactions with a coordinated substrate.[1,2] Such complexes have been employed as hosts for a variety of neutral, cationic, and anionic guests. We have described a series of organopalladium based metalloreceptors capable of hydrogen-bonding and/or π-stacking interactions. These receptors have been applied to the molecular recognition of aliphatic amines,[2] aromatic amines,[3] hydrazines[4] and DNA nucleobases.[5]

Calix[4]arenes[6,7] are known to act as receptors for cationic,[7,8] anionic,[9-11] or neutral[12,13] substrates by providing a platform for the attachment of convergent binding groups, at the upper[15,12] or lower rim[8,9,13] or by utilizing the bowl shaped arrangement of the four aromatic groups as a hydrophobic cavity.[14] Our design strategy was to build a metalloreceptor by constructing a "handle" on the top of the calixarene "basket." This would allow for a substrate to be *both* coordinated to the metal center (handle) *and* situated inside the hydrophobic cavity of the calix[4]arene (basket). Ideally, metal coordination would act to anchor the substrate to the receptor while the best fit of size and shape to the calixarene cavity would allow for molecular recognition and determine selectivity. The synthesis and characterization of two organopalladium-based metalloreceptors are reported herein, along with selectivity studies on some model aromatic amines to demonstrate the potential of these systems for chemical separations.

Experimental

All starting materials were purchased from Aldrich Chemicals and used without further purification, except acetonitrile which was distilled from CaH_2 under $N_2(g)$. All reactions were performed under an atmosphere of $N_2(g)$. 5,17-Bis(2-chloroacetamido)-25,26,27,28-tetrapropoxycalix[4]arene and 5,17-bis[2-(4-chloromethylphenoxy)acetamido]-25,26,27,28-tetrapropoxycalix[4]arene were prepared by literature methods.[15,16] ^{1}H NMR and ^{13}C NMR spectra were recorded on a Bruker AC300 spectrometer locked to the deuterated solvent at 300.1 and 75.5 MHz respectively. Liquid secondary-ion and electron-impact mass spectra were recorded on a Kratos Profile mass spectrometer. Elemental analyses were performed by Canadian Microanalytical Service, Delta, British Columbia. Molecular mechanics calculations and computer modelling were performed on a 400 MHz Pentium II PC employing Oxford Molecular's Personal CAChe Software Package.

Preparation of Macrobicyclic Calix[4]arene Ligand HL^1

m-Xylene-α,α'-dithiol (57.8 μL, 0.39 mmol) was dissolved in a solution of sodium ethoxide freshly prepared using Na (0.018 g, 0.78 mmol) and ethanol (100 mL). 5,17-bis(2-chloroacetamido)-25,26,27,28-tetrapropoxycalix[4]arene (0.303 g, 0.39 mmol) in ethanol (50 mL) was added dropwise over a 4 h period. The reaction mixture was stirred an additional 12h at room temperature. The solvent was removed and the residue dissolved in CH_2Cl_2 and washed successively with 1.0 M HCl, H_2O and $NaHCO_3$ solutions. The organic extracts were dried over anhydrous $MgSO_4$ and the solvent removed. The crude product was recrystallized from CH_2Cl_2/CH_3CN. Yield 0.336 g (86 %). ^{1}H NMR (CDCl$_3$): δ (ppm) 7.79 (s, NH, 2H), 7.28-7.11 (m, ArH, 8H), 6.93 (t, ArH, 2H), 5.96 (s, ArH, 4H), 4.42 (d, ArCH$_2$, 4H), 4.03 (t, OCH$_2$, 4H), 3.68-3.59 (m, OCH$_2$ + SCH$_2$, 8H), 3.42 (s, 3H, CH$_3$OH), 3.14 (d, ArCH$_2$, 4H), 2.90 (s, CH$_2$S, 4H), 1.99-1.84 (m, CH$_2$, 8H), 1.09 (t, CH$_3$, 6H), 0.87 (t, CH$_3$, 6H). ^{13}C NMR (CDCl$_3$): δ (ppm) 167.00 (CO), 157.92, 153.78, 137.12, 136.72, 133.79, 130.36, 130.00, 129.58, 129.05, 128.26, 124.44, 122.33 (ArC), 77.33, 76.58 (OCH$_2$), 47.54 (CH$_3$OH), 35.52, 34.03 (CH$_2$S), 31.12 (ArCH$_2$), 23.60, 22.98 (CH$_2$), 10.93, 9.87 (CH$_3$). LSI-MS: [M+1]$^+$ = 873. Anal. Calcd. for C$_{52}$H$_{60}$N$_2$O$_6$S$_2$·CH$_3$OH: C, 70.32; H, 7.13; N, 3.10. Found: C, 70.48; H, 6.75; N, 3.48.

Preparation of Metalloreceptor [Pd(L^1)(MeCN)][BF$_4$]

Calixarene HL^1 (0.1 g, 0.12 mmol) was suspended in acetonitrile (50 mL) and heated to reflux. After complete dissolution, a solution of [Pd(CH$_3$CN)$_4$][BF$_4$]$_2$ in acetonitrile was added dropwise over a period of 15 min. The reaction mixture was refluxed 2h, cooled to room temperature and the solvent removed. Yield: 0.12 g (93 %). ^{1}H NMR (CD$_3$CN): δ (ppm) 8.29 (s, NH, 2H), 7.69 (br s, ArH, 2H), 7.23 (d, ArH, 4H) 7.05 (s, ArH, 3H), 6.95 (t, ArH, 2H), 6.54 (br s, ArH, 2H), 4.48 (d, ArCH$_2$, 4H), 4.38 (br s, CH$_2$S, 4H), 4.07 (m, CH$_2$S + OCH$_2$, 8H), 3.69 (t, OCH$_2$, 4H), 3.29 (d, ArCH$_2$, 4H), 2.12 (m, CH$_2$, 4H), 1.95 (m, CH$_2$, 4H), 1.02 (t, CH$_3$, 6H), 0.94 (t, CH$_3$, 6H). ^{1}H NMR (CD$_3$NO$_2$): δ (ppm) 8.09 (s, NH, 2H), 7.72 (br s, ArH, 2H), 7.28 (d, ArH, 4H), 7.08 (m, ArH, 3H), 7.00 (t, ArCH$_2$, 2H), 6.67 (br s, ArCH$_2$, 2H), 4.59 (d, ArCH$_2$, 4H), 4.45 (br s, CH$_2$S, 4H), 4.21 (s, CH$_2$S, 4H), 4.18 (t, OCH$_2$, 4H), 3.78 (t, OCH$_2$, 4H), 3.33 (d, ArCH$_2$, 4H), 2.21 (m, CH$_2$, 4H), 1.98 (m, CH$_2$, 4H), 1.09 (t, CH$_3$, 6H), 1.01 (t, CH$_3$, 6H), -1.80 (br s, MeCN, 3H). ^{13}C NMR (CD$_3$NO$_2$): δ (ppm) 164.90 (br), 158.58, 154.92, 148.93 (br), 137.44, 136.11, 134.16 (br), 130.95 (br), 127.76 (br), 124.58 (br), 121.48 (br), 79.82, 77.82, 45.91 (br), 45.25 (br), 31.96, 24.61, 24.28, 11.23, 10.27. LSI-MS: [M-CH$_3$CN-BF$_4$]$^+$ = 978. Anal. Calcd. for C$_{54}$H$_{62}$BF$_4$N$_3$O$_6$PdS$_2$·CHCl$_3$·CH$_3$OH: C 53.53; H, 5.44; N, 3.35. Found: C, 53.56; H, 5.49; N, 2.92.

Preparation of Macrobicyclic Calixarene Ligand HL^2

Ethanol solutions (50 mL) of α,α'-*meta*-xylene dithiol (0.0427 g, 0.25 mmol) and 5,17-bis[2-(4-chloromethylphenoxy)acetamido]-25,26,27,28-tetrapropoxycalix[4]-arene (0.248 g, 0.25 mmol) were added dropwise to a solution of Na (0.0115 g, 0.5 mmol) in EtOH (100 mL) over a 12 h period at room temperature. The reaction mixture was stirred an additional 12h after which time the solvent was removed and the residue dissolved in ethyl acetate and washed with H$_2$O. The organic extracts were dried over anhydrous MgSO$_4$ and the solvent removed. The crude product was purified by chromatography (SiO$_2$; eluent: 1% MeOH/CH$_2$Cl$_2$). Yield 0.045 g (17 %). ^{1}H NMR (CDCl$_3$): δ (ppm) 7.80 (s, NH, 2H), 7.28 (m, ArH, 2H), 7.14 (m, ArH, 8H), 7.00 (m, ArH, 3H), 6.81 (s, ArH, 1H), 6.64 (d, ArH, 4H), 6.31 (s, ArH, 4H), 4.47 (d, ArCH$_2$, 4H), 4.20 (s, CH$_2$CO, 4H), 4.08 (t, OCH$_2$, 4H), 3.80 (s, CH$_3$OH, 3H), 3.65 (t, OCH$_2$, 4H), 3.53 (s, CH$_2$S, 4H), 3.50 (s, CH$_2$S, 4H), 3.17 (d, ArCH$_2$, 4H), 2.00 (m, CH$_2$, 4H), 1.88 (m, CH$_2$, 4H), 1.09 (t, CH$_3$, 6H), 0.89 (t, CH$_3$, 6H). ^{13}C NMR (CDCl$_3$): δ (ppm) 165.22 (CO), 157.75, 155.71, 152.86, 137.48, 136.56, 133.55, 131.48, 130.68, 130.32, 130.15, 129.02, 127.57, 122.26, 121.55, 114.53 (ArC), 77.18, 76.45 (ArOCH$_2$), 67.07 (CH$_2$CO), 44.40 (CH$_3$OH), 35.02, 34.31 (CH$_2$S), 31.04 (ArCH$_2$), 23.44, 22.85 (CH$_2$), 10.78, 9.75 (CH$_3$). LSI-MS: [M+1] = 1085. Anal. Calcd. for C$_{66}$H$_{72}$N$_2$O$_8$S$_2$·CH$_3$OH: C, 72.01; H, 6.86; N, 2.51. Found: C, 71.96; H, 6.54; N, 2.48.

Preparation of Metalloreceptor [Pd(L^2)(MeCN)][BF$_4$]

Calixarene HL^2 (0.100 g, 0.092 mmol) was suspended in CH$_3$CN (100 mL) and refluxed until completely dissolved. Once dissolved, a solution of [Pd(CH$_3$CN)$_4$][BF$_4$]$_2$ (0.041 g, 0.092 mmol) in CH$_3$CN was added dropwise. The reaction mixture was refluxed for 6 h after which time the solvent was removed.

Yield: 0.117 g (96 %). ^{1}H NMR (CD$_3$CN): δ (ppm) 8.19 (s, NH, 2H), 7.38 (d, ArH, 4H), 7.15 (d, ArH, 4H), 6.89 (m, ArH, 9H), 6.61 (s, ArH, 4H), 4.47 (d, ArCH$_2$, 4H), 4.37 (s, CH$_2$CO, 4H), 4.33 (br s, CH$_2$S, 4H), 4.28 (s, CH$_2$S, 4H), 4.08 (t, OCH$_2$, 4H), 3.66 (t, OCH$_2$, 4H), 3.20 (d, ArCH$_2$, 4H), 2.05 (m, CH$_2$, 4H), 1.90 (m, CH$_2$, 4H), 1.07 (t, CH$_3$, 6H), 0.91 (t, CH$_3$, 6H). ^{13}C NMR (CD$_3$CN): δ (ppm) 166.91 (CO), 158.43, 153.68, 151.38, 137.43, 134.88, 132.53, 130.05, 128.58, 123.66, 121.65, 115.93 (ArC), 78.52, 77.31 (ArOCH$_2$), 67.84 (CH$_2$CO), 45.14, 43.49 (CH$_2$S), 31.87 (ArCH$_2$), 24.24, 23.79 (CH$_2$), 11.14, 10.13 (CH$_3$). LSI-MS: [M-CH$_3$CN-BF$_4$+H]$^+$ = 1190. Anal. Calcd. for C$_{68}$H$_{74}$BF$_4$N$_3$O$_8$PdS$_2$·CH$_2$Cl$_2$: C, 59.09; H, 5.47; N, 3.00. Found: C, 59.77, H, 5.51, N, 2.67.

Preparation of [Pd(L^1)(pyridine)][BF$_4$]

Metalloreceptor [Pd(L^1)(MeCN)][BF$_4$] (0.05g, 0.045 mmol) was dissolved in CH$_3$CN (50 mL) and an excess of pyridine was added. The reaction mixture was warmed to 50°C and stirred for 2 h. The solvent was reduced, and the product precipitated with the addition of diethyl ether. The crude product was recrystallized from CH$_2$Cl$_2$/Et$_2$O. Yield: 0.040 g (77 %). ^{1}H NMR (CD$_3$NO$_2$): δ (ppm) 8.86 (m, py, 1H), 8.75 (t, py, 1H), 8.17 (t, py, 1H), 7.52 (br s, NH + py, 4H), 7.37 (d, ArH, 4H), 7.31 (br s, ArH, 2H), 7.17 (t, ArH, 2H), 7.07 (s, ArH, 3H), 6.01 (br s, ArH, 2H), 5.44 (s, 2H, CH$_2$Cl$_2$), 4.65 (d, ArCH$_2$, 4H), 4.52 (br s, CH$_2$S, 4H), 4.35 (br s, CH$_2$S, 4H), 3.93 (m, OCH$_2$, 4H), 3.72 (t, OCH$_2$, 4H), 3.32 (d, ArCH$_2$, 4H), 2.24 (m, CH$_2$, 4H), 1.99 (m, CH$_2$, 4H), 1.11 (t, CH$_3$, 6H), 0.99 (t, CH$_3$, 6H). ^{13}C NMR (CD$_3$NO$_2$): δ (ppm) 163.34, 159.44, 154.92, 150.27, 148.34, 138.98, 137.84, 134.67, 134.16, 131.83, 127.72, 124.81, 124.20, 121.38 (br), 79.62, 77.82, 52.02, 47.72, 44.66, 32.51 (br), 24.64, 24.14, 11.20, 10.22. LSI-MS: [M-py-BF$_4$]$^+$ = 978. Anal. Calcd. for C$_{57}$H$_{64}$N$_3$O$_6$S$_2$PdBF$_4$·CH$_2$Cl$_2$: C, 56.66; H, 5.41; N, 3.42. Found: C, 56.83; H, 5.47; N, 3.46.

Preparation of [Pd(L^2)(4-Phpy)][BF$_4$]

Metalloreceptor [Pd(L^2)(MeCN)][BF$_4$] (0.085 g, 0.064 mmol) was dissolved in CHCl$_3$ and warmed to 50 °C. 4-Phenylpyridine (0.01 g, 0.064 mmol) was added and the reaction mixture stirred an additional 2 h at 50°C. The solvent was reduced and product precipitated upon addition of diethyl ether. The isolated solid was recrystallized from CHCl$_3$/Et$_2$O. Yield 0.091 g (99 %). ^{1}H NMR (CD$_3$NO$_2$): δ (ppm)7.49 (d, ArH, 2H), 7.28 (d, ArH, 4H), 7.11-6.96 (m, ArH, 13H), 6.88 (d, ArH, 2H), 6.37 (s, ArH, 2H), 6.27 (d, ArH, 4H), 4.73 (m, m-PhPy, 2H), 4.58 (d, ArCH$_2$, 4H), 4.36 (s, OCH$_2$CO, 4H), 4.24 (br s, CH$_2$S, 4H), 4.19 (t, OCH$_2$, 4H), 3.74 (s, CH$_2$S, 4H), 3.67 (t, OCH$_2$, 4H), 3.56 (t, p-Phpy, 1H), 3.26 (d, ArCH$_2$, 4H), 2.25 (m, CH$_2$, 4H), 1.92 (m, CH$_2$, 4H), 1.02 (t, CH$_3$, 6H), 0.99 (t, CH$_3$, 6H). ^{13}C NMR (CD$_3$NO$_2$): δ (ppm) 165.38, 158.88, 157.96, 153.90, 153.41, 153.20, 138.09, 134.78, 134.14, 133.51, 133.16, 132.26, 130.88, 127.37, 126.83, 124.88, 124.32, 124.02, 122.45, 122.28, 115.83, 79.51, 77.85, 67.04, 47.14, 46.02, 32.06, 24.69, 24.40, 11.29,

10.35. LSI-MS: [M -BF$_4$]$^+$ = 1345. Anal. Calcd. for C$_{77}$H$_{80}$BF$_4$N$_3$O$_8$PdS$_2$·CH$_2$Cl$_2$: C, 61.67; H, 5.45; N, 2.78. Found: C, 61.49; H, 5.50; N, 2.94.

Preparation of Model Ligand, HL^3

Na (0.20 g, 8.8 mmol) was dissolved in EtOH (80 mL). α,α'-*meta*-Xylene dithiol (0.75 g, 4.4 mmol) was added to the NaOEt solution and stirred for 1 h at room temperature. Chloroacetanilide (1.5 g, 8.8 mmol) was added to the reaction mixture and the solution was stirred for 12 h. Yield: 1.85 g, (96 %). ^{1}H NMR (CDCl$_3$): δ 8.59 (s, NH, 2H), 7.42 (d, ArH, 4H), 7.26-7.09 (m, ArH, 10H), 3.68 (s, CH$_2$S, 4H), 3.20 (s, CH$_2$S, 4H). ^{13}C NMR (CDCl$_3$): δ (ppm) 167.14, 137.62, 137.40, 129.56, 128.92, 128.14, 124.59, 119.86, 36.80, 36.17. Anal. Calcd. for C$_{24}$H$_{24}$N$_2$O$_2$S$_2$: C, 66.04; H, 5.55; N, 6.42. Found: C, 65.87; H, 5.55; N, 6.31.

Preparation of Model Metalloreceptor [Pd(L^3)(MeCN)][BF$_4$]

HL^3 (0.98 g, 2.25 mmol) and [Pd(CH$_3$CN)$_4$][BF$_4$]$_2$ (1.0 g, 2.25 mmol) were gently refluxed in CH$_3$CN (50 mL) for 12 hours. The solution was concentrated under vacuum and pure product was obtained from crystallization at 4°C. Yield: 1.37 g (91 %). ^{1}H NMR (CD$_3$CN): δ (ppm) 8.74 (s, NH, 2H), 7.52 (d, ArH, 4H), 7.33 (t, ArH, 2H), 7.13 (t, ArH, 1H), 6.98 (m, ArH, 6H), 4.51 (br s, CH$_2$S, 4H), 4.01 (s, CH$_2$S, 4H), 2.25 (s, H$_2$O, 2H). ^{13}C NMR (CDCl$_3$): δ (ppm) 164.06, 137.65, 129.07, 126.33, 124.90, 123.50, 120.02, 45.71 (br), 42.03. Anal. Calcd. for C$_{26}$H$_{26}$BF$_4$N$_3$O$_2$PdS$_2$·H$_2$O: C, 45.40; H, 4.10; N, 6.11. Found: C, 45.70; H, 3.93; N, 6.51.

Results

Synthesis and Characterization Macrobicyclic Calix[4]arene Ligands

The macrobicyclic calix[4]arene based ligands HL^1 and HL^2 were prepared by building upon a known calix[4]arene platform. HL^1 and HL^2 were prepared in 86 % and 17 % yield respectively under high dilution conditions. The difference in yields reflects the much higher degree of flexibility and size of the dithioether ring of HL^2 as compared to HL^1. Both HL^1 and HL^2 have the lower rim substituted with *n*-propoxy groups in order to inhibit interconversion between the four possible conformations of the calix[4]arene (cone, partial cone, 1,3-alternate and 1,2-alternate). Once the cone conformation is isolated, the bulky nature of the propoxy groups insures retention of the desired cone conformation throughout the synthetic sequence. The ^{1}H NMR spectra of HL^1 and HL^2 show an AB quartet for the methylene Ar*CH$_2$*Ar protons, consistent with calix[4]arenes in the cone conformation.[7] and the Ar*CH$_2$*Ar peak at ~ δ 31 ppm in the ^{13}C NMR spectrum is also indicative of the cone conformation.[18]

HL¹ **HL²**

Synthesis and Characterization of the Metalloreceptors [Pd(L¹)(MeCN)][BF₄] and [Pd(L²)(MeCN)][BF₄]

Palladation of HL¹ and HL² employing [Pd(CH$_3$CN)$_4$][BF$_4$]$_2$ in acetonitrile solution yielded the metalloreceptors Pd(L¹)(MeCN)][BF$_4$] and [Pd(L²)(MeCN)][BF$_4$] in 93 and 96 % yield respectively. Both complexes were isolated as pale yellow solids that could be recrystallized from CH$_3$CN/diethyl ether solutions. All spectroscopic and analytical data are consistent with palladation and a formula, [Pd(L)(MeCN)][BF$_4$]. Evidence of metallation is observed in the ^{1}H NMR spectra as benzylic CH$_2$S protons are shifted downfield and broadened compared to that for the free ligand HL. A strong ion peak for [Pd(L)]$^+$ was also observed in the LSI mass spectrum for both compounds.

Interestingly, the ^{1}H NMR spectrum of [Pd(L¹)(MeCN)][BF$_4$] in CD$_3$NO$_2$ shows the coordinated MeCN group is shifted significantly *upfield* to δ -1.80 ppm. This is consistent with the methyl group of the MeCN ligand being trapped inside the calix[4]arene cavity.

Synthesis and Characterization of [Pd(L¹)(py)][BF₄] and [Pd(L²)(4-Phpy)][BF₄]

Examination of CPK and computer generated models suggested that pyridine (py) would be a suitable substrate for [Pd(L¹)(MeCN)]$^+$ since formation of a Pd-N bond might orient the bound aromatic moiety *into* the hydrophobic cavity provided by the calix[4]arene. Similarly, it was estimated that since [Pd(L²)(MeCN)]$^+$ contains a much larger cavity, 4-phenylpyridine (4-Phpy) might be a good candidate for a substrate that could interact with both the Pd coordination site and the cavity of the

[Pd(L¹)(MeCN)]⁺   [Pd(L²)(MeCN)]⁺

calix[4]arene. In this case, additional π-π stacking interactions could also occur between the pyridine ring and the aromatic spacer of the metalloreceptor if the phenyl substituent were indeed oriented into the calixarene cavity.

Both [Pd(L¹)(py)]⁺ and [Pd(L²)(4-Phpy)]⁺, were easily prepared in quantitative yield by displacing the labile acetonitrile ligand with one equivalent of the substrate molecule. For both adducts, ¹H NMR spectral data exhibited i) complexation shifts

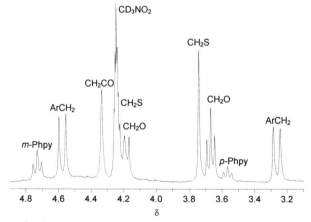

Figure 1. ¹H NMR spectrum (300 MHz, CD₂Cl₂/CD₃NO₂ (3:1)), in the region δ 3.2 - 4.8 ppm identifying the upfield shifted resonances, m-Phpy and p-Phpy, indicative of binding 4-Phpy inside the calix[4]arene cavity.

for the pyridine protons consistent with binding to the palladium center via σ-donation[3] and ii) distinct features indicative of substrate interaction with the calix[4]arene moiety. The most dramatic effects were observed for [Pd(L^2)(4-Phpy)]$^+$ as seen in Fig. 1 The protons of the aromatic spacer on the metalloreceptor are shifted *upfield* approximately 0.8 ppm, indicative of a π-stacking interaction[19] while a dramatic change is observed for the *meta* and *para* protons on the phenyl ring of the 4-Phpy substrate. Resonances at 3.59 and 4.71 ppm belong to the aromatic protons of the phenyl group of 4-Phpy. These protons are shifted *upfield* by 3.90 and 2.78 ppm relative to free substrate. These very large shifts almost certainly arise from inclusion of the phenyl ring *inside* the calix[4]arene cavity.[20]

For [Pd(L^1)(py)]$^+$, the pyridine protons are shifted downfield as a result of coordination but the only change to the calixarene portion of the molecule is a splitting of the calixarene aromatic protons on the substituted rings (labeled *a* in Fig 2). This splitting is consistent with an asymmetric coordination of the pyridine unit and suggests that py is too large for the receptor cavity. That is, in solution a conformation is adopted which coordinates py to the Pd center but *not* inside the

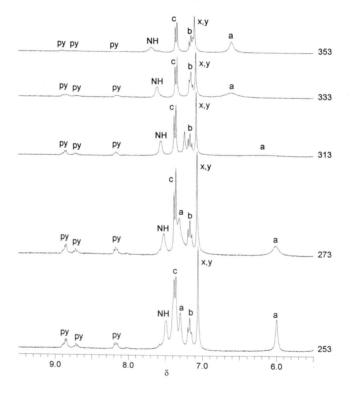

Figure 2. Variable temperature ^{1}H NMR spectra of [Pd(L^1)(py)]$^+$ in CD$_3$NO$_2$.

calixarene cavity. A variable temperature ^{1}H NMR study on this compound, shown in Fig 2, demonstrates that the asymmetry observed at room temperature disappears with increasing temperature until a symmetrical spectrum is observed. This is indicative of a fluxional process in which the pyridine moves "in and out of" or "through" the cavity. It should be noted that when the bound ligand is small such as in the complex [Pd(L^1)(Cl)], only a single symmetrical species is observed which is not temperature dependent.[16] Fig 3 shows results of a molecular mechanics study on [Pd(L^1)(py)]$^+$, which infers that the most stable conformation is that in which the pyridine ligand is oriented outside the cavity consistent with the described NMR spectroscopic results. The symmetrical high temperature structure observed would simply be an average of the two conformations in which the pyridine is coordinated to Pd but oriented outside the cavity. The lack of upfield shifts similar to that observed for [Pd(L^2)(4-Phpy)]$^+$, is also evidence for this dynamic behavior and positioning of the ligand outside the cavity.

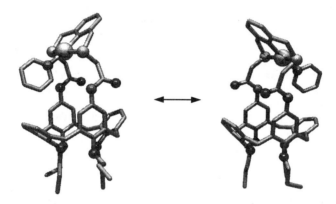

Figure 3. Although pyridine coordinates to the palladium centre it is too large for the calix[4]arene cavity. ^{1}H NMR experiments are consistent with a fluxional process in which the substrate moves through the cavity but prefers to reside outside due to the poor fit.

Molecular Recognition of Phenyl-Substituted Pyridines

Since solution and modeling studies show that 4-Phpy is an excellent fit for the receptor [Pd(L^2)(MeCN)], it was reasoned that this complimentarity could be used to effect molecular recognition of this species over related phenylpyridine compounds. Accordingly, competition studies were performed between 4-Phpy and 2-Phpy or 3-Phpy. In separate reactions, one equivalent of [Pd(L^2)(MeCN)]$^+$ was mixed with one equivalent of 4-Phpy and one equivalent of either 2-Phpy or 3-Phpy. The ^{1}H NMR spectra were recorded and the ratio of complexed to free 4-Phpy determined by integration of the *p*-Ph proton. As a comparison, the same competition experiments

[Pd(L³)(MeCN)]⁺

were carried out using [Pd(L³)(MeCN)]⁺, as a model receptor containing no calixarene unit. The results are summarized in Table 1. A comparison of the results for the model receptor with those for the metalloreceptor [Pd(L²)(MeCN)]⁺ provides a measure of the ability of [Pd(L²)(MeCN)]⁺ to selectively bind 4-Phpy. Thus, [Pd(L²)(MeCN)]⁺ exhibited a selectivity of *4x* for 4-Phpy over 2-Phpy as the ratio of 4-Phpy:2-Phpy changed from 50:50 (1:1) for the model to 80:20 (4:1) for the metalloreceptor and a remarkable selectivity of *36x* for 4-Phpy over 3-Ph-py as the ratio of 4-Phpy:3-Phpy reversed from 4:96 (1:24) for the model to 60:40 (3:2) for the receptor. Molecular mechanics calculations were performed for all three isomers in this study. Fig. 4 shows how extreme the structural differences are and lends further support to the idea that the selection of 4-phenylpyridine must be attribute to the encapsulation of the 4-substituted phenyl ring *inside* the cavity of the calix[4]arene.

Discussion

The nature of the binding sites in these new metalloreceptors was investigated by ¹H NMR spectroscopy and molecular mechanics calculations. For [Pd(L²)(4-Phpy)]⁺, the marked upfield shifts of the phenyl protons must surely result from shielding of these hydrogens by an aromatic group *inside* the calixarene cavity.[20] This type of non-covalent interaction occurs as a result of the strong, oriented binding provided by the organopalladium center. In these new metalloreceptor, the Pd atom provides two important attributes: 1) it acts as a binding site for coordination of the substrate, and 2) it directs the substrate into the hydrophobic cavity of the calix[4]arene. The first-sphere coordination via σ-donation to the organopalladium center anchors the substrate in place. The non-covalent, second-sphere interactions provide a source for selectivity based on complimentary size and shape.

Table I. Competition Reactions Showing Molecular Recognition of 4-Phpy

Substrate Pairs	$[Pd(L^3)(MeCN)]^+$	$[Pd(L^2)(MeCN)]^+$	Selectivity[b]
4-Phpy / 2-Phpy	50/50[a]	80/20	4
4-Phpy / 3-Phpy	4/96	60/40	36

[a] Ratio of receptor bound substrates in a 1:1:1 mixture of the two competing substrates and the receptor. [b] The ratio for metalloreceptor $[Pd(L^2)(MeCN)]^+$ over the ratio for model receptor $[Pd(L^3)(MeCN)]^+$.

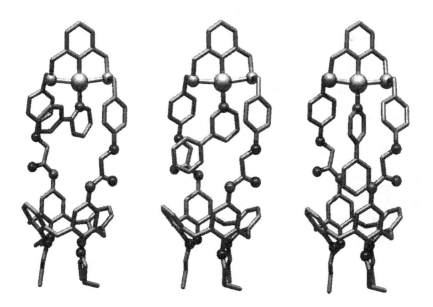

Figure 4. The orientations of coordinated 2-phenylpyridine, 3-phenylpyridine and 4-phenylpyridine differ significantly. Only the 4-phenylpyridine is positioned inside the cavity of the calix[4]arene giving rise to selectivity.

References

1. (a) van Staveren, C. J. M.; van Veggel, J. F. C.; Harkema, S.; Reinhoudt, D. N. *J. Am. Chem. Soc.* **1988**, *110*, 4994. (b) van Doorn, A. R.; Rushton, D. J.; van Straaten-Nijenhuis, W. F.; Verboom, W.; Reinhoudt, D. N. *Recl. Trav. Chim. Pays-Bas* **1992**, *111*, 421. (c) Reetz, M. T.; Niemeyer, C. M.; Hermes, M.; Goddard, M. *Angew. Chem., Int. Ed. Engl.* **1992**, *31*, 135. (d) Aoyama, Y.; Yamagishi, A.; Asagawa, M.; Toi. H.; Ogoshi, H. *J. Am. Chem. Soc.* **1988**, *110*, 4076. (e) Mizutani, T.; Ema, T.; Yoshida, T.; Kurado, Y.; Ogoshi, H. *Inorg. Chem.* **1993**, *32*, 2072. (f) Shionoya, M.; Kimura, E.; Shiro, S. *J. Am. Chem. Soc.* **1993**, *115*, 6730. (f) Shionoya, M.; Ikeda, T.; Kimura, E.; Shiro, S. *J. Am. Chem. Soc.* **1994**, *116*, 3848.
2. Kickham, J. E.; Loeb, S. J. *Inorg. Chem.* **1995**, *34*, 5656.
3. Kickham, J. E.; Loeb, S. J. *Inorg. Chem.* **1994**, *33*, 4351.
4. Kickham, J. E.; Loeb, S. J. *J. Chem. Soc., Chem. Commun.* **1993**, 1848.
5. (a) Kickham, J. E.; Loeb, S. J.; Murphy, S. L. *J. Am. Chem. Soc.* **1993**, *115*, 7031. (b) Kickham, J. E.; Loeb, S. J.; Murphy, S. L. *Chem. Eur. J.* **1997**, *3*, 1203.
6. Cameron, B. R.; Loeb, S. J. *J. Chem. Soc., Chem. Commun.* **1996**, 2003.
7. (a) Gutsche, C. D. *Calixarenes, Monographs in Supramolecular Chemistry*, vol. 1, ed. Stoddart, J. F. Royal Society of Chemistry; Cambridge, **1989**. (b) Vicens, J.; Böhmer, V. eds, *Calixarenes: A Versatile Class of Macrocyclic Compounds*; Kluwer Academic Publisher: Dordecht, **1991**. (c) Weiser, C.; Matt, D; Toupet, L.; Bourgeois, H.; Kintzinger, J. P. *J. Chem. Soc., Chem. Commun.* **1993**, 604. (d) Matt, D.; Loeber, C.; Vicens, J.; Asfari, Z. *J. Chem. Soc., Dalton Trans.* **1996**, 4041 and references therein.
8. (a) van Dienst, E.; Iwerna-Bakker, W. I.; Engbersen, J. F. J.; Verboom, W.; Reinhoudt, D. N. *Pure Appl. Chem.* **1993**, *65*, 387 and references therein. (b) Cobben, P. L. H. M.; Egberink, R. J. M.; Bomer, J. G.; Bergveld, P.; Verboom, W.; Reinhoudt, D. N. *J. Am. Chem. Soc.* **1992**, *114*, 10573. (c) Malone, J. F.; Marrs, D. J.; McKervey, M. A.; O'Hagan, P.; Thompson, N.; Walker, A.; Arnaud-Neu, F.; Mauprivez, O.; Schwing-Weill, M-J.; Dozol, J-F.; Rouquette, H.; Simon, N. *J. Chem. Soc., Chem. Commun.* **1995**, 2151.
9. (a) Scheerder, J.; Fochi, M.; Engbersen, J. F. J.; Reinhoudt, D. N. *J. Org. Chem.* **1994**, *59*, 7815. (b) Morzherin, Y.; Rudkevich, D. M.; Verboom, W.; Reinhoudt, D. N. *J. Org. Chem.* **1993**, *58*, 7602.
10. Beer, P. D.; Drew, M. G. B.; Hazelwood, C.; Hesek, D.; Hodacova, J.; Stokes, S. E.; *J. Chem. Soc., Chem. Commun.* **1993**, 229.
11. Xu, W.; Vittal, J. J.; Puddephatt, R. J. *J. Am. Chem. Soc.* **1993**, *115*, 6456.
12. (a) Kubo, Y.; Maruyama, S.; Ohhara, N.; Nakamura, M.; Tokita, S. *J. Chem. Soc., Chem. Commun.* **1995**, 1727. (b) van Loon, J-D.; Heida, J. F.; Verboom, W.; Reinhoudt, D. N. *Recl. Trav. Chim. Pays-Bas* **1992**, *111*, 353. (c) Gutsche, C. D.; See, K. A. *J. Org. Chem.* **1992**, *57*, 4527.
13. (a) Beer, P. D.; Chen, Z.; Goulden, A. J.; Grieve, A.; Hesek, D.; Szemes, F.; Wear, T. *J. Chem. Soc., Chem. Commun.* **1994**, 1269. (b) Giannini, L.; Solari, E.; Zanotti-Gerosa, A.; Floriani, C.; Chiesi-Villa, A.; Rizzoli, C. *Angew. Chem., Int. Ed. Engl.* **1996**, *35*, 2825. (c) Xu, B.; Carroll, P. J.; Swager, T. M. *Angew.*

Chem., Int. Ed. Engl. **1996**, *35*, 2094. (d) Giannini, L.; Solari, E.; Zanotti-Gerosa, A.; Floriani, C.; Chiesi-Villa, A.; Rizzoli, C. *Angew. Chem., Int. Ed. Engl.* **1996**, *35*, 85. (e) Stolmar, M.; Floriani, C.; Chiesi-Villa, A.; Rizzoli, C. *Inorg. Chem.* **1997**, *36*, 1074.
14. (a) Atwood, J. L.; Orr, G. W.; Bott, S. G.; Robinson, K. D. *Angew. Chem., Int. Ed. Engl.* **1993**, *32*, 1093. (b) McKervey, M. A.; Seward, E. M.; Ferguson, G.; Ruhl, B. L. *J. Org. Chem.* **1986**, *51*, 133. (c) Andreetti, G. O.; Ungaro, R.; Pochini, A. *J. Chem. Soc., Chem. Commun.* **1979**, 1005. (d) Gutsche, C. D.; Bauer, L. J.; *J. Am. Chem. Soc.* **1985**, *107*, 6052. (e) Bott, S. G.; Coleman, A. W.; Atwood, J. L. *J. Am. Chem. Soc.* **1988**, *110*, 610.
15. Rudkevich, D. M.; Verboom, W.; Reinhoudt, D. N. *J. Org. Chem.* **1994**, *59*, 3683.
16. B. R. Cameron, S. J. Loeb, G. P. A. Yap, *Inorg. Chem.* **1997**, *36*, 5498.
17. Arduini, A.; Fabbi, M.; Mantovani, M.; Mirone, L.; Pochini, A.; Secchi, A.; Ungaro, R. *J. Org. Chem.* **1995**, *60*, 1454 and references therein.
18. Jaime, C.; de Mendoza, J.; Prados, P.; Neito, P. M.; Sanchez, C. *J. Org. Chem.* **1991**, *56*, 3372.
19. (a) Hunter, C. A.; Sanders, J. K. M.; *J. Am. Chem. Soc.* **1990**, *112*, 5525. (b) Burley, S. K.; Petsko, G. A. *J. Am. Chem. Soc.* **1986**, *108*, 7995.
20. Komoto, T.; Ando, I.; Nakamoto, Y.; Ishida, S-I. *J. Chem. Soc., Chem. Commun.* **1988**, 135.

Chapter 22

Synthesis and Structural Analysis of Thiacalixarene Derivatives

Mir Wais Hosseini

Laboratoire de Chimie de Coordination Organique,
Université Louis Pasteur, Institut Le Bel, Strasbourg, France

A new family of calixarenes composed of thiacalix[4]arene derivatives, in which the methylene junctions between the aromatic moieties were replaced by sulfur atoms, was developed. Partial, as well as complete oxidation of the sulfur atoms to the tetrasulfoxide and tetrasulfone derivatives respectively was achieved. All compounds were characterised by single-crystal X-ray diffraction. The coordination ability of thiacalix[4]arene derivatives towards transition, metals was explored. Doubly fused calix dimers were obtained in the presence of copper, zinc and cobalt and characterised by X-ray diffraction techniques. In the case of copper complex, a tetranuclear core holding the two calix units with strong anti ferromagnetic coupling between metallic centres was obtained.

Introduction

Molecular solids are defined by the chemical nature of their molecular components and by their interactions with respect to each other in the crystalline phase. With our present level of knowledge the complete understanding, and therefore prediction, of all intermolecular interactions in the crystalline phase seems unreachable (*1*). However, using appropriate molecular modules, one may predict some of the inter-component's interactions (*2*). Molecular networks are defined in the solid state as supramolecular structures composed of an infinite number of molecules interacting with each other in a specific manner (*3*). Whereas molecules are described as assemblies of atoms interconnected by covalent bonds, by analogy, molecular networks are hypermolecules in which the connectivity between elementary components (molecular modules) is ensured by non-covalent inter-component interactions (*4*). Thus, the design of molecular networks requires both molecular recognition between molecular modules composing the crystal and iteration of the recognition processes.

In terms of the energy of interactions between molecular modules governing the recognition and assembling processes, in addition to hydrogen

(*5,6*) and coordination (*7,8*) bonds, we proposed some time ago to use rather weak van der Waals interactions responsible for the inclusion processes between concave and convex molecules (*9,10*).

Design of Koilands

The chemistry of inclusion complexes which is based on concave and convex molecules, *i.e.* the inclusion of a substrate within the cavity of a receptor molecule, is an established area. We have extended the concept of inclusion in solution to the construction of inclusion networks in the solid state. Koilands (*9,10*) (from Greek *koilos* : hollow) were defined as multicavity receptor molecules composed of at least two cavities arranged in a divergent fashion. Since each individual cavity offers the possibility of forming inclusion complexes with convex molecules, fusing two or more such cavities in the presence of appropriate rigid connector leads to non-covalently assembled polymeric species which was referred to as koilate. For example, in the solid state linear *koilates* (one dimensional linear molecular arrays or a-networks) or linear inclusion networks were obtained using non-covalent van der Waals interactions between rigid and compact direceptors possessing two divergent cavities with an angle of 180° between them (linear *koiland*) and a linear connector, possessing two extremities capable each to be included within the cavities of the direceptor (Fig. 1).

For the design of koilands, calix[4]arene derivatives (*11*) appeared to be candidates of choice (Fig. 2). Indeed, these compounds offer a preorganised and tuneable hydrophobic pocket surrounded by four aryl moieties as well as four hydroxy groups for further functionalisation.
Furthermore, both the entrance and the depth of the preorganised cavity of calixarenes may be controlled by the nature of the substituent R at the *para* position (Scheme 1), *i.e.* C(CH$_3$)$_3$ (**1**), H (**2**), CH$_3$ (**3**), Ph (**4**). Whereas for both parent compounds **1**, **2** and **4** the cone conformation was established in the solid state by X-ray studies, to our knowledge, for compound **3** no X-ray data is available so far. The basket-type cavity of calix[4]arene derivatives such as **1** resulting from the cone conformation has been shown to accommodate in the solid state a variety of neutral guests (*11*). The design of koilands was based on the fusion of two calix[4]arenes derivatives in the cone conformation by two silicon atoms (Fig. 2) (*10*). Examples of fused calix[4]arene using other metals have been reported (*12*).

Design of New Koilands

Calixarenes are among the most used macrocyclic frameworks. The increasing interest over the last two decades in this class of molecules results, on one hand, from the ease and reproducibility of synthetic procedures reported by

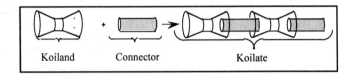

Figure 1 : *schematic representation of the formation of a centrosymmetric koilate by centrosymmetric koilands and connectors.*

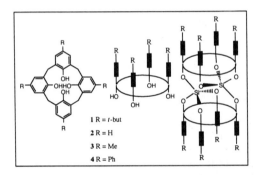

Figure 2 : *Calix[4]arene derivatives (left) and schematic representations of their cone conformation (middle) and of linear koilands obtained by fusion of two such units by two silicon atoms (right).*

Gutsche *et al* (*13*) and on the other hand, from synthetically achievable structural as well as functional modifications. Compounds such as **1** and **2** mainly bind metal cations *via* the OH groups (*12*), although in some cases the contribution of the aromatic π clouds was demonstrated (*14*).

However, the replacement of CH_2 junctions by S atoms in **5** and **6**, introduces some additional coordination sites and, due to the size of sulphur atom, modifies the dimension of the cavity of the calixarene. Thus, using thiacalixarenes such as **5** and **6**, we thought we might extend our original design of koilands, which was based on the fusion of two calix units by two silicon atoms, to transition metals as connecting points. In particular, using paramagnetic metals, one may design magnetically active koilands which may lead to magnetic inclusion networks. Recently an efficient procedure for preparing thiacalixarenes **5** (*15*) and **6** (*16*) was reported (scheme 1). Whereas **5** (*15*) was obtained upon direct treatment of *p-tert*-butylphenol by S_8, **6** (*16*) was obtained by aluminium chloride de-*tert*-butylation of **5** using the procedure developed for the synthesis of calix[4]arene **2**. Furthermore, partial oxidation of the thio junctions affording the tetrasulfoxide derivatives **7** and **8** (*17,18*) or complete oxidation of the sulfur atoms leading to the tetrasulfone derivative **9** (*19*) have been recently achieved. The O-alkylation of **5** leading to the methylether derivative **10** (*19*) or to the ester compound **12** (*20-22*) has been also recently reported. The methyether tetrasulfone **11** was also reported (*19*). Dealing with other sulfur containing calixarenes, one may mention the synthesis of di- (**14**) (*23*) and tetra-(**15**) mercaptocalix[4]arenes (*24,25*) in which two or four OH groups were replaced by SH groups. Another example of tetrathiacalix[4]arene based on four thiophen moieties has been published (*26*). Recently an example of substitution of CH_2 groups by Me_2Si was also reported (*27*).

Structural analysis of thiacalix[4]arenes

The inclusion ability of **5** towards small solvent molecules such as CH_2Cl_2, $CHCl_3$ and MeOH was investigated by single-crystal X-ray diffraction (*16*). For both CH_2Cl_2 (Fig. 3a) and $CHCl_3$ (Fig. 3b) suitable crystals were obtained upon slow liquid-liquid diffusion of MeOH into a dichloromethane or chloroform solution of **5** respectively, whereas for the methanol inclusion complex, crystals were grown by slow liquid-liquid diffusion of MeOH into a *p*-xylene solution of **5** (Fig. 3c).

In all three cases, the following common structural features were observed : i) the thiacalix **5** adopts a cone conformation leading to inclusion complexes with all three guest molecules; ii) in all three cases, the substrate penetrates deeply into the cavity of the calix; iii) both CH_2Cl_2 and $CHCl_3$ substrates as well as the *p-tert*-butyl groups were found to be disordered in the crystalline phase; iv) the average distance between two adjacent oxygen atoms was 2.85 Å. Although one can not exclude the stabilizing role of the included substrate, the rather short distance between adjacent oxygen atoms for all three structures may be responsible for the existence of an intramolecular H-bonds array stabilizing the cone conformation. In the case of **1** (average distance

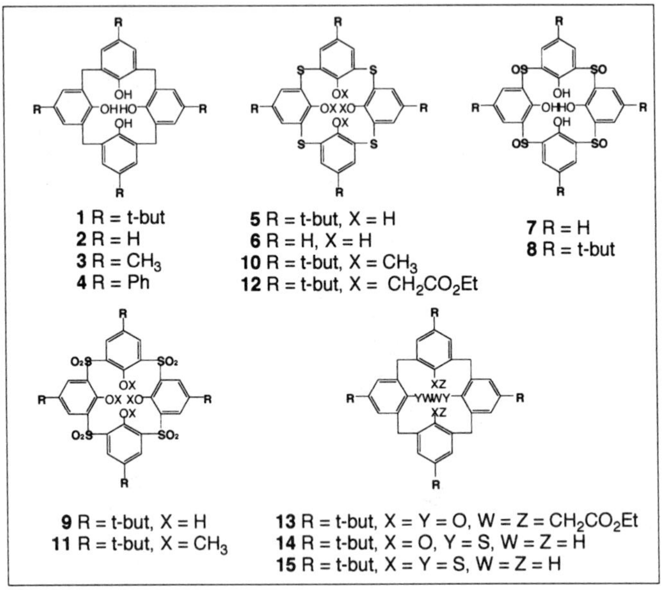

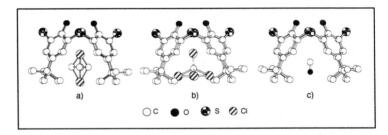

Scheme 1

Figure 3 : *Lateral views of the crystal structures of the inclusion complexes formed by compound 5 with by CH_2Cl_2 (a), $CHCl_3$ (b), and MeOH (c). For clarity, H atoms are not represented. Both CH_2Cl_2 and $CHCl_3$ substrate as well as the p-tert-butyl groups of the receptor molecule were found to be disordered in the crystalline phase.*

between two adjacent oxygen atoms of *ca* 2.70 Å), the same argument has been previously employed to justify the cone conformation at low temperature and in the solid state *(11)*. The dimension of the thiacalix **5** was found to be, as expected, larger than for the parent calix **1**.

In the case of dichloromethane inclusion complex, the crystal (tetragonal crystal system with P4/n as the space group) is formed, in a centrosymmetric mode, by alternate columns composed of inclusion complexes (Fig. 4a). Interestingly, within each infinite column, the inclusion complexes were packed one on the top of the other leading thus to a distance of 3.31 Å between the chlorine atom of the substrate pointing towards the exterior of the cavity and all four oxygen atoms belonging to the consecutive thiacalix unit (Fig. 4b). For the other two cases, the same type of packing was observed.

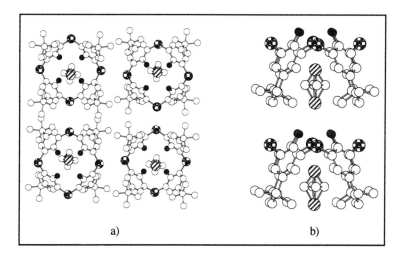

Figure 4 : *Portions of the structure of the inclusion complex formed by compound* **5** *with CH$_2$Cl$_2$. The top view (a) shows the alternate packing of columns composed of the inclusion complexes. The lateral view (b) shows the packing of complexes within the same column. For clarity, H atoms are not represented.*

The solid state structure of **6** was also investigated by X-ray analysis *(16)*. Suitable single-crystals were obtained upon slow liquid-liquid diffusion of hexane into a chloroform solution of **6**. The study revealed the following features (Fig. 5): i) **6** crystallized in the hexagonal crystal system with P6$_3$/m as the space group and the unit cell contained both **6** and 2 water molecules which were not localized within the cavity of the host molecule; ii) **6** adopts a cone conformation; iii) the average C-S and C-O distances were ca 1.78 Å and 1.37 Å respectively; iv) the average distance between two adjacent sulfur atoms was 5.51 Å; v) the average distance between two adjacent oxygen atoms was 2.64 Å which may again be responsible for the formation of intramolecular H-bonded array; vi) in marked contrast with **5** which was found to form inclusion complexes with small molecules (see above), **6** forms, by self-inclusion, trimeric units (Fig. 5). In the case of calix[4]arene **2**, the same particular behaviour was observed. Suitable monocrystals of **2** were grown from CH_2Cl_2 solution and studied by X-ray diffraction (Fig. 5) (*28*). Crystals (hexagonal crystal system with P6$_3$/m as the space group) were composed of **2** and CH_2Cl_2 molecules. Although in the lattice CH_2Cl_2 molecules were present, none of them were located within the cavity of the calix adopting a cone conformation. Interestingly, again similar formation of trimeric inclusion complex was observed. The formation of the same type of trimeric complex in the solid state has been previously observed with calix[4]arene acetone clathrates (*29*). These observations may be rationalized in the following manner. In the case of compounds **1** and **5**, due to the presence of bulky groups at the upper rim, self-inclusion seems to be unfavourable. Consequently, both compounds form inclusion complexes with appropriate substrates. In marked contrast, in the case of both compounds **2** and **6**, since the formation of self-inclusion complexes may not be excluded for steric reasons, both compounds form trimeric complexes by self-inclusion and therefore, no inclusion complexes with small guests may take place.

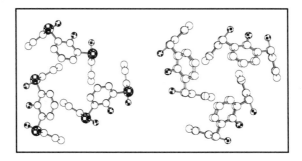

Figure 5 : *Crystal structures of the compounds* **6** *(left) and* **2** *(right) : lateral views of the trimeric inclusion complex. For clarity, solvent molecules and H atoms are not represented.*

Synthesis and structural analysis of tetrasulfinylcalix[4]arenes

The oxidation of sulfur atoms in **5** and **6** would generate either sulfoxide or sulfone type derivatives which may show extended coordination ability. The partial oxidation of the sulfur atoms in **5** and **6** leads to new sulfinyl type ligands **7** and **8** which are based on four SO groups linking the phenolic groups (Scheme 1). Compounds **7** and **8**, obtained by partial oxidation using H_2O_2 in glacial acetic acid and *m*-chlorobenzoic acid in CH_2Cl_2 of the precursors **5** and **6** respectively, were structurally characterized in the crystalline phase by X-ray crystallography on single-crystals (*18*).

Although at the earlier stage of development of calixarenes, their conformational lability appeared as a severe limitation, over the last two decades, due to considerable synthetic as well as structural achievements, the conformational issue is currently positively explored in the design of receptors, catalysts, and building blocks. For the tetrasulfoxide calix **7** and **8**, in addition to classical conformational mobility, due to the relative orientation of the oxygen atoms and consequently the lone pairs of the sulfur atoms an impressive set of isomers may be obtained (*18*). Indeed, based on the relative orientation of the aromatic moieties one may envisage four limit conformers (Cone (C), Partial cone (PC), 1,2- and 1,3- alternate) as in the case of the parent compounds **1, 2, 5** and **6**. On the other hand, for each sulfoxide group the oxygen atom may occupy either axial (a) or equatorial (e) position with axial-equatorial exchange by a flip-flop process. It is worth noting that, whereas the four conformers may interconvert, as it was demonstrated in the case of compounds **1, 2, 5** and **6**, the isomers defined by the relative orientation of the SO groups can not be interconverted without breaking covalent S-C bonds.

The solid state structure of **7** was investigated by X-ray analysis (Fig. 6). Compound **7** crystallized in the tetragonal crystal system with $P\ 4_2/n$ as the space group. **7** adopts a 1,3-alternate conformation for which all four oxygen atoms of the SO groups are oriented above and below the main plane defined by four S atoms leading to the aeae isomer stabilized by four strong hydrogen bonds with a OH···OS bond distance of 2.64 Å between the OH and SO groups. Interestingly, the packing pattern showed a 3-D network obtained by strong stacking interactions between all four aromatic groups of the molecular units (the distance between centroids of the aromatic rings was 3.49 Å).

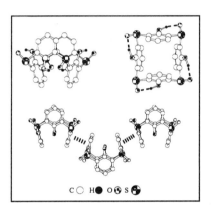

Figure 6 : *Lateral (left) and top (right) views of the solid state structure of **7** showing the 1,3-alternate conformer of the calix unit and the alternate orientation of the SO groups (1,3aeae), as well as a portion of the structure (bottom) showing the formation of a 3-D network through stacking of the aromatic groups (3.49 Å). For clarity H atoms are not represented.*

The solid state structure of **8** was also investigated by X-ray analysis (Fig. 7). Compound **8** crystallized in the monoclinic crystal system with $P\ 2_1/n$ as the space group. Again, **8** adopts a 1,3-alternate conformation with all four oxygen atoms of the SO groups (d_{SO} = 2.62-2.72 Å) oriented above and below the main plane defined by four S atoms leading again to the aeae isomer again stabilized, as in the above mentioned case of **7**, by four strong intramolecular hydrogen bonds with an average OH···OS bond distance of 2.67 Å between the OH and SO groups. In marked contrast with compound **7** which formed a 3-D network based on stacking interactions between the aromatic groups, in the case of **8**, due to the presence of bulky *tert*-butyl groups no such a network was observed.

Synthesis and structural analysis of tetrasulfonylcalix[4]arenes

The complete oxidation of the thiacalixarene derivatives into their sulfone analogues was achieved so far only in the case of the thiacalixarene **5** (*19*). The synthetic strategy for the preparation of the tetrasulfone **9** was based on the complete oxidation of thiacalix **5**. In a first attempt, all four OH groups of **5** were protected as methoxy groups (compound **10**) prior to the oxidation of the sulfide linkages to sulfones. The complete oxidation of **10** to the tetrasulfone **11** was first attempted by refluxing the compound **10** in a 30% H_2O_2/glacial acetic acid mixture for eight days. Probably due to the low solubility of **10**, the reaction isolated yield was ca 10 %. The yield increased to 88 % when performing the oxidation with *m*-chloroperbenzoic acid. Since the removal of the protecting groups by treatment with a large excess of BBr_3 in CH_2Cl_2 afforded the final compound **9** in only ca 30 % yields, the direct oxidation of the tetrasulfide **5** was carried out in 59 % yield using 30% H_2O_2/glacial acetic acid.

In solution, the ^{1}H- and ^{13}C-NMR spectra for compounds **5-11** were extremely simple and in agreement with the proposed structure. In marked contrast with calix derivatives **1** and **2** for which both the ^{1}H- and ^{13}C-NMR signals corresponding to the methylene groups are usually used for conformational assignment in solution, due to the absence of such a NMR probes for compounds **5-11**, their conformation in solution could not be established unambiguously.

In the solid state, all three compounds **9-11** were studied by X-ray diffraction methods on single-crystals. In marked contrast with the tetramethoxy derivative of the parent compound **1** which was shown to adopt the partial cone conformation (*30*), and with the tetrathiacalix[4]arene parent compounds **5** and **6** which were present in cone conformation (*30*), the methyl protected thia analogue **10** was found to be in the 1,3-alternate conformation (Fig. 8).

For the sulfone derivatives **9** (Fig. 9), again the conformation was found to be 1,3-alternate with oxygen atoms of the sulfones pointing outwardly. The average SO and CS bond distance were *ca* 1.43 Å *ca* 1.77 Å respectively. The average distance between two adjacent sulfur atoms was *ca* 5.52 Å.

Based on the observed average O···O distance between two adjacent oxygen atoms of *ca* 2.70 Å, 2.85 Å and 2.64 Å for **1**, **2** and **5** respectively, as previously proposed, the cone conformation was, at least partially stabilized, by an array of intramolecular H-bonds. Interestingly, the 1,3-alternate conformation for compound **9** may be also rationalized in terms of H-bonding between the OH and SO_2 groups (Fig. 9). Indeed, O···OS distances varying from 2.81 to 3.02 Å were observed. Furthermore, compound **9** was found to form a 3-D network through intermolecular H-bonding between the OH and SO_2 groups belonging to adjacent units with an average O···OS distance of *ca* 2.87 Å.

305

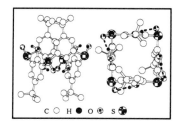

Figure 7 : *Lateral (left) and top (right) views of the solid state structure of **8** showing the 1,3-alternate conformer of the calix unit and the alternate orientation of the SO groups (1,3aeae).*

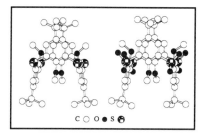

Figure 8 : *X-ray structure of **10** (left) and **11** (right) demonstrating the adopted 1,3-alternate conformation in the crystalline phase. For sake of clarity, the H atoms are not represented.*

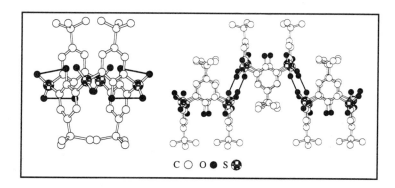

Figure 9 : *A portion of the X-ray structure of the 3-D network formed between consecutive compounds **9**. The 3-D network is obtained by intermolecular H-bonding between OH and SO groups. For sake of clarity, the H atoms are not represented.*

Functionalization of thiacalixarnes at the lower rim

The ionophoric properties of calix[4]arene derivatives bearing four ester groups have been established (*31*). For the reaction of **1** with ethyl bromoacetate, the role of alkalin metal cations, used as metal carbonates, on the isomers distribution of **13** has been studied (*32*). Using the same methodology, a systematic synthetic as well as structural studies dealing with the conformer distribution of compound **12** (Scheme 1) was achieved (*22*). The same study has been also reported by another group (*20,21*).

The role played by alkali metal cations in the synthesis as well as in the distribution of conformers of **12** was studied by treatment of tetrathiacalix[4]arene **5** in refluxing acetone by 8 eq. of ethylbromoacetate in the presence of 6 eq. of M_2CO_3. The highest overall yield of *ca* 76-80 % (all yields are based on isolated compounds, however, when the yields were less than *ca* 3 %, they were not considered) was obtained after 7 days. Except for the **12**$_{1,2-A}$ (1,2-A and 1,3-A refer to 1,2- and 1,3-alternate conformers, whereas C and PC refer to cone and partial cone conformers) isomer which was not isolated, the other three isomers were purified and structurally studied both in solution and in the solid state.

It has been reported that in the case of **1**, in the presence of alkali metal carbonate no trace of **13**$_{1,2-A}$ nor of **13**$_{1,3-A}$ conformers could be detected (32). The results obtained with thiacalix **12** were rather different. Indeed, although no **12**$_{1,2-A}$ isomer could be isolated, in marked contrast with **1**, up to 85 % of the **12**$_{1,3-A}$ isomer was obtained. When using Li_2CO_3, in agreement with the observation for **1** (*32*), no trace of the tetra substituted **12** was detected. Again, as observed for **1** (*32*), in the presence of Na_2CO_3 the **12**$_C$ conformer was exclusively formed although with higher overall yield of 80 %. In the case of K_2CO_3, in marked contrast with the reported results for **1** (96 and 3 % of **13**$_C$ and **13**$_{PC}$, respectively) (*32*), for **5**, although no **12**$_C$ conformer was isolated, in addition to 70 % of **12**$_{PC}$, 30 % of **12**$_{1,3-A}$ were formed. Finally, in the presence of Cs_2CO_3, again in marked contrast with **1** for which the **13**$_{PC}$ conformer was found to be exclusively formed, in the case of **5**, although the **12**$_{PC}$ conformer was obtained in 15 % yield, the major isomer obtained was the **12**$_{1,3-A}$ (85 % yield).

In solution, all three conformers isolated (**12**$_C$, **12**$_{PC}$ and **12**$_{1,3-A}$) were studied by both ^1H and ^{13}C NMR spectroscopy which showed sharp signals indicating the presence of conformationally blocked isomers. As expected for symmetry reasons, for both **12**$_C$ and **12**$_{1,3-A}$ conformers, the same number of signals as well as the same pattern were observed thus precluding precise assignments. On the other hand, for the **12**$_{PC}$ isomer, possessing a lower symmetry, both the proton as well as the carbon signals could be assigned. In order to precisely assign all three isomers, a structural analysis in the solid state based on X-ray diffraction on single-crystals was achieved. **12**$_C$ crystallized in a monoclinic form, space group P 1 2$_1$/n 1 (Fig. 10). No solvent molecules were present in the lattice. The average CS distance and CSC angle were 1.785 Å and 100.1°, respectively. The average C=O distance was 1.190 Å whereas the average C-O(arom) and C-O (ethyl) distances were found to be 1.376 and 1.326 Å, respectively.

12$_{PC}$ crystallized in monoclinic system with P1 2$_1$/n as the space group. In the unit cell two crystallographically slightly different molecules of 12$_{PC}$ as well as a CH$_2$Cl$_2$ molecule (not included in the cavity of the calix) were present. Only one of the two molecules are shown in Figure 10. The average CS distance and CSC angle were 1.784 Å and 100.5°, respectively. The average C=O distance was 1.191 Å whereas the average C-O(arom) and C-O (ethyl) distances were found to be 1.377 and 1.333 Å, respectively.

12$_{1,3-A}$ crystallized in monoclinic form, space group I 1 2/a 1 (Fig. 10). No solvent molecules were present in the lattice. The average CS distance and CSC angle were 1.777 Å and 102.25°, respectively. The average C=O distance was 1.169 Å whereas the average C-O (arom) and C-O (ethyl) distances were found to be 1.377 and 1.268 Å, respectively.

The use of thiacalixarenes in the synthesis of koilands

We have shown that inclusion based molecular networks may be formed in the solid state using hollow molecular units possessing at least two divergent cavities and connector molecules capable of undergoing inclusion processes (*10*). However, our approach was mainly oriented so far towards the understanding of structural aspects governing the formation of molecular inclusion networks. A further development, which remain to be achieved, is concerned with the design and preparation of functional inclusion molecular networks. Towards this goal, one may imagine that by using magnetically active building blocks, if the distance and orientation requirements were filled, magnetic networks could be obtained.

The replacement of CH$_2$ junctions by S atoms leading to thia calix derivatives **5** and **6** and their further partial oxidation to tetrasulfoxide derivatives **7** and **8** as well as their complete oxidation tetrasulfone **9** should enhance the number of coordination sites. As stated above, this feature may be exploited for the design new koilands and polynuclear transition metal complexes. In particular using paramagnetic transition metals such as Cu(II), a magnetically active koiland may be formed. This aspect was investigated using the tetrathiacalix **5** and Cu^{II} (*33*).

Upon heating a dimethylformamide solution of compound **5**, hydrated copper acetate and Et$_3$N, a dark blue-black, crystalline complex was obtained. The ^{1}H-NMR spectrum of a solution of the complex in CDCl$_3$ at 298 K showed only a single, broad resonance, indicating that the complex was paramagnetic. Under the same conditions, an electron spin resonance could not be detected and measurements of magnetic susceptibility (4 - 298 K) could be interpreted, in the light of the structural information described below, in terms of anti ferromagnetic coupling (J = -103±1 cm^{-1}) of four equivalent Cu(II) ions arranged in a square.

The solid state structure of the Cu$_4$(**5**)$_2$ complex was elucidated by X-ray crystallography on single-crystals. Two unequivalent Cu$_4$(**5**)$_2$ units with subtle differences are found within the unit cell. In both, the Cu$_4$ array, sandwiched between two thiacalix entities in a "cone" conformation similar to that of the free ligand (*16*), is close to exactly square, with each copper in a six-coordinate O$_4$S$_2$ donor-atom environment (Fig. 11). The coordination sphere is far from regular, with one Cu-O bond *ca* 0.5 Å longer than the other three and one Cu-S bond *ca* 0.3 Å longer than the other.

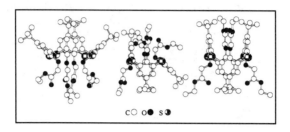

Figure 10 : *X-ray structure of 12_C (left), 12_{PC} (middle) and $12_{1,3-A}$ confortmers of 12.*

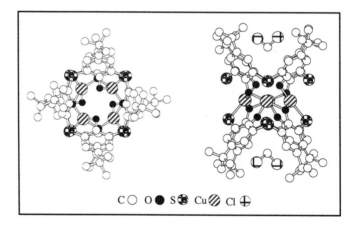

Figure 11 : *Top (left) and lateral (right) views of the crystal structure of the tetranuclear copper complex showing the square arrangement of the metal centres and their bridging by phenoxide groups (left) and the inclusion of solvent molecules (right). For clarity, H atoms and solvent molecules are not represented.*

Interestingly, the copper koiland, as in the case of silicon koilands, forms a binuclear inclusion complex with two molecules of CH_2Cl_2. Each one of the solvent molecule penetrates deeply into the cavity of the koiland (Fig. 11).

Other structures obtained using the same strategy showed that this mode of binding is certainly not restricted to Cu(II). For example, solid state structural analysis of the Zn(II) and Co(II) complexes formed with the tetrathiacalix **5** revealed the formation of koilands in which the two calix units are fused by three metal centres (Fig. 12). In both cases, again, both cavities were occupied by solvent molecules demonstrating the inclusion ability of the new koilands (*34*).

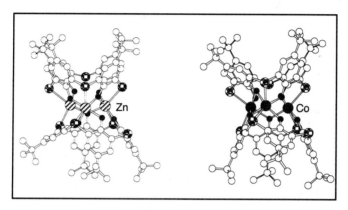

***Figure 12** : lateral views of the solid state structure of the trinuclear zinc (left) and cobalt (right) complexes possessing two cavities occupied each by one CH_2Cl_2 molecule. For clarity, H atoms are not represented.*

Conclusion

In conclusion, pursuing our investigations on hollow molecular building blocks which may be used for the design of solid state inclusion networks, the synthesis as well as structural features of a new class of calixarenes based on sulfide (S), sulfoxide (SO) and sulfone (SO_2) groups connecting the aromatic moieties were investigated. It was demonstrated that upon addition of supplementary coordination sites (sulfur atoms) within the framework of calix[4]arene derivatives, a rather rich coordination chemistry could be opened. In particular, a variety of thiacalix dimers were obtained with transition metal cations, thus opening the way to the possible design of magnetically active koilands and koilates. The formation of inclusion molecular networks in the crystalline phase using doubly fused thiacalixarenes by a large number of transition metals and a variety of connectors us currently under investigation. Furthermore, the use thiacalixarenes (S) as well as partially (SO) or completely (SO_2) derivatives in the design of new receptor, catalysts as well as extractants is under currently under study

References

1. A. Gavezzotti, *Acc. Chem. Res.*, **1994**, *27*, 309.
2. Desiraju, G. R., *Angew. Chem. Int. Ed. Engl.*, **1995**, *34*, 2311.
3. Dunitz, J. D., *Pure & Appl. Chem.*, **1991**, *63*, 177; Lehn, J.-M., in *Supramolecular Chemistry, Concepts and Perspectives,* VCH, Weinheim, 1995.
4. Etter, M.C., *Acc. Chem. Res.*, **1990**, *23*, 120; Whitesides, G.M.; Mathias, J.P.; Seto, T., *Science*, **1991**, *254*, 1312; Fowler, F. W.; Lauher, J. W., *J. Amer. Chem. Soc.*, **1993**, *115*, 5991; Delaigue, X., Graf, E.; Hajek, F.; Hosseini, M. W.; Planeix, J.-M., in *Crystallography of Supramolecular Compounds*, G. Tsoucaris, J. L. Atwood, J. Lipkowski, Eds., NATO ASI, Kluwer, **1996**, *C480*, 159; Brand, G.; Hosseini, M. W.; Félix, O.; Schaeffer, P.; Ruppert, R., in *Magnetism a Supramolecular Function*, O. Kahn, Ed., NATO ASI, Kluwer, **1996**, *C484*, 129..
5. Aakeröy, C. B.; Seddon, K. R.; *Chem. Soc. Rev.*, **1993**, *22*, 397; Subramanian, S.; Zaworotko, M. J., *Coord. Chem. Rev.*, **1994**, *137*, 357; Braga, D.; Grepioni, F., *Acc. Chem. Res.*, **1994**, *27*, 51; Russell, V. A.; Ward, M. D., *Chem. Mater.*, **1996**, *8*, 1654; Simard, M.; Su, D.; Wuest, J.D., *J. Amer. Chem. Soc.*, **1991**, *113*, 4696; Lawrence, D. S.; Jiang, T.; Levett, M., *Chem. Rev.*, **1995**, *95*, 2229; Philp, D.; Stoddart, J. F., *Angew. Chem. Int. Ed. Engl.*, **1996**, *35*, 1155.
6. Brand, G.; Hosseini, M. W.; Ruppert, R.; De Cian, A.; Fischer, J.; Kyritsakas, N., *New J. Chem.*, **1995**, *19*, 9; Hosseini, M. W.; Ruppert, R.; Schaeffer, P.; De Cian, A.; Kyritsakas, N.; Fischer, J., *J. Chem. Soc. Chem. Comm.*, **1994**, 2135; Hosseini, M. W.; Brand, G.; Schaeffer, P.; Ruppert, R.; De Cian, A.; Fischer, J., *Tetrahedron Lett.* **1996**, *37*, 1405; Félix, O.; Hosseini, M. W.; De Cian, A.; Fischer, J., *Angew. Chem. Int. Ed. Engl.*, **1997**, *36*, 102; Félix, O.; Hosseini, M. W.; De Cian, A.; Fischer, J., *Tetrahedron Lett.* **1997**, *38*, 1755; Félix, O.; Hosseini, M. W.; De Cian, A.; Fischer, J., *Tetrahedron Lett.* **1997**, *38*, 1933; Félix, O.; Hosseini, M. W.; De Cian, A.; Fischer, J., *New J. Chem.* **1997**, *21*, 285; Félix, O.; Hosseini, M. W.; De Cian, A.; Fischer, J., *New J. Chem.*, **1998**, *22*, 1389.
7. R. Robson in *Comprehensive Supramolecular Chemistry*, Eds; J. L. Atwood, J. E. D. Davies, D. D. Macnicol, F. Vögtle, Pergamon, Vol. 6 (Eds. D. D. Macnicol, F. Toda, R. Bishop), 1996, p. 733.
8. Kaes, C.; Hosseini, M. W.; Rickard, C. E. F.; Skelton, B. W.; White, A. H.; *Angew. Chem. Int. Natl. Ed. Engl.* **1998**, *37*, 920; Mislin, G.; Graf, E.; Hosseini, M. W.; De Cian, A.; Kyritsakas, N.; Fischer, J., *J. C. S. Chem Comm.*, **1998**, 2545; Loï, M.; Graf, E.; Hosseini, M. W.; De Cian, A.; Fischer, J., *J. C. S., Chem. Comm.*, **1999**, 603.
9. Delaigue, X.; Hosseini, M. W.; De Cian, A.; Fischer, J.; Leize, E.; Kieffer, S.; van Dorsselaer, A., *Tetrahedron Lett.*, **1993**, *34*, 3285.Delaigue, X.; Hosseini, M. W.; Leize, E.; Kieffer, S.; Van Dorsselaer, A., *Tetrahedron Lett.*, **1993**, *34*, 7561; Delaigue, X.; Hosseini, M. W.; Graff, R.; Kintzinger, J.-P.; Raya, J., *Tetrahedron Lett.*, **1994**, *35*, 1711; Hajek, F.; Graf, E.; Hosseini, M. W., *Tetrahedron*

Lett., **1996**, *37*, 1409; Hajek, F.; Graf, E.; Hosseini, M. W.; Delaigue, X.; De Cian, A.; Fischer, J., *Tetrahedron Lett.* **1996**, *37*, 1401; Hajek, F.; Graf, E.; Hosseini, M. W.; De Cian, A.; Fischer, J., *Tetrahedron Lett.*, **1997**, *38*, 4555.; Hajek, F.; Graf, E.; Hosseini, M. W.; De Cian, A.; Fischer, J., *Angew. Chem. Int. Ed. Engl.*, **1997**, *36*, 1760; Martz, J.; Graf, E.; Hosseini, M. W.; De Cian, A.; Kyritsakas, N.; Fischer, J., *J. Mat. Chem.*, **1998**, *8*, 2331; Hajek, F.; Graf, E.; Hosseini, M. W.; De Cian, A.; Fischer, J., *Crystal Engineering*, **1998**, *1*, 79.

10. Hosseini, M. W.; De Cian, A., *J. C. S., Chem. Comm.*, **1998**, 727;
11. C. D. Gutsche, *"Calixarenes"*, Monographs in Supramolecular Chemistry, Ed. J. F. Stoddart, R.S.C., London, 1989; C. D. Gutsche, *"Calixarenes revisited"*, Monographs in Supramolecular Chemistry, Ed. J. F. Stoddart, R.S.C., London, 1998; *Calixarenes. A Versatile Class of Macrocyclic Compounds*, Ed. J. Vicens and V. Böhmer, Kluwer Academic Publishers, Dordrecht, 1991; V. Böhmer, *Angew. Chem. Int. Ed. Engl.*, **1995**, *34*, 713.
12. Wieser, C.; Dieleman, C.; Matt, D., *Coord. Chem. Rev.*, **1997**, *165*, 93; Olmstead, M. M.; Sigel, G.; Hope, H.; Xu, X.; Power, P., *J. Amer. Chem. Soc.*, **1985**, *107*, 8087; Corazza, F.; Floriani, C.; Chiesti-Villa, A.; Guastini, C., *J. C. S, Chem. Comm.*, **1990**, 1083; Atwood, J. L.; Bott, S. G.; Jones, C.; Raston, C. L., *J. C. S., Chem.Comm.*, **1992**, 1349; Atwood, J. L.; Junk, P. C.; Lawrence, S. M.; Raston, C. L., *Supramolecular Chem.*, **1996**, *7*, 15; Acho, J. A.; Ren, T.; Yun, J. W.; Lippard, S. J., Inorg. Chem., 1995, 34, 5226; Chisholm, M. H.; Folting, K.; Streib, W. E.; Wu D-D, *J. C. S, Chem. Comm.*, **1998**, 379.
13. Gutsche, C. D.; Iqbal, M., *Organic Syntheses* **1989**, *68*, 234.
14. Harrowfield, J. MBc.; Ogden, M. I.; Richmond, W. R.; White, A. H., *J. C. S., Chem. Comm.*, **1991**, 1159; Asmuss, R.; Böhmer, V.; Harrowfield, J. MBc.; Ogden, M. I.; Richmond, W. R.; Skelton, B. W.; White, A. H., *J. C. S., Dalton Trans*, **1993**, 2427.
15. Kumagai, H.; Hasegawa, M.; Miyanari, S.; Sugawa, Y.; Sato, Y.; Hori, T.; Ueda, S.; Kamiyama, H.; Miyano, S., *Tetrahedron Lett.*, **1997**, *38*, 3971; Sone, T.; Ohba, Y.; Moriya, K.; Kumada, H.; Ito, K., *Tetrahedron Lett*, **1997**, *38*, 10689.
16. Akdas, H.; Bringel, L.; Graf, E.; Hosseini, M. W.; Mislin, G.; Pansanel, J.; De Cian, A.; Fischer, J., *Tetrahedron Lett.*, **1998**, *39*, 2311.
17. Iki, N.; Kumagai, H.; Morohashi, N.; Ajima, K.; Hasegawa, M.; Miyano, S., *Tetrahedron Lett*, **1998**, *39*, 7559.
18. Mislin, G.; Graf, E.; Hosseini, M. W.; De Cian, A.; Fischer, J., *Tetrahedron Lett.*, **1999**, *40*, 1129.
19. Mislin, G.; Graf, E.; Hosseini, M. W.; De Cian, A.; Fischer, J., *J. C. S., Chem. Comm.*, **1998**, 1345
20. Lhotak, P.; Himl, M.; Pakhomova, S.; Stibor I., *Tetrahedron Lett.*, **1998**, *39*, 8915.
21. Iki, N.; Narumi, F.; Fujimoto, T.; Morohashi, N.; Miyano, S., *J. C. S. Perkin Trans. 2*, **1998**, 2745.
22. Akdas, H.; Mislin, G.; Graf, E.; Hosseini, M. W.; De Cian, A.; Fischer, J., Tetrahedron Lett., **1999**, *40*, 2113.
23. Delaigue, X.; Hosseini, M. W.; De Cian, A.; Kyritsakas, N.; Fischer, J., *J. C. S., Chem. Comm.* **1995**, 609.

24. Delaigue, X.; Hosseini, M. W., *Tetrahedron Lett.* **1993**, *34*, 8111; Delaigue, X.; Harrowfield, J. McB.; Hosseini, M. W.; De Cian, A.; Fischer, J.; Kyritsakas, N., *J. C. S., Chem. Comm.* **1994**, 1579.
25. Gibbs, C. G.; Gutsche, C. D.; *J. Amer. Chem. Soc.*, **1993**, *115*, 5338; Gibbs, C. G.; Sujeeth, P. K.; Rogers, J. S.; Stanley, G. G.; Krawiec, M.; Watson, W. H.; Gutsche, C. D., *J Org. Chem.*, **1995**, *60*, 8394.
26. König, B.; Rödel, M.; Dix, I.; Jones, P. G., *J. Chem. Res.*, **1997**, 0555.
27. König, B.; Rödel, M.; Bubenitschek, P.; Jones, P. G., *Angewandte Chem. Int. Ed. Engl.*, **1995**, *34*, 661.
28. Hajek, F.; Graf, E.; Hosseini, M. W.; Decian, A.; Fischer, J., unpublished.
29. Ungaro, R.; Pochini, A.; Andreetti, G. D.; Sangermano, V., *J. C. S., Perkin Trans. 2*, **1984**, 1979.
30. Ungaro, R.; Pochini, A.; Andreetti, G. D.; Sangermano, V., *J. C. S., Perkin Trans. 2*, **1984**, 1979.
31. McKervey, M. A.; Seward, E. M.; Ferguson, G.; Ruhl, BV.; Harris, S., *J. C. S., Chem. Comm.* **1985**, 388; Arduini, A.; Pochini, A.; Reverberi, S.; Ungaro, R.; Andreetti, G. D.; Ugozzoli, F., *Tetrahedron*, **1986**, *42*, 2089; Chang, S.-K.; Cho, I., *J. C. S, Perkin Trans. 1*, **1986**, 211.
32. Iwamoto, K.; Shinkai, S., *J. Org. Chem.*, **1992**, *57*, 7066.
33. Mislin, G.; Graf, E.; Hosseini, M. W.; Bilyk, A.; Hall, A. K.; Harrowfield, J. McB.; Skelton, B. W.; White, A. H., *J. C. S., Chem. Comm.*, **1999**, 373.
34. Mislin, G.; Graf, E.; Hosseini, M. W.; Bilyk, A.; Hall, A. K.; Harrowfield, J. McB.; Skelton, B. W.; White, A. H., *to be published*.

Chapter 23

Synthesis and Fullerene Complexation Studies of *p*-Allylcalix[5]arenes

Charles G. Gibbs, Jian-she Wang, and C. David Gutsche[1]

Department of Chemistry, Texas Christian University, Fort Worth, TX 76129

The synthesis of calix[5]arenes substituted on the upper rim by one, two, three, four, and five allyl groups has been carried out using a modification of the Claisen rearrangement that employs a silylating agent. The allyl ethers of the calixarenes required for the Claisen rearrangements were prepared by selective silylation using the hindered, hydrolytically stable triisopropylsilyl moiety followed by allylation of the unsubstituted OH groups. The complexation constants of the fullerenes C_{60} and C_{70} with the members of this family of calix[5]arenes were measured and were found to increase in a more or less regular fashion as the number of *p*-allyl groups increases, with the ratios of the complexation constants for C_{60}/C_{70} also increasing.

The formation of complexes of several bis-calixarenes (*1*) with fullerenes C_{60} and C_{70} has been reported from this laboratory (*2*). Included among these were the bis-calix[5]arene **1** and its octaallyl counterpart **2**. The K_{assoc} values of **1** for C_{60} and C_{70} are 93 M^{-1} and 119 M^{-1}, respectively, whereas those for **2** are 1300 M^{-1} and 625 M^{-1}, considerably larger. For comparison, the complexation constants of the simple calixarenes calix[5]arene (**12**) and its pentaallyl counterpart (**19**) were also measured, and the allyl groups were seen to have a significant complexation-enhancing effect in this case as well. The purpose of the present paper, therefore, is to explore the family of *p*-allylcalix[5]arenes to determine the regularity, or lack thereof, with which the *p*-allyl groups contribute to the magnitudes and ratios of K_{assoc} for C_{60} and C_{70}.

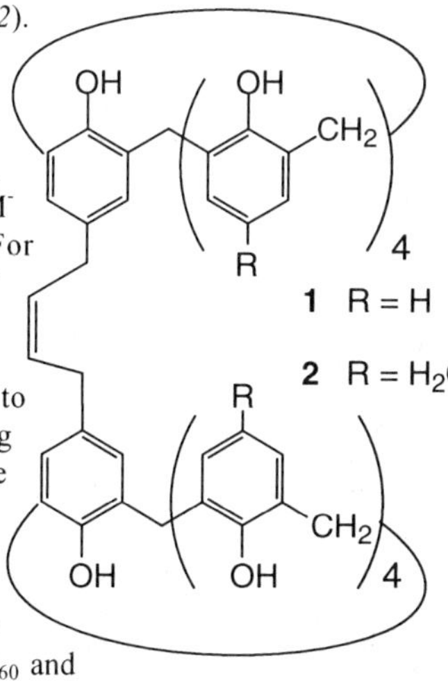

1 R = H

2 R = H$_2$C=CHCH$_2$

[1]Corresponding author.

Synthesis of *p*-Allylcalix[5]arenes.

Tetraallylcalix[4]arene, prepared by Gutsche and Levine (*3*) in 1982 by the Claisen rearrangement of the tetraallyl ether of calix[4]arene, was one of the earliest methods for introducing a moiety into the *p*-position, providing a route to a variety of *p*-functionalized calix[4]arenes (*4*). However, extension of the procedure to larger calixarenes has been disappointing, the yield of hexaallylcalix[6]arene from the hexaallyl ether being only *ca* 20%, and the yields of *p*-allylcalixarenes from the pentallyl ether of calix[5]arene and the octaallyl ether of calix[8]arene being zero. A solution to this problem of low or no yields for the Claisen rearrangement with the larger calixarenes has now been found. By carrying out the reaction in refluxing N,N-diethylaniline (DEA) containing greater than stoichiometric amounts of 1,3-bis-(trimethylsilyl)urea (BTMSU) the products are the silyl ethers rather than the free phenols and are more resistant to decomposition. Without isolation of the silyl ethers, the silyl groups are removed under mild conditions either with MeOH-H_2O-HCl (if only Me_3Si groups are present) or tetrabutylammonium fluoride (TBAF) (if *i*-Pr_3Si groups are present) to yield the rearranged products as the free phenols, as shown in Scheme 1. The choice of BTMSU was based on its high melting point, its stability under the reaction conditions, and its relatively low price. By means of this procedure *p*-tetraallylcalix[4]arene can be obtained in 98% yield, *p*-pentaallylcalix[5]arene in 67% yield, *p*-hexaallylcalix[6]arene in 92% yield, and *p*-octaallylcalix[8]arene in 75% yield. It is this modification of the Claisen rearrangement that made possible the preparation of the bis-calixarenes reported earlier (*2*), and it promises to expand the utility of this functionalization procedure within the calixarene family and, perhaps, to other areas of organic synthesis as well.

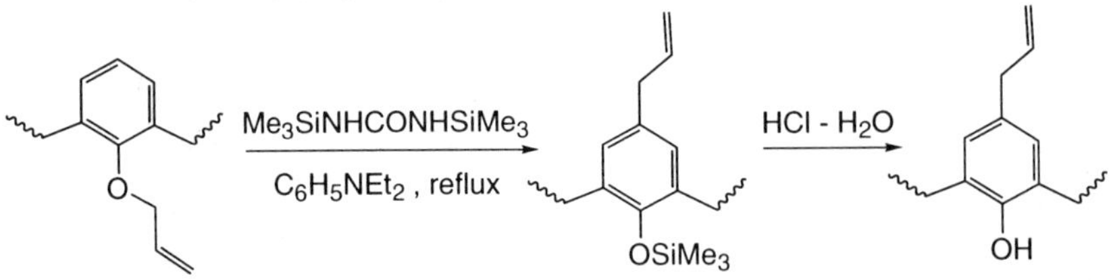

Scheme 1. Silylation modification of Claisen rearrangement

As portrayed in Scheme 2, the allyl ethers necessary for the Claisen rearrangement were obtained either by direct allylation or by allylation of appropriately substituted silyl ethers. Calix[5]arene (**12**) was directly converted to the pentaallyl ether **11** with an excess of allyl bromide or to the monoallyl ether **3** with a limiting amount of allyl tosylate. Treatment of **12** with triisopropylsilyl chloride yielded the monoisopropylsilyloxy ether **9**, the 1,3-bis(triisopropylsilyloxy) ether **4** or the 1,2,4-tris-(triisopropyl-silyloxy) ether **6** depending on the ratio of reagents and the base used. Treatment of the 1,3-diether **4** with allyl bromide in the presence of the weak base K_2CO_3 placed allyl groups only on the two adjacent free OH groups and not on the more hindered OH between the silyl ethers, affording the 1,2-diallyl ether **5**. With the stronger base NaH, however, all three free OH groups were allylated to produce the triallyl ether **8**. Treatment of the mono ether **9** with an excess of allyl bromide afforded the tetraallyl ether **10**. Employing the modified Claisen rearrangement, as described above, the allyl ethers **3, 5, 7, 8, 10**, and **11** were converted to the mono (**14**), 1,2-di (**15**), 1,3-di (**16**), 1,2,4-tri (**17**), tetra (**18**), and pentaallyl (**19**) calix[5]arenes in yields ranging from 50% for **14** to 81% for **15**.

TIPS = triisopropylsilyl

3 R^1 = CH$_2$CH=CH$_2$; $R^{2,3,4,5}$ = H
4 $R^{1,2,4}$ = H; $R^{3,5}$ = TIPS
5 $R^{1,2}$ = CH$_2$CH=CH$_2$; $R^{3,5}$ = TIPS; R^4 = H
6 $R^{1,3}$ = H; $R^{2,4,5}$ = TIPS
7 $R^{1,3}$ = CH$_2$CH=CH$_2$; $R^{2,4,5}$ = TIPS
8 $R^{1,2,4}$ = CH$_2$CH=CH$_2$; $R^{3,5}$ = TIPS
9 $R^{1,2,3,4}$ = H; R^5 = TIPS
10 $R^{1,2,3,4}$ = CH$_2$CH=CH$_2$; R^5 = TIPS
11 $R^{1,2,3,4,5}$ = CH$_2$CH=CH$_2$

12 $R^{1,2,3,4,5}$ = H
13 $R^{1,2,3,4,5}$ = *tert*-Butyl
14 R^1 = CH$_2$CH=CH$_2$; $R^{2,3,4,5}$ = H
15 $R^{1,2}$ = CH$_2$CH=CH$_2$; $R^{3,4,5}$ = H
16 $R^{1,3}$ = CH$_2$CH=CH$_2$; $R^{2,4,5}$ = H
17 $R^{1,2,4}$ = CH$_2$CH=CH$_2$; $R^{3,5}$ = H
18 $R^{1,2,3,4}$ = CH$_2$CH=CH$_2$; R^5 = H
19 $R^{1,2,3,4,5}$ = CH$_2$CH=CH$_2$

Complex Formation with C_{60} and C_{70}

The complexation constants of the *p*-allylcalix[5]arenes **14-19** with the fullerenes C_{60} and C_{70} were measured in toluene at 25 °C, using the intensity of the UV-vis absorption at 430 nm for C_{60} and 420 nm for C_{70}. As noted in other studies of the fullerenes with cyclodextrins *(5,6,7)*, azacrowns *(8)*, cyclotriveratrylenes *(9)*, and calixarenes *(10-34)*, complexation induces a color change from magenta to red for C_{60} and from red to colorless for C_{70}. Application of the Benesi-Hildebrand treatment of the spectrophotometric measurements *(35)* produced the complexation constants shown in Table 1

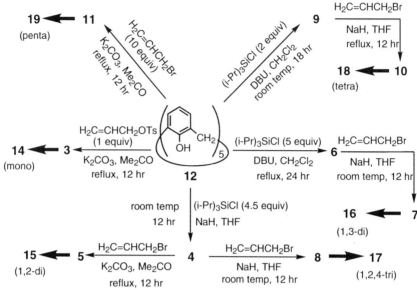

Scheme 2. Pathways for the synthesis of p-allylcalix[5]arenes

Compound	K_{assoc}, M^{-1} (C_{60})	K_{assoc}, M^{-1} (C_{70})	$K_{assoc}C_{60}/K_{assoc}C_{70}$
p-tert-Butylcalix[5]arene (13)	9 ± 1	2.3 ± 0.2	0.26
Calix[5]arene (12)	30 ± 2	51 ± 3	0.59
Mono-p-allylcalix[5]arene (14)	78 ± 9	106 ± 10	0.74
1,2-Di-p-allylcalix[5]arene (15)	140 ± 13	113 ± 11	1.24
1,3-Di-p-allylcalix[5]arene (16)	150 ± 1	120 ± 13	1.25
1,2,4-Tri-p-allylcalix[5]arene (17)	223 ± 2	166 ± 20	1.34
Tetra-p-allylcalix[5]arene (18)	290 ± 25	220 ± 24	1.32
Penta-p-allylcalix[5]arene (19)	530 ± 63	270 ± 37	1.96
Bis-calixarene 1	93 ± 5	119 ± 6	0.78
Bis-calixarene 2	1300 ± 65	625 ± 32	2.08

Table 1. Complexation constants for calix[5]arenes and fullerenes C_{60} and C_{70}.

Inspection of these data shows that there is an almost 60-fold difference between the association constants for penta-tert-butylcalix[5]arene and pentaallylcalix[5]arene, suggesting that a steric factor probably plays a part in the formation of the complex. The fairly regular upward trend in the magnitude of the complexation constants as allyl groups are added to the upper rim of calix[5]arene is also commensurate with a steric component, the complexation constant increasing as the depth and circumference of the cavity increase. A similar response to the p-substituent was observed in the earlier study (2) where the C_{60} complexation constant of simple bis-calixarene 1 is almost 14 times smaller than that of the fully allylated bis-calixarene 2.

The incremental increase resulting from the addition of the first allyl group to **12** is 2.3, while those for the additions of the remaining allyl groups are somewhat less, ranging between 1.3 and 1.8. Since the allyl groups may also exert an electronic influence, these incremental changes may be the result of a combination of both steric and electronic contributions. The electronic influence, however, will be reported in a separate study. It is interesting to note that the ratio of the C_{60} to C_{70} complexation also tends to increase as allyl groups are added to the upper rim, viz. C_{70} becoming almost three times less well complexed with the pentaallyl compound than with the parent compound, a phenomenon that was also noted with the bis-calixarenes **1** and **2** where a 2.67 difference in ratio between the two was observed. This, likewise, can probably be attributed to a steric factor, C_{70} being somewhat larger and less symmetrical than C_{60}.

Job plots of $\Delta_{absorbance}$ vs $[C_{60}]/([C_{60}] + [Calixarene])$ for the complexes of the *p*-allylcalix[5]arenes show a maximum $\Delta_{absorbance}$ at$[C_{60}]/([C_{60}] + [Calixarene]) = 0.5$, indicating calixarene:fullerene ratios of 1:1 in all cases. However, in the solid state complexes that have been studied (*31-34*) the calixarene to fullerene ratio is 2:1. Thus, at least some of the 2:1 complex presumably is present in solution, and a recent study by Fukazawa and coworkers has assigned K_{assoc} constants of 450 M^{-1} and 110 M^{-1} for the formation of the 1:1 complex and 2:1 complex, respectively, for a calix[5]arene bearing a functionality that can engage in intermolecular bridging via hydrogen bonding. This same group of workers (*34*) has reported a bis-calixarene, containing a 7-carbon bridge between the upper rims, that forms 1:1 complexes with C_{60} and C_{70} with K_{assoc} of 76,000 M^{-1} and 163,000 M^{-1}, respectively, by far the highest values yet reported for calixarene-fullerene complexes in solution. Also reported is a bis-calixarene carrying a 2-carbon alkyne bridge which has a considerably lower K_{assoc} of 8300 M^{-1} for C_{60} but, nevertheless, more than six times higher than the 1300 M^{-1} value for the bis-calixarene **2**. On the other hand, Shinkai's capsule-like cage molecule (*26*) comprising a pair of calixarenes linked via chelation with Pd has a K_{assoc} of only 56 M^{-1}, illustrating the great sensitivity of the strength of complexation to the particular host assembly. In the case of the complex of the bis-calixarene **2** with C_{60} a Job plot (see Figure 1) shows a maximum $\Delta_{absorbance}$ at $[C_{60}]/([C_{60}] + [Calixarene]) = 0.4$, suggesting the possibility of still another calixarene:fullerene ratio, viz a 3:2 complex such as that depicted in Figure 2. Why **2** behaves so differently from the bis-calixarenes reported by Fukazawa and coworkers and Shinkai and coworkers is not clear but may arise from the greater flexibility of the linking chain in **2** as compared with those in the compounds studied by these other workers.

Experimental Section (*36*)

5-Allylcalix[5]arene-31,32,33,34,35-pentol (14). A stirred suspension of 0.8 g (1.5 mmol) of calix[5]arene (**12**), 0.32 g (1.5 mmol) of allyl tosylate (*37*) and 0.21 g (1.5 mmol) of K_2CO_3 in 60 mL of dry acetone was heated at reflux for 12 h. The reaction mixture was subjected to flash chromatography (*38*) (petroleum ether-CH_2Cl_2 eluent) to give 0.75 g (88%) of the monoallyloxycalix[5]arene (**3**) which was used without further purification (*39*). A 0.64 g (1.22 mmol) sample of **3** was heated in 20 mL of DEA containing 1.8 g (8.8 mmol) of BTMSU for 3 h, diluted with CH_2Cl_2, washed with 2N HCl and dried. To this solution was added MeOH and 3 mL of 2N HCl, and the mixture was stirred 4 h. Evaporation and flash chromatography (petroleum ether-CH_2Cl_2) gave

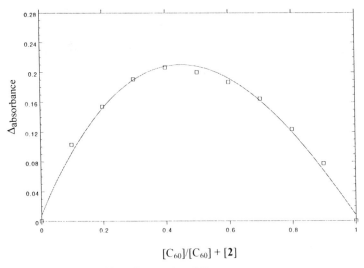

Figure 1. Job plot for bis-calixarene **2** and C$_{60}$

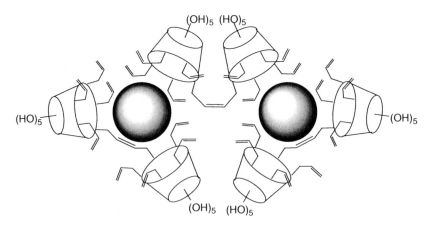

Figure 2. 3:2 Complex of bis-calixarene **2** and C$_{60}$

0.32 g (50%) of **14** as a white solid after trituration with MeOH. An analytical sample was obtained by recrystallization from CHCl$_3$ - MeOH: mp 232-234 °C with preliminary shrinking; ^{1}H NMR (CDCl$_3$) δ 8.95 (s, 2), 8.93 (s, 2), 8.83 (s, 1), 7.20 (d, 8, J = 7.5 Hz), 7.00, (s, 2), 6.84 (t, 4, J = 7.5 Hz), 5.96-5.83 (m, 1), 5.06-5.00 (m, 2), 3.84 (br s, 10), 3.25 (d, 2, J = 6.6 Hz); ^{13}C NMR (CDCl$_3$) δ 150.01, 149.97, 148.28, 137.69, 132.95, 129.14, 126.73, 126.67, 126.55, 121.55, 115.55, 39.20, 31.32, 31.23. Anal. Cacld for C$_{38}$H$_{34}$O$_5$·1/2 CH$_3$OH: C, 78.60; H, 6.251. Found: C, 78.94; H, 5.91.

5,11-Diallylcalix[5]arene-31,32,33,34,35-pentol (15). To a suspension of 0.35 g (8.8 mmol) of 60% NaH in 60 mL of THF was added 1.06 g (2 mmol) of calix[5]arene (**12**), and the mixture was stirred for 1 h at room temperature. A 1.7 g (8.8 mmol) sample of triisopropylsilyl chloride was added, and the mixture was stirred for 12 h. The solution was concentrated by partial removal of the THF on a Rotovap, CH$_2$Cl$_2$ was added, and the solution was washed with 1N HCl and H$_2$O. After drying over Na$_2$SO$_4$ the solvent was removed, and the residue was flash chromatographed (petroleum ether-CH$_2$Cl$_2$ eluent) to give 1.28 g (76%) of 33,35-bis(triisopropylsilyloxy)calix[5]arene-31,32,34-triol (**4**) which was used without further purification (*40*). A stirred suspension of 1.7 g (2 mmol) of 33,35-bis-(triisopropylsilyloxy)calix[5]arene-31,32,34-triol (**4**), 1.21 g (10 mmol) of allyl bromide, and 1.38 g (10 mmol) of K$_2$CO$_3$ in 60 mL of acetone was heated at reflux for 12 h. The solution was concentrated by partial removal of the THF on a Rotovap, H$_2$O was added, and the mixture was extracted with CH$_2$Cl$_2$. A white solid was obtained after flash chromatography (petroleum ether-CH$_2$Cl$_2$ eluent) and trituration with MeOH (*41*). A 1.42 g (1.54 mmol) sample of this material was heated in 30 mL of DEA containing 1.9 g (9.3 mmol) of BTMSU for 5 h, diluted with CH$_2$Cl$_2$, washed with 2N HCl, and dried. Concentration of the solution and trituration with MeOH gave a tan solid. This material was dissolved in 50 mL of THF, 4.0 g (15.4 mmol) of TBAF was added, and the mixture was stirred at room temperature for 18 h. The solution was concentrated by partial removal of the THF on a Rotovap, CH$_2$Cl$_2$ was added, and the solution was washed with 1N HCl and H$_2$O. After drying over Na$_2$SO$_4$ the solvent was removed, and flash chromatography (petroleum ether - CH$_2$Cl$_2$ eluent) gave 0.76 g (81%) of **15** after trituration with MeOH. An analytical sample was obtained by recrystallization from CHCl$_3$-MeOH: mp 192-194 °C with preliminary shrinking; ^{1}H NMR (CDCl$_3$) δ 8.94 (s, 2), 8.93 (s, 1), 8.84 (s, 2), 7.20 (d, 6, J = 7.5 Hz), 7.00 (s, 4), 6.84 (t, 3, J = 7.5 Hz), 5.94-5.85 (m, 2), 5.07-5.00 (m, 4), 3.83 (br s, 10), 3.25 (d, 4, J = 6.6 Hz); ^{13}C NMR (CDCl$_3$) δ 150.00, 149.95, 148.23, 137.72, 132.91, 129.23, 129.13, 126.72, 126.65, 126.59, 126.52, 121.52, 115.48, 39.18, 31.31. Anal. Calcd for C$_{41}$H$_{38}$O$_5$: C, 80.63; H, 6.27. Found: C, 80.78; H, 6.44.

5,17-Diallylcalix[5]arene-31,32,33,34,35-pentol (16). A stirred solution of 1.06 g (2 mmol) of calix[5]arene (**12**), 1.52 g (10 mmol) of 1,8-diazobicyclo[5.4.0]undec-7-ene (DBU), and 1.93 g (10 mmol) of triisopropylsilyl chloride in 75 mL of CH$_2$Cl$_2$ was heated at reflux for 24 h. The reaction mixture was washed with 1N HCl and H$_2$O. After drying, the solvent was removed and the residue was flash chromatographed (petroleum ether-CH$_2$Cl$_2$ eluent) to give 1.4 g (70%) of 32,34,35-tris-(triisopropyl-silyloxy)calix[5]arene-31,33-diol (**6**) which was used without further purification (*42*). To a suspension of 0.22 g (5.5 mmol) of 60% NaH in 60 mL of THF was added 1.38 g (1.38 mmol) of **6**, and the mixture was stirred for 1 h at room temperature.Then 0.67 g (5.5 mmol) of allyl bromide was added, and the mixture was stirred for 12 h. The solution was concentrated by partial removal of the THF on a Rotovap, CH_2Cl_2 was added, and the solution was washed with 1N HCl and H_2O. After drying over Na_2SO_4 the solvent was removed to yield 7 as a white solid after flash chromatography (petroleum-ether CH_2Cl_2 eluent) and trituration with MeOH (*41*). A 1.39 g sample of this material was heated in 25 mL of DEA containing 1.12 g (5.5 mmol) of BTMSU for 4 h, diluted with CH_2Cl_2, washed with 2N HCl, and dried. Concentration of the solution and trituration with MeOH gave a tan solid. This material was dissolved in 40 mL of THF, 3.9 g (15.1 mmol) of TBAF was added, and the mixture was stirred at room temperature for 18 h. The solution was concentrated, diluted with CH_2Cl_2, washed with 1N HCl and H_2O, and dried. Evaporation and flash chromatography (petroleum ether - CH_2Cl_2 eluent) gave 0.46 g (60%) of **16** after trituration with MeOH. An analytical sample was obtained by recrystallization from $CHCl_3$-MeOH: mp 200-202 °C with preliminary shrinking; ^1H NMR ($CDCl_3$) δ 8.94 (s, 2), 8.93 (s, 1), 8.84 (s, 2), 7.20 (d, 6, J = 7.5 Hz), 7.00 (s, 4), 6.84 (t, 3, J = 7.5 Hz), 5.96-5.82 (m, 2), 5.06- 5.00 (m, 4), 3.82 (br s, 10), 3.25 (d, 4, J = 5.5 Hz); ^{13}C NMR ($CDCl_3$) δ 150.02, 148.28, 137.74, 137.70, 132.95, 129.81, 126.74, 126.68, 126.61, 126.56, 121.54, 115.54, 115.49, 39.20, 31.34. Anal. Calcd. for $C_{41}H_{38}O_5 \cdot 1/2CH_3OH$: C, 79.53; H, 6.43. Found: C, 79.83; H, 6.18.

5,11,23-Triallylcalix[5]arene-31,32,33,34,35-pentol (17). To a suspension of 0.2 g (5.2 mmol) of 60% NaH in 50 mL of THF was added 0.74 g (0.86 mmol) of **4**, prepared as described above, and the mixture was stirred for 1 h at room temperature. A 0.63 g (5.2 mmol) sample of allyl bromide was added, and the mixture was stirred for 12 h. The solution was concentrated by partial removal of the THF on a Rotovap, CH_2Cl_2 was added, and the solution was washed with 1N HCl and H_2O. After drying over Na_2SO_4 the solvent was removed to yield **8** as a white solid after flash chromatography (petroleum-ether CH_2Cl_2 eluent) and trituration with MeOH (*41*). A 0.75 g (0.78 mmol) sample of this material was heated in 20 mL of DEA containing 0.95 g (4.68 mmol) of BTMSU for 5 h, diluted with CH_2Cl_2, washed with 2N HCl, and dried. Concentration of the solution and trituration with MeOH gave a tan solid. This material was dissolved in 30 mL of THF, 0.22 g (8.6 mmol) of TBAF was added, and the mixture was stirred at room temperature for 18 h. The solution was concentrated, diluted with CH_2Cl_2, washed with 1N HCl and H_2O, and dried. Evaporation and flash chromatography (petroleum ether - CH_2Cl_2 eluent) gave 0.32 g (57%) of **17** after trituration with MeOH. An analytical sample was obtained by recrystallization from $CHCl_3$-MeOH: mp 205-207 °C with preliminary shrinking; ^1H NMR ($CDCl_3$) δ 8.96 (s, 2), 8.83 (s, 2), 7.20 (d, 4, J = 7.5Hz), 7.00 (s, 6), 6.85 (t, 2, J = 7.4 Hz), 5.91-5.86 (m, 3), 5.06-5.01 (m, 6), 3.82 (br s, 10), 3.26 (d, 6, J = 6.5 Hz); ^{13}C NMR ($CDCl_3$) δ 150.02, 148.28, 137.74, 137.70, 132.95, 129.81, 126.74, 126.68, 126.61, 126.56, 121.54, 115.54, 115.49, 39.20, 31.34. Anal. Calcd. For $C_{44}H_{42}O_5 \cdot 3/4$ MeOH: C, 79.89; H, 6.44. Found: C, 79.59; H, 6.37.

5,11,17,23-Tetraallylcalix[5]arene-31,32,33,34,35-pentol (18). The precursor 30-triisopropylsilyl-oxycalix[5]arene-31,32,33,34-tetrol (**9**) was prepared by stirring a solution of 1.06 g (2 mmol) of calix[5]arene (**12**), 0.67 g (4.4 mmol) of DBU, and 0.85 g (4.4 mmol) of triisopropylsilyl chloride in 75 mL of CH_2Cl_2 at room temperature for 18 h. The reaction mixture was washed with 1N HCl and H_2O. After drying over Na_2SO_4 the solvent was removed, and the residue was flash chromatographed (petroleum ether-CH_2Cl_2 eluent) to give 1.1 g (80%) of **9** which was used without further purification (*43*). To a suspension of 0.65 g (16.2 mmol) of 60% NaH in 50 mL of THF was added 1.86 g

(2.7 mmol) of **9**, and the mixture was stirred for 1 h at room temperature. A 1.96 g (16.2 mmol) sample of allyl bromide was added, and the mixture was heated at reflux for 12 h. The solution was concentrated, diluted with CH_2Cl_2, and washed with 1N HCl and H_2O. The tetraallyl ether **10** was obtained as a white solid after flash chromatography (petroleum ether-CH_2Cl_2 eluent) and trituration with MeOH (*41*). A 1.2 g (1.44 mmol) sample of this material was heated in 35 mL of DEA containing 2.9 g (14.4 mmol) of BTMSU for 6 h, diluted with CH_2Cl_2, washed with 2N HCl, and dried. Concentration of the solution and trituration with MeOH gave a tan solid. This material was dissolved in 40 mL of THF, 3.8 g (14.4 mmol) of TBAF was added, and the mixture was stirred at room temperature for 18 h. The solution was concentrated, diluted with CH_2Cl_2, washed with 1N HCl and H_2O, and dried. Evaporation and flash chromatography (petroleum ether - CH_2Cl_2 eluent) gave 0.60 g (61%) of **18** after trituration with MeOH. An analytical sample was obtained by recrystallization from $CHCl_3$-MeOH: mp 225-227 °C with preliminary shrinking; ^{1}H NMR ($CDCl_3$) δ 8.95 (s, 1), 8.84 (s, 4), 7.19 (d, 2, J = 7.5 Hz), 7.00 (s, 8), 6.84 (t, 1, J = 7.4 Hz), 5.96-5.84 (m, 4), 5.07-5.01 (m, 8), 3.80 (br s, 10), 3.27 (d, 8, J = 6.5 Hz); ^{13}C ($CDCl_3$) δ 148.34, 137.77, 132.83, 129.20, 126.58, 115.42, 39.17, 31.37. Anal. Calcd for $C_{47}H_{46}O_5$: C, 81.71; H, 6.71. Found: C, 81.48; H, 6.76.

5,11,17,23,29-Pentaallylcalix[5]arene-31,32,33,34,35-pentol (19). A stirred suspension of 5.5 g (10.4 mmol) of calix[5]arene (**12**), 12.6 g (104 mmol) of allyl bromide, and 14.4 g (104 mmol) of K_2CO_3 in 300 mL of dry acetone was heated at reflux for 12 h. The solution was concentrated by partial removal of the THF on a Rotovap, H_2O was added, and the mixture was extracted with CH_2Cl_2. After drying over Na_2SO_4 the solvent was removed, and the residue was triturated with MeOH to give 6.5 g (85%) of 31,32,33,34,35-pentaallyloxycalix[5]arene (**11**) as a white solid. An analytical sample was obtained by recrystallization from hexane: mp 143-144 °C; ^{1}H NMR ($CDCl_3$) δ 7.08 (d, 10, J = 7.5 Hz), 6.84 (t, 5, J = 7.5 Hz), 5.59 (m, 5), 5.06 (d, 5, J = 17.1 Hz), 4.91 (d, 5, J = 10.2 Hz), 3.78 (s, 10), 3.70 (s, 10); ^{13}C NMR ($CDCl_3$) δ 155.90, 134.34, 134.17, 129.62, 123.16, 116.21, 73.24, 31.95. Anal. Calcd for $C_{50}H_{50}O_5$: C, 82.16; H, 6.89. Found: C, 82.17; H, 6.93.

The procedure for the Claisen rearrangement previously described (*2*) gave 0.6 g (67%) of **19** from 0.91 g of **11** after flash chromatography (petroleum ether-CH_2Cl_2 eluent) as a white solid. An analytical sample was obtained by recrystallization from $CHCl_3$ - MeOH: mp 235-236 °C; ^{1}H NMR ($CDCl_3$) δ 8.83 (s, 5), 6.99 (s, 10), 5.91-5.86 (m, 5), 5.07-5.01 (m, 10), 3.79 (br s, 10), 3.27 (d, 10, J = 6.5 Hz); ^{13}C NMR ($CDCl_3$) δ 148.36, 137.78, 132.86, 129.22, 126.56, 115.42, 39.19, 31.38. Anal. Calcd for $C_{50}H_{50}O_5$: C, 82.16; H, 6.89. Found: C, 82.08; H, 7.09.

Measurement of Complexation Constants. Complexation constants (K_{assoc}, M^{-1}) were determined in toluene solution with a Cary 3 UV-vis spectrometer at 25 ± 0.5 °C. One set of toluene stock solutions was prepared with a 2.0 x 10^{-4} M concentration of C_{60} and C_{70}. Another set of toluene stock solutions was prepared containing five different concentrations (2.0, 4.0, 6.0, 8.0, and 10.0 x 10^{-3} M) of the particular calixarene being measured. For each spectrophotometric determination, identical volumes of the fullerene and calixarene stock solutions were placed in the sample cell, and identical volumes of pure toluene and the calixarene stock solution were placed in the reference cell. The absorption spectra were measured for both C_{60} and C_{70} for each of the five calixarene concentrations. The absorption of a solution of C_{60} in toluene increased in the 400-450 nm region upon the addition of p-allyllcalix[5]arene, and the color changed from magenta to red, indicating the formation of a complex. Similarly, the absorption of a solution of C_{70} increased in the 400-450 nm region upon the addition of p-allylcalix[5]arene, and the color changed from red to colorless, again indicating the formation of a complex. The wavelengths for the maximum intensity changes were selected as 430 nm for C_{60} and 420 nm for C_{70}. From the absorption intensities at these wavelengths, along with the concentrations of the host, the complexation constants of Table 1 were calculated using the Benesi-Hildebrand equation (35).

Job plots were made by measuring the change in absorbence ($\Delta_{absorbance}$) as a function of the ratio of the concentration of C_{60} to the sum of the concentrations of C_{60} and the calixarene ([C_{60}]/[C_{60}] + [calixarene]), as illustrated for the bis-calixarene 2 in Figure 2.

Acknowledgment. We are indebted to the Robert A. Welch Foundation and the National Science Foundation for generous support of this research.

References

1. For general reviews of calixarene chemistry *cf* (a) Gutsche, C. D. *Calixarenes Revisited* in "Monographs in Supramolecular Chemistry", Stoddart, J. F. Ed., Royal Society of Chemistry, London, **1998**. (b) Böhmer, V., "Calixarenes, Macrocycles with (Almost) Unlimited Possibilties", *Angew. Chem. Int. Ed. Engl.* **1995**, *34*, 713 - 745. (c) Gutsche, C. D. "Calixarenes", *Aldrichimica Acta*, **1995**, *28*, 3-9; (d) "Calixarenes, A Versatile Class of Macrocyclic Compounds", Vicens, J.; Böhmer, V. eds, Kluwer, Dordrecht, **1991**; (e) Gutsche, C. D. *Calixarenes* in "Monographs in Supramolecular Chemistry", Stoddart, J. F. Ed., Royal Society of Chemistry, London, **1989**.
2. Wang, J.; Gutsche, C. D. *J. Am. Chem. Soc.* **1998**, *120*, 12226.
3. Gutsche, C. D.; Levine, J. A. *J. Am. Chem. Soc.* **1982**, *104*, 2652.
4. Gutsche, C. D.; Levine, J. A.; Sujeeth, P. K. *J. Org. Chem.* **1985**, *50*, 5802.
5. Andersson, T.; Nilsson, K.; Sundahl, M.; Westman, G.; Wennerström, O. *J. Chem Soc Chem.Commun.* **1992**, 604.
6. Yoshida, Z.-i.; Takekuma, H.; Takekuma, S.-i.; Matsubara, Y. *Angew. Chem. Int. Ed. Engl.* **1994**, *33*, 1597.
7. Andersson, T.; Westman, G.; Stenhagen, G; Sundahl, M.; Wennerström, O.*Tetrahedron Lett.*, **1995**, 36, 597.
8. Diederich, F.; Effing, J.; Jonas, U.; Jullien, L.; Plesnivy, T.; Ringsdorf, H.; Thilgen, C.; Weinstien, D. *Angew. Chem., Int. Ed. Engl.*, **1992**, 31, 1599.

9. Steed, J. W.; Junk, P. C.; Atwood, J. L.; Barnes, M. J.; Raston, C. L.; Burkhalter, R. S. *J. Am. Chem. Soc.* **1994**, 116, 10346.
10. Williams, R. M.; Verhoeven, J. W. *Rec. Trav. Chim. Pays-Bas* **1992**, *111*, 531. Williams, R. M.; Zwier, J. M.; Verhoeven, J. W.; Nachtegaal, G. H.; Kentgens, A. P. M. *J. Am. Chem. Soc.* **1994**, *116*, 6965.
12. Atwood, J. L.; Koutsantonis, G. A.; Raston, C. L. *Nature* **1994**, *368*, 229.
13. Raston, C. L.; Atwood, J. L.; Nichols, P. J.; Sudria, I. B. N. *J. Chem. Soc. Chem. Commun.* **1996**, 2615.
14. Barbour, L. J.; Orr, G. W.; Atwood, J. L. *Chem. Commun.* **1997**, 1439.
15. Atwood, J. L.; Barbour, L. J.; Nichols, P. J.; Raston, C. L.; Sandoval, C. A. *Chem. Eur. J.* **1999**, *5*, 990.
16. Isaacs, N. S.; Nichols, P. J.; Raston, C. L.; Sandova, C.A.; Young, D. J. *Chem.Commun.* **1997**, 1839.
17. Croucher, P. D.; Marshall, J. M. E.; Nichols, P. J.; Raston, C. L. *Chem. Commun.* **1999**, 193.
18. Olsen, S. A.; Bond, A. M.; Compton, R. G.; Lazarov. G.; Mahon, P. J.; Marken, F.; Raston, C. L.; Tedesco, V.; Webster, R. D. *J. Phys. Chem. A.* **1998**, *102*, 2641.
19. Susuki, T.; Nakashima, K.; Shinkai, S. *Chem. Lett.* **1994**, 699.
20. Suzuki, T.; Nakashima, K.; Shinkai, S. *Tetrahedron Lett.* **1995**, *36*, 249.
21. Ikeda, A.; Yoshimura, M.; Shinkai, S. *Tetrahedron Lett.* **1997**, *38*, 2107.
22. Paci, B.; Amoretti, G.; Arduini, G.; Ruani, G.; Shinkai, S.; Suzuki, T.; Ugozzoli, F.; Caciuffo, R. *Phys. Rev.* **1997**, *55*, 5566.
23. Shinkai, S.; Ikeda, A. *Gazz. Chim. Ital.* **1997**, *127*, 657.
24. Ikeda, A.; Suzuki, Y.; Yoshimura, M.; Shinkai, S. *Tetrahedron* **1998**, *54*, 2497.
25. Cliffel, D. E.; Bard, A. J.; Shinkai, S. *Anal. Chem.* **1998**, *70*, 4146.
26. Ikeda, A.; Yoshimura, M.; Udzu, H.; Fukuhara, C.; Shinkai, S. *J. Am.Chem. Soc.* **1999**, *121*, 4296.
27. Castillo, R.; Ramos, S.; Cruz, R.; Martinez, M.; Lara, F.; Ruiz-Garcia, J. *J. Phys.Chem.* **1996**, *100*, 709.
28. Lo Nostro, P.; Casnati, A.; Bossoletti, L.; Dei, L.; Baglioni, P. *Colloids and Surfaces* **1996,** *116*, 203.
29. K.; Tanaka, K.; Kinoshita, T.; Fuji, K. *Chem.Comm.* **1998**, 895.
30. Georghiou, P. E.; Mizyed, S.; Chowdhury, S. *Tetrahedron Lett.* **1999**, *40*, 611.
31. Haino, T.; Yanase, M.; Fukazawa, Y. *Angew. Chem. Int. Ed. Engl.* **1997**, *36*, 259.
32. Haino, T.; Yanase, M.; Fukazawa, Y. *Tetrahedron Lett.* **1997**, *38*, 3739.
33. Haino, T.; Yanase, M.; Fukazawa, Y. *Angew. Chem Intl. Ed. Engl.* **1998**, *37*, 997.
34. Haino, T.; Yanase, M.; Fukazawa, Y.*Angew. Chem Intl. Ed. Engl.*. **1999**, 997.
35. Benesi, H. A.; Hildebrand, J. H. *J. Am. Chem. Soc.* **1949**, *71*, 2703.
36. Unless otherwise noted, starting materials were obtained from commercial suppliers and used without further purification. THF was freshly distilled from Na-benzophenone. All reactions were carried out in an inert atmosphere of nitrogen. The melting points of all compounds were taken in sealed evacuated capillary tubes on a Mel-Temp apparatus (Laboratory Devices, Cambridge, MA) using a 500°C thermometer calibrated against a thermocouple, and they are uncorrected. ^{1}H and ^{13}C NMR spectra were recorded on a Varian XL-300 spectrometer at 300 and 75 MHz,

respectively. Analytical samples were dried in a drying pistol under vacuum for 24 hrs. Microanalyses were performed by Desert Analytics, Tucson, AZ. TLC analyses were carried out on Analtech silica gel plates (absorbent thickness 250 μm) containing a fluorescent indicator. Flash chromatography was carried out with J. T. Baker silica gel No. JT7042-2 (40 μm particle diameter).

37. Johnson, C. R.; Dutra, G. A. *J. Am. Chem. Soc.*, **1973**, *95*, 7777.
38. As previosly described (Gibbs, C. G.; Sujeeth, P. K.; Rogers, J. S.; Stanley, G. G.; Krawiec, M.; Watson, W. H.; Gutsche, C. D. *J. Org. Chem.* **1995**, *60*, 8394) flash chromatography was carried out by dissolving the crude solid in CH_2Cl_2, adding 10-15 times its weight of silica gel, and evaporating to dryness. This material was slurried in a small amount of petroleum ether- CH_2Cl_2 (90:10) and added to a prepared column made up of the same solvent system. Elution was carried out with a gradient solvent system (petroleum ether- CH_2Cl_2, increasing the CH_2Cl_2 in 10% incremental steps).
39. ^{1}H NMR ($CDCl_3$) δ 8.00 (s, 4), 7.19- 6.78 (m, 15), 6.17-6.04 (m, 1), 5.56 (d, 1, J = 17.2 Hz), 5.34 (d, 1, J = 10.6 Hz), 4.49 (d, 2, J = 5.7 Hz), 4.32- 3.56 (m, 10); NMR ($CDCl_3$) δ 152.5, 151.79, 150.16, 133.00, 132.44, 129.43, 129.26, 129.22 129.07, 129.00, 128.86, 127.31, 126.97, 126.91, 125.37, 121.43, 120.29, 119.01, 76.38, 31.64, 31.08.
40. ^{1}H NMR ($CDCl_3$) δ 7.27-6.51 (m, 17), 5.29 (s, 1), 4.47 (d, 2, J = 14.8 Hz), 4.44 (d 2, J = 14.5 Hz), 4.14 (d, 1, J = 14.0 Hz), 3.47 (d, 1, J = 14.0 Hz), 3.40 (d, 2, J = 15.0 Hz), 3.38 (d, 2, J = 14.5 Hz), 1.47-1.35 (m, 6), 1.54 (d, 36, J = 7.5 Hz).
41. This material has a very complex ^{1}H NMR spectrum.
42. ^{1}H NMR ($CDCl_3$) δ 7.25-6.20 (m, 12), 4.43 (d, 2, J = 14.7 Hz), 4.35 (d, 1, J = 15.6 Hz), 4.21 (d, 2, J = 14.9 Hz), 3.52 (d, 2, J = 15.2 Hz), 3.47 (d, 1, J = 17.0 Hz), 3.29, (d, 2, J = 15.1 Hz), 1.45-1.07 (m, 63).
43. ^{1}H NMR ($CDCl_3$) δ 8.12 (s, 2), 6.57-7.21 (m, 17), 4.52 (d, 2, J = 14.5 Hz), 4.16 (d, 2, J = 14.2 Hz), 3.94 (d, 1, J = 14.3 Hz), 3.55-3.43 (m, 5), 1.52-1.40 (m, 3), 1.22 (d, 18, J = 7.2 Hz).

Chapter 24

Hydrogen-Bonded Cavities Based upon Resorcin[4]arenes by Design

Leonard R. MacGillivray[1] and Jerry L. Atwood[2]

[1]Steacie Institute for Molecular Sciences, National Research Council of Canada, Ottawa, Ontario K1A 0R6, Canada
[2]Department of Chemistry, University of Missouri at Columbia, Columbia, MO 65211

Strategies for the design of multi-molecular hosts that display recognition properties analogous to monomolecular predecessors are described. In particular, the bowl-shaped cavities of molecular receptors known as resorcin[4]arenes have been elaborated, using hydrogen bonds, for the construction of open and closed cavities that accommodate single and multiple guest species. Both chemical and geometric considerations derived from these studies have led us to identify host frameworks, based upon the five Platonic and 13 Archimedean solids, which have yet to be synthesized or discovered and we propose such systems as targets in chemical synthesis.

Introduction

Chemistry has witnessed the emergence of an approach to chemical synthesis that focuses on the exploitation of non-covalent forces (*e.g.* hydrogen bonds, π-π interactions) for the design of multi-component supramolecular frameworks.[1] Our fascination with Nature's ability to direct the self-assembly of small subunits into large superstructures (*e.g.* viruses, fullerenes), coupled with a desire to construct materials possessing unique bulk physical properties (*e.g.* optical, magnetic), has undoubtedly provided a major impetus for their design.[2]

Multi-Component Hosts

Along these lines, there is much interest in utilizing non-covalent forces, particularly in the form of hydrogen bonds, for the construction of multi-component hosts (*e.g.* molecular capsules **1**) that display recognition properties analogous to their monomolecular predecessors (*e.g.* carcerands **2**).[3-4] Such frameworks typically

© 2000 American Chemical Society

involve replacing covalent bonds with supramolecular synthons that retain the structural integrity of the parent host molecule.

Notably, in addition to providing access to systems which are difficult to obtain using conventional covalent approaches to molecular synthesis,[5] such frameworks can display properties not found in the molecular analog (*e.g.* reversible formation) which, in some instances, can bear relevance in understanding related biological phenomena (*e.g.* virus formation).[6]

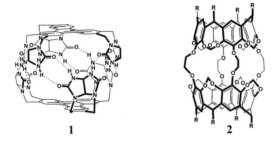

Resorcin[4]arenes for Multi-Component Host Design

With this in mind, we recently initiated a program of study aimed at extending the cavities of resorin[4]arenes supramolecularly.[7] We chose the readily available *C*-methylcalix[4]resorcinarene **3** as a platform for the assembly process.[8] Indeed, solid state studies revealed the ability of **3** to adopt a bowl-like conformation with C_{2v} symmetry in which four upper rim hydroxyl hydrogen atoms of **3** are pointed upward above its cavity which, in turn, effectively make **3** a quadruple hydrogen bond donor.[9] Using a resorcinol-based supramolecular synthon[10] **4** for host design, we reasoned that co-crystallization of **3** with hydrogen bond acceptors such as pyridines would result in formation of four O-H···N hydrogen bonds between the upper rim of **3** and four pyridine units which would extend the cavity of **3** and yield a discrete, multi-component host, **3**·4(pyridine) **5** (where pyridine = pyridine and derivatives), capable of entrapping a guest, **3**·4(pyridine)·guest.

Notably, from these studies, we also discovered the ability of **3** to self-assemble in the solid state as a spherical hexamer, along with eight water molecules, to form, in a way similar to **1**, a spherical container assembly held together by 60 O-H···O hydrogen bonds **6** (Fig. 1a).[11]

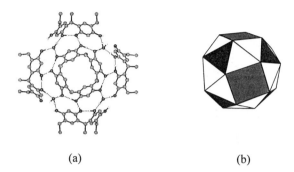

(a)   (b)

*Figure 1. (a) Cross-sectional view of X-ray crystal structure of **6** and (b) snub cube. Reproduced with permission from reference 11.*

The spherical assembly, which is chiral, was found to possess a well-defined central cavity with a maximum diameter of 1.8 nm and an internal volume of about 1.4 nm^3, a volume five times larger than the largest molecular capsule previously reported.[12] Consultation of polyhedron models also revealed the structure of **6** to conform to a snub cube, one of the 13 Archimedean solids (Fig. 1b).[13] Indeed, from these observations we concluded that to design related spherical containers, one must consider the limited topologies available in space for such frameworks (*i.e.* the five Platonic and 13 Archimedean solids).[11]

Elaborating Resorcin[4]arenes Supramolecularly

Cavities Based Upon Rigid Extenders

The product of the co-crystallization of **3** with pyridine from boiling pyridine is shown in Fig. 2.[7] The assembly is bisected by a crystallographic mirror plane and consists of **3** and five molecules of pyridine, four of which form four O-H···N hydrogen bonds, as two face-to-face dimers, such that they adopt an orthogonal orientation, in a similar way to **4**, with respect to the upper rim of **3**. As a consequence of the assembly process, a cavity has formed inside which a disordered molecule of pyridine is located, interacting with **3** by way of C-H···π-arene interactions.[14] Notably, the remaining hydroxyl hydrogen atoms of the six-component assembly form four intramolecular O-H···O hydrogen bonds along the upper rim of **3** resulting in a total of eight structure-determining O-H···X (X = N, O) forces. Indeed, the inclusion of an aromatic such as pyridine within 3·4(pyridine) is reminiscent of the ability of covalently modified calix[4]arenes, such as *p-tert*-butylcalix[4]arene, to form molecular complexes with aromatics such as benzene and toluene.[15]

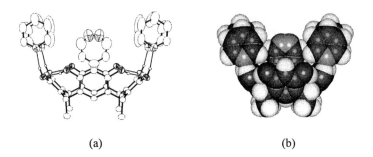

(a) (b)

Figure 2. X-ray crystal structure of 3·4(pyridine)·pyridine: (a) ORTEP perspective and (b) space-filling view. The pyridine molecule located within the cavity lies disordered across a crystallographic mirror plane.

To determine whether it is possible to isolate a guest within **5** which, unlike 3·4(pyridine), is chemically different than the 'substituents' hydrogen bonded to the upper rim of **3**, we turned to pyridine derivatives, namely, 4-picoline (monopyridine) and 1,10-phenanthroline (bipyridine).[16] In a way similar to pyridine, both molecules possess hydrogen bond acceptors along their surfaces and π-rich exteriors which we anticipated would allow these units to assemble along the upper rim of **3** as stacked dimers. As shown in Fig. 3, co-crystallization of **3** with either 4-picoline or 1,10-phenanthroline from MeNO$_2$ and MeCN, respectively, yielded six-component

(a) (b)

Figure 3. ORTEP perspective of: (a) 3·4(4-picoline)·MeNO$_2$ and (b) 3·4(1,10-phenanthroline)·MeCN. The included MeNO$_2$ molecule lies disordered on a crystallographic two-fold rotation axis. Reproduced with permission from reference 16.

complexes, **3**·4(4-picoline)·MeNO$_2$ and **3**·4(1,10-phenanthroline)·MeCN, which are topologically equivalent to the parent pyridine system.[7] Unlike the parent assembly, however, the cavities created by the five molecules were occupied by guests different than the walls of the host. Indeed, this observation illustrated that this approach to discrete, extended frameworks based upon **3** is not limited to two different components. Notably, whereas the 4-picoline moiety was observed to interact with **3** by way of conventional O-H···N hydrogen bonds, the 1,10-phenanthroline extender was also observed to interact with **3** by way of a bifurcated O-H···N force.

A Cavity Based Upon a Flexible Extender

With the realization that **5** may be exploited for the inclusion of guests different than the supramolecular extenders of **3** achieved, we shifted our focus to pyridines that possess flexible substituents. In addition to introducing issues of stereochemistry, we anticipated that this approach would allow us to further address the robustness and structural parameters which define **4** and thereby aide the future design of analogous host-guest systems based upon **3**.

Our first study in this context has involved 4-vinylpyridine.[17] As shown in Fig. 4, in a similar way to the discrete systems described above, four 4-vinylpyridines assemble along the upper rim of **3**, as two face-to-face stacked dimers, in **3**·4(4-vinyl-pyridine)·MeNO$_2$, to form a six-component assembly. Interestingly, the olefins of this system, in contrast to resorcinol·4-vinylpyridine, adopt a parallel orientation in the crystalline state, the bonds being separated by a distance of 4.18 Å. Indeed, approaches that utilize host frameworks to promote alignment of olefins in the solid

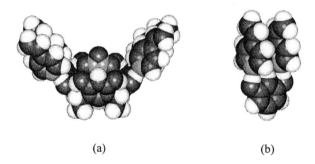

(a)   (b)

*Figure 4. Space-filling view of the X-ray crystal structure of (a) **3**·4(4-vinylpyridine)· MeNO$_2$ and (b) resorcinol·2(4-vinylpyridine). The included MeNO$_2$ molecule lies disordered on a crystallographic two-fold rotation axis. Reproduced with permission from reference 17.*

state, for conducting [2+2] photochemical reactions, for example, are rare[18] and these observations suggest that similar complexes based upon **3** may provide a route to achieving this goal.

Spherical Molecular Assemblies

It is clear that co-crystallization of **3** with pyridine and its derivatives in the presence of a suitable guest results in the elaboration of the cavity of **3** in which four pyridine units assemble along the upper rim of **3**, as two stacked dimers, to yield a multi-component complex **5** in which **3** serves as a quadruple hydrogen bond donor.[7,16-17] During studies aimed at co-crystallizing **3** with pyridines from aromatic solvents - potential guests for **5** (*e.g.* nitrobenzene) - we discovered the ability of **3** to self-assemble in the solid state, along with eight water molecules, to form a chiral spherical container assembly, with idealized octahedral symmetry, that is held together by 60 O-H···O hydrogen bonds **6** (Fig. 1).[11] Although guest species could be located within **6** (*i.e.* electron density maxima), it was not possible to determine their identity from the X-ray experiment, presumably owing to the high symmetry of the host and high thermal motion within the cavity. Notably, solution studies also revealed the ability of *C*-undecylcalix[4]resorcinarene to maintain the structure of **6** in apolar organic solvents such as benzene.[11]

Polyhedron Model - Snub Cube

With a rational approach to elaborating the cavity of **3** supramolecularly achieved, we turned to determine whether the structure of **6** could be rationalized according to principles of solid geometry. Indeed, to us, the ability of six molecules of **3** to self-assemble to form **6** was reminiscent of spherical viruses[6] (*e.g.* hepatitis B) and fullerenes[19] (*e.g.* C_{60}) in which identical copies of proteins and carbon atoms self-assemble to form spherical molecular structures having icosahedral symmetry and a shell-like enclosure. Consultation of polyhedron models revealed the structure of **6** to conform to a snub cube, one of the 13 Archimedean solids,[13] in which the vertices of the square faces correspond to the corners of **3** and the centroids of the eight triangles that adjoin three squares correspond to the eight water molecules. Moreover, owing to its classification as an Archimedean solid, we concluded that to rationally construct similar hosts, one must consider the limited possibilities available in space for such spherical frameworks, those being the five Platonic and 13 Archimedean solids.

Platonic Solids

The Platonic solids comprise a family of five convex uniform polyhedra which possess cubic symmetry (*i.e. 32, 432,* or *532* symmetry)[20] and are made of the same regular polygons (*e.g.* equilateral triangle, square) arranged in space such that the vertices, edges, and three coordinate directions of each solid, are equivalent (Fig. 5).[13] That there is a finite number of such polyhedra is due to the fact that there exists a

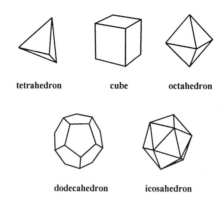

Figure 5. The five Platonic solids.

limited number of ways in which identical regular polygons may be adjoined to construct a convex corner. Equilateral triangles may be adjoined in three ways while squares and pentagons may be adjoined in only a single manner. Moreover, it is impossible to create a convex corner using regular polygons with six or more sides since the sum of the angles around each vertex would be greater than or equal to 360 degrees.[13] These principles give rise to five isometric polyhedra which are achiral and whose polygons are related by combinations of *n*-fold rotation axes. The Platonic solids include the tetrahedron, which belongs to the point group T_d and possesses *32* symmetry; the cube and octahedron, which belong to the point group O_h and possess *432* symmetry; and the dodecahedron and icosahedron, which belong to the point group I_h and possess *532* symmetry.

Archimedean Solids

In addition to the Platonic solids, there exists a family of 13 convex uniform polyhedra known as the Archimedean solids (Fig. 6). Each member of this family is made up of at least two different regular polygons and may be derived from at least one Platonic solid through either trunction or the twisting of faces.[13] In the case of the latter, two chiral members, the snub cube and the snub dodecahedron, are realized. The remaining Archimedean solids are achiral. Like the Platonic solids, the Archimedean solids possess identical vertices and exhibit either *32*, *432*, or *532* symmetry, respectively. The Archimedean solids possess a wider variety of polygons than the Platonic solids. These include the equilateral triangle, square, pentagon, hexagon, octagon, and decagon.

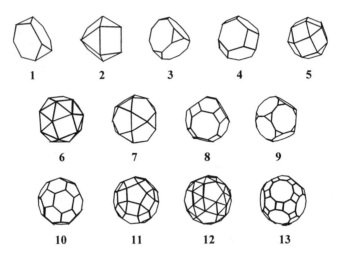

Figure 6. The 13 Archimedean solids, in order of increasing number of vertices. Truncated tetrahedron (1), Cuboctahedron (2), Truncated cube (3), Truncated octahedron (4), Rhombicuboctahedron (5), Snub cube (6), Icosidodecahedron (7), Rhombitruncated cuboctahedron (8), Truncated dodecahedron (9), Truncated icosahedron (10), Rhombicosidodecahedron (11), Snub dodecahedron (12), Rhombitruncated icosidodecahedron (13).

Spherical Organic Hosts from the Laboratory and from Nature

Our contention here that spherical molecular hosts may be constructed, in a way similar to **6**, according to principles of solid geometry renders a variety of organic components (*i.e.* subunits that correspond to regular polygons) viable for their design and permits the individual components of the host to be held together by covalent and/or noncovalent bonds. To demonstrate the utility of this symmetry-based approach, we now present selected examples of spherical hosts composed of organic components from both the laboratory and nature. We believe that this method allows one to identify similarities at the structural level which, at the chemical level, may not seem obvious and may be used to design assemblies similar to **6**.[21]

Platonic Solids
As stated, the Platonic solid constitute a family of five convex uniform polyhedra made up of the same regular polygons and possess either *32*, *432*, or *532*, symmetry.

Tetrahedral Systems

The macrotricyclic spherand designed by Lehn et al. was the first tetrahedral host (Fig. 7a).[22] The bridgehead nitrogen atoms, located at the corners of the tetrahedron, and ethyleneoxy units, the edges, supply the three-fold and two-fold rotation axes, respectively. Notably, this molecule and its tetraprotonated form has been shown to bind an ammonium and chloride ion, respectively. Schmidtchen et al. have introduced similar tetrahedral cages with edges comprised entirely of methylene bridges[23] while Vögtle et al. have demonstrated the synthesis of a hollow hydrocarbon called spheriphane (Fig. 7b).[24]

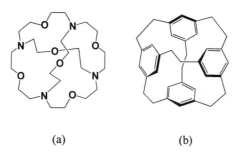

(a)   (b)

Figure 7. Tetrahedral hosts with T_d symmetry, (a) Lehn's spherand, (b) Vögtle's spheriphane.

Octahedral Systems

We present two octahedral hosts related to the Platonic solids. Both are based upon the cube.

The first is a cyclophane-based system reported by Murakami et al. (Fig. 8).[25] The sides of the host consist of tetraaza-[3.3.3.3]paracyclophane units and its octaprotonated cation has been shown to bind anionic guests.

The second is a cube synthesized by Chen and Seeman the components of which are based upon DNA.[26] The directionality and ability of the double helix to form branched junctions are exploited for the edges and vertices, respectively. Interestingly, each face of this DNA-based system forms a cyclic strand which is catenated with strands of adjacent faces. Molecular modeling experiments indicate the length of each edge to be approximately 6.8 nm.

Figure 8. Murakami's cyclophane-based cube.

Icosahedral Systems

Spherical viruses are icosahedral molecular hosts related to the Platonic solids (Fig. 9a).[6] Consisting of identical copies of proteins which assemble, in a similar way to 6, by way of non-covalent forces, these hosts range from 15 to 90 nm in diameter and encapsulate strands of ribonucleic acid (RNA). Although the shells of spherical viruses require a minimum of 60 subunits, most are made up of 60n (n = 1, 3, 4...)

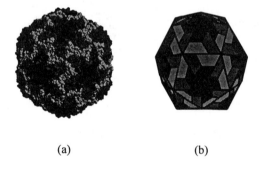

Figure. 9 An icosahedral host, (a) X-ray crystal structure of the rhinovirus, a spherical virus linked to the common cold, (b) a schematic representation of the rhinovirus displaying triangulation. Reproduced with permission from reference 21.

subunits owing to a reduction in symmetry of their polygons (Fig. 9b). This process, known as triangulation, gives rise to quasi-equivalent positions along the surface of the shell which enable the virus particle to cover the RNA with the largest number of subunits. Since only certain triangulations are permitted by symmetry, viruses may be classified into a coherent system.[6]

Archimedean Solids

As stated, the Archimedean solids constitute a family of 13 convex uniform polyhedra made up of two or more regular polygons and, like the Platonic solids, possess either *32*, *432*, or *532* symmetry.

Cuboctahedron (2)

Ross *et al.*, using MM2 molecular model simulations, have considered the existence of a molecule containing eight benzene rings and either 12 oxygen or sulfur atoms which they refer to as heterospherophane (Fig. 10).[27] Although it is not mentioned in the original report, the shell exhibits a topology identical to a cuboctahedron.

Figure 10. A theoretical organic shell based upon the cuboctahedron (X = O, S).

Truncated Octahedron (4)

Seeman and Zhang have constructed a DNA-based shell analogous to a truncated octahedron.[28] The edges of this molecule, each of which contains two turns of the double helix, contain 1440 nucleotides and the molecular weight of the structure, which is an overall 14-catenane, is 790 000 Daltons. Interestingly, the design strategy relies on a solid support approach in which a net of squares is ligated to give the polyhedron. It is currently unclear what shape the molecule adopts in various media.

Snub Cube (6)

As stated, we have demonstrated the ability of six molecules of **3** and eight molecules of H_2O to assemble in apolar media to form a spherical molecular assembly which conforms to a snub cube (Fig. 11).[11] Each resorcin[4]arene lies on a four-fold rotation axis and each H_2O molecule on a three-fold axis. The assembly, which exhibits an external diameter of 2.4 nm, possesses an internal volume of about 1.4 nm^3 and is held together by 60 O-H···O hydrogen bonds.

336

Figure 11. Space-filling view of the X-ray crystal structure of the cavity of 6. Reproduced with permission from reference 11.

Truncated Icosahedron (10)

Buckminsterfullerene, an allotrope of carbon, is topologically equivalent to a truncated icosahedron, an Archimedean solid that possesses 12 pentagons and 20 hexagons (Fig. 12).[19] Each carbon atom of this fullerene corresponds to a vertex of the polyhedron. As a result, C_{60} is held together by 90 covalent bonds, the number of edges of the solid.

Figure 12. X-ray crystal structure of buckminsterfullerene, C_{60}, a shell based upon the truncated icosahedron. Reproduced with permission from reference 21.

Archimedean Duals and Irregular Polygons

As stated, the Platonic and Archimedean solids comprise two finite families of polyhedra in which each solid consists of identical vertices, edges, and either a single or two or more regular polygons. It is of interest to note, however, that there exists a family of spherical solids which are made up of irregular polygons which may also be used as models for spheroid design. Known as Archimedean duals,[13] these polyhedra are constructed by simply connecting the midpoints of the faces of an Archimedean solid. Such a treatment gives rise to 13 polyhedra which possess two or more different vertices and identical irregular polygon faces (Fig. 13). As a result, chemical subunits used to construct hosts which conform to these polyhedra cannot be based upon regular polygons.

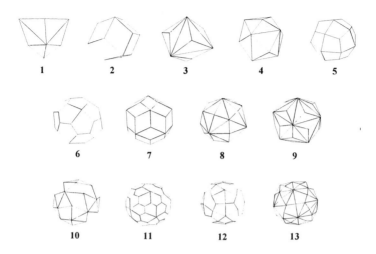

Figure 13. The 13 Archimedean duals derived from corresponding Archimedean solids. Triakis tetrahedron (1), Rhombic dodecahedron (2), Triakis octahedron (3), Tetrakis hexahedron (4), Deltoidal icositetrahedron (5), Pentagonal icositetrahedron (6), Rhombic tricontahedron (7), Disdyakis dodecahedron (8), Triakis icosahedron (9), Pentakis dodecahedron (10), Deltoidal hexecontahedron (11), Pentagonal hexecontahedron (12), Disdyakis triacontahedron (13).

Rhombic Dodecahedron (2)

To the best of our knowledge, there is one host which conforms to the structure of an Archimedean dual. Harrison was the first to point out that the quaternary structure of ferritin, a major iron storage protein in animals, bacteria, and plants, corresponds to the structure of a rhombic dodecahedron.[29] This protein, which is approximately 12.5 nm in diameter, consists of 24 identical polypeptide subunits (Fig. 14a), and holds up to 4500 iron atoms in the form of hydrated ferric oxide with varying amounts of phosphate $[Fe_2O_3(H_2O/H_3PO_4)_n]$.[30] The polypeptides, which consist of four helix bundles, assemble by way of non-covalent forces and form dimers which correspond to the faces of the solid (Fig. 14b).

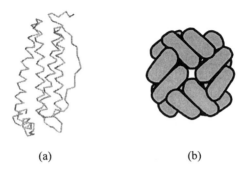

(a)   (b)

Figure 14. Structure of ferritin. A spherical host based upon a rhombic dodecahedron, (a) carbon trace of the polypeptide subunit, (b) the assembly displayed by the subunits. Reproduced with permission from reference 21.

Irregular Polygons

It is important to point out that if partial truncation is applied to the Platonic solids such that Archimedean solids are not realized, or if truncation is applied to the Archimedean solids, then the resulting polyhedra will not possess regular faces but, like the Archimedean duals, may be used as models for spheroid design as a consequence of their cubic symmetries. Indeed, of the spherical hosts synthesized to date, all have been constructed using chemical subunits which either correspond to regular polygons (*e.g.* calix[4]arenes) or form regular polygons (*e.g.* carbon-based hexagons and pentagons). Moreover, that spherical shells may be constructed using polyhedra with irregular faces, as in the case of ferritin, implies that spherical shells based upon irregular polygons may be rationally designed.

Conclusions

We have presented here approaches for rationally designing hydrogen bonded cavity-containing assemblies based upon **3**. By utilizing **3** as a quadruple hydrogen bond donor, we have illustrated the ability of **3** to assemble with either itself or aromatic-based hydrogen bond acceptors to form multi-component hosts that display molecular recognition properties analogous to mono-molecular predecessors. Specifically, the ability of pyridine and its derivatives, to deepen the cavity of **3** has been demonstrated[7,16-17] while **3** has also been observed to self-assemble as a spherical hexamer, along with eight water molecules, to form a container assembly held together by 60 hydrogen bonds.[11] In the case of the latter, the host conforms to the structure of a snub cube, one of the 13 Archimedean solids, and a general strategy for the design of similar host systems has been developed.[21] With such observations realized, we anticipate that, in addition to expanding the library of components used to construct these frameworks, focus will be placed upon incorporating increasing levels of function (*e.g.* selectivity, chirality) within these materials.

Acknowledgements. We are grateful for funding from the Natural Sciences and Engineering Research Council of Canada (NSERC) and the National Science Foundation (NSF).

References

1. Lehn, J. -M. (1995) *Supramolecular Chemistry, Concepts and Perspectives*, VCH, Weinheim.
2. Philip, D.; Stoddart, J. F. *Angew. Chem., Int. Ed. Engl.* **1996**, *35*, 1155.
3. Heinz, T. H.; Rudkewich, D. M.; Rebek, J., Jr. *Nature* **1998**, *394*, 764.
4. Jetti, R. K. R.; Kuduva, S. S.; Reddy, D. S.; Xue, F.; Mak, T. C. W.; Nangia, A.; Desiraju, G. R. *Tetrahedron Lett.* **1998**, *39*, 913.
5. de Mendoza, J. *Chem. Eur. J.* **1998**, *4*, 1373.
6. Caspar, D.; Klug, A. *Cold Spring Harb. Symp. Quant. Biol.* **1962**, *27*, 1.
7. MacGillivray, L. R. and Atwood, J. L. *J. Am. Chem. Soc.* **1997**, *119*, 6931.
8. Högberg, A. G. S. *J. Am. Chem. Soc.* **1980**, *102*, 6046.
9. Murayama, L.; Aoki, K. *Chem. Commun.* **1997**, 119, and references therein.
10. Desiraju, G. R. *Angew. Chem., Int. Ed. Engl.* **1995**, *34*, 2311.
11. MacGillivray, L. R.; Atwood, J. L. *Nature* **1997**, *389*, 469.
12. Kang, J.; Rebek, J., Jr. *Nature* **1997**, *385*, 50.
13. Wenniger, M. J. *Polyhedron Models*, Cambridge Univ. Press, New York, 1971.
14. Leigh, D. A., Linnane, P., Pritchard, R. G. and Jackson, G. *J. Chem. Soc., Chem. Commun.* **1994**, 389.
15. Andretti, G. D., Ugozzoli, F., Ungaro, R. and Pochini, A. in *Inclusion Compounds*, Oxford University Press, Oxford, 1991; Vol. 4, pp. 64.
16. MacGillivray, L. R.; Atwood, J. L. *Chem. Commun.*, **1999**, 181.
17. MacGillivray, L. R.; Reid, J. L.; Atwood, J. L.; Ripmeester, J. A. *Cryst. Eng.*, in press.
18. Toda, F. *Top. Curr. Chem.* **1988**, *149*, 211.
19. Kroto, H. W.; Heath, J. R.; O'Brien, S. C.; Curl, R. F.; Smalley, R. E. *Nature* **1985**, *318*, 162.
20. The phrases *32*, *432*, and *532* refer to tetrahedral, octahedral, and icosahedral symmetries, respectively, and denote the presence of 2-, 3-, 4-, and/or 5- fold rotation axes.
21. A detailed account of this design strategy which also includes inorganic frameworks can be found in: MacGillivray, L. R.; Atwood, J. L. *Angew. Chem., Int. Ed. Engl.* **1999**, *38*, 1018.
22. Graf, E.; Lehn, J. -M. *J. Am. Chem. Soc.* **1975**, 97, 5022.
23. Schmidtchen, F. P.; Müller, G. *J. Chem. Soc., Chem. Commun.* **1984**, 1115.
24. Vögtle, F.; Seel, C.; Windscheif, P. -M. In *Comprehensive Supramolecular Chemistry, Cyclophane Hosts: Endoacidic, Endobasic, and Endolipophilic Large Cavities, Vol. II*; Lehn, J. -M.; Atwood, J. L.; Davies, J. E. D.; MacNicol, D. D.; Vögtle, F., Ed.; Pergamon, New York, 1996.
25. Murakami, Y.; Kikuchi, J.; Hirayama, T. *Chem. Lett.* **1987**, 161.
26. Chen, J.; Seeman, N. C. *Nature* **1991**, *350*, 631.

27. Ross, R. S.; Pincus, P.; Wudl, F. *J. Phys. Chem.* **1992**, *96*, 6169.
28. Zhang, Y.; Seeman, N. C. *J. Am.Chem. Soc.* **1994**, *116*, 1661.
29. Smith, J. M. A.; Stansfield, R. F. D.; Ford, G. C.; White, J. L.; Harrison, P. M. *J. Chem. Edu.* **1988**, *65*, 1083.
30. Trikha, J.; Theil, E. C.; Allewell, N. M. *J. Mol. Biol.* B, **1995**, 949.

Author Index

Allain, F., 56
Arnaud-Neu, F., 150
Atwood, Jerry L., 325
Baaden, M., 71
Babain, Vasily A., 107
Barboso, S., 150
Bartsch, Richard A., 112, 125
Bencze, Zsolt, 195
Berny, F., 71
Besançon, Frédéric 179
Böhmer, Volker, 135
Bonnesen, Peter V., 26
Bryan, Jeffrey C., 86
Bünzli, Jean-Claude G., 179
Byrne, D., 150
Cameron, Beth R., 283
Casnati, A., 12
Charbonnière, L. J., 150
Chirakul, Panadda, 195
Chmutova, Marina K., 107
Crawford, C. L. 45
Delmau, Laetitia H., 86
Desreux, J. F., 165
Dozol, J.-F., 12, 56
Duesler, Eileen N., 195
Elkarim, Nazar S. A., 125
Engle, Nancy L., 26, 86
Fondeur, F. F., 45
Gale, Philip A., 238
Genge, John W., 238
Gibbs, Charles G., 313
Gopalan, Aravamudan S., 208
Gutsche, C. David, 2, 313
Hampton, Philip D., 195
Hartman, Deborah H., 223
Haverlock, Tamara J., 26
Hay, Benjamin, P., 86
Hosseini, Mir Wais, 296
Huddleston, Jonathan G., 223
Hwang, Hong-Sik, 112, 125
Ihringer, Frédéric 179

Jacques, V., 165
Jankowski, C., 56
Jarvinen, Gordon D., 208
Kal'tchenko, Vitaly I., 107
Koch, H. Fred, 255
Král, Vladimír, 238
Lamare, V., 12, 56
Lambert, B., 165
Lambert, Timothy N., 208
Loeb, Stephen J., 283
Logunov, Mikhail V., 107
MacGillivray, Leonard R., 325
Moulin, C., 56
Moyer, Bruce A., 26, 86
Muzet, N., 71
Park, Chunkyung, 112
Peterson, R. A., 45
Rebek, Julius, Jr., 270
Rogers, Robin D., 223
Roundhill, D. Max, 255
Rudkevich, Dmitry M., 270
Sachleben, Richard A., 26, 86
Schwing-Weill, M. J., 150
Sessler, Jonathan L., 238
Shadrin, Andrey Yu., 107
Simon, N., 12
Swatloski, Richard P., 223
Tabet, J.-C., 56
Talanov, Vladimir S., 112, 125
Talanova, Galina G., 112, 125
Tallant, Matthew D., 208
Troxler, L., 71
Ulrich, G., 150
Ungaro, R., 12
Virelizier, H., 56
Visser, Ann E., 223
Wang, Jian-she, 313
White, T. L., 45
Wipff, G., 71
Wu, Si, 195

Subject Index

A

Acidic media, cesium extraction, 18, 20f
Acidic nuclear waste. *See* Cesium extraction from acidic nuclear waste
ACTINEX (actinide extraction) program, recovery of minor actinides, 13
Actinides
  extractants, 135
  See also Acyclic analogs of *tert*-butyl-calix[4]arene hydroxamate extractants; Carbamoylmethylphosphine oxide-substituted calixarenes
Acyclic analogs of *tert*-butyl-calix[4]arene hydroxamate extractants
  calixarene-based extractants for actinide (III or IV) ions, 210f
  calixarene-based extractants for tetra- and trivalent actinides, 209, 210f
  calix[4]arene tetrahydroxamates and acyclic analogs, 212f
  competitive extraction at pH 1 Th(IV) versus Fe(III), 215t
  competitive extraction at pH 2 Th(IV) versus Fe(III), 216t
  competitive extraction studies– Th(IV) versus Fe(III), 214–217
  considerations when designing chelators, 208–209
  extraction preference of trihydroxamate for Th(IV) cation, 217, 218f, 220
  goal of research program, 209, 211
  ligand molar variation studies, 217
  pH dependence of extraction capabilities, 214
  resorcin[4]arene cavitands for Eu(III) extraction, 209
  single metal ion extraction studies, 211, 214
  spectroscopic analyses, 217
  syntheses, 211, 213
  synthetic route to tetrahydroxamate extractant, 213
  Th(IV) and Fe(III) single metal extraction studies, 214t
  UV–VIS spectra of extracted species of Fe(III) complexes at pH 2 and trihydroxamate-Fe(III) complexes from ligand molar variation study, 219f
Adenosine nucleotides, separation on HPLC-modified column, 246–247
Alkaline radioactive wastes
  batch contact flow diagram, 111f
  chemical composition of feed and resultant solutions, 110t
  extraction of Cs, Sr, Tc, and Am from alkaline media by *tert*-butyl-calix[6]arene, 108t
  extraction tests from simulated solutions, 108–109
  mixture containing t-butyl-calix[6]arene, 107–108
  radiochemical composition of feed and resultant solutions, 110t
  single extractant for long-lived radionuclides, 107–108
  static test on actual high-level waste, 109, 111
p-Allylcalix[5]arenes
  5-allylcalix[5]arene-

31,32,33,34,35-pentol, 317, 319
Benesi–Hildebrand treatment for complexation constants, 315, 322
complexation constants for calix[5]arenes and fullerenes C_{60} and C_{70}, 316t
complex formation from bis-calixarenes and fullerenes, 313
complex formation with C_{60} and C_{70}, 315–317
complex of bis-calixarene and C_{60}, 318f
5,11-diallylcalix[5]arene-31,32,33,34,35-pentol, 319
5,17-diallylcalix[5]arene-31,32,33,34,35-pentol, 319–320
Job plot for bis-calixarene and C_{60}, 318f
Job plots, 317
measurement of complexation constants, 321–322
pathways for synthesis of, 316
5,11,17,23,29-pentaallylcalix[5]arene-31,32,33,34,35-pentol, 321
silylation modification of Claisen rearrangement, 314
substituted calix[5]arenes under study, 315
synthesis, 314–315
synthetic procedures and results, 317, 319–321
5,11,17,23-tetraallylcalix[5]arene-31,32,33,34,35-pentol, 320–321
5,11,23-triallylcalix[5]arene-31,32,33,34,35-pentol, 320
Americium
extraction from aqueous nitric acid, 144t
See also Alkaline radioactive wastes
Amide substituted calix[4]arenes. See Phase transfer extractants for oxyions

Amine substituted calix[4]arenes. See Phase transfer extractants for oxyions
Anion separations
separation on HPLC-modified column, 244–245
See also Silica gels, calix[4]pyrrole-functionalized
Aqueous biphasic systems (ABS)
alternative to volatile organic compounds (VOCs), 224
See also Calixarene partitioning
Archimedean solids
Archimedean duals and irregular polygons, 336–337
cuboctahedron, 335
duals, 337f
rhombic dodecahedron, 337, 338f
snub cube, 335, 336f
spherical molecular assemblies, 331–332
structures, 332f
theoretical organic shell based on cuboctahedron, 335f
truncated icosahedron, 336
truncated octahedron, 335
Assisted liquid-liquid ion extraction
adsorption of Cs^+ calixarene complexes at interface, 76–78
cations approaching interface, 82–83
complexation and ion recognition at liquid-liquid interface, 82
computational issues concerning energy representation of system, 83–84
computer demixing experiment on water-chloroform mixture, 72
Cs^+ calixarene complex simulations at tributyl phosphate (TBP)/water interface, 78f, 79f
demixing of "perfectly mixed" water/chloroform binary mixture containing 30 TBPs and 5

$UO_2(NO_3)_2$ salts, 80f
demixing of ternary water/chloroform/TBP solution, 79–81
effect of concentration and synergism with TBP, 76–78
extracting complexes from interface to organic phase, 83
molecular dynamics (MD) simulations for cesium and uranyl cations extraction from water, 72
role of interfacial region in cation capture, 71–72
simulated species, 72
simulating synergistic and salting out effects using TBP, 78
simulation conditions at interface, 73t
simulation of Cs^+ calixarene complex at interface, 77f
simulation of models of $10[UO_2(NO_3)_2]$ and $10(Cs^+Pic^-)$ salts at interface, 75f
simulations methods, 73–74
snapshot of $[UO_2(NO_3)_2]$ tetramer in water phase, 76f
snapshot of Cs^+Pic^- salt at interface, 76f
synergistic effects with addition of TBP, 83
TBP•water "supermolecule" in chloroform and 1:1 and 1:2 complexes at TBP-water interface, 81f
uncomplexed salts at water/chloroform interface, 74–76
water-chloroform interface, 74
water/organic liquid separation at interface, 82
Azacalix[3]arenes
alkali-metal and alkylammonium ion extraction, 204
alkali metal and ammonium picrate extraction studies, 206
convergent approach, 198–199
crystal structures of $Y(H_3L)Cl_3$ and La(L) complexes (3 and 4), 200f, 203f
early synthetic work, 195–197
experimental section, 204–206
isolation of group 3 metal-ion complexes of, 199–202
structural data for $Y(H_3L)Cl_3$ and La(L) complexes (3 and 4), 202t
structures, 196f
synthesis by O-methylation, 205–206
synthesis of La(L) (4), 205
synthesis of precursors, 198f
synthesis of $Y(H_3L)Cl_3$ (3), 204–205
synthetic routes, 197–199
X-ray crystallographic data for complexes 3 and 4, 201t
X-ray crystallographic studies on complexes 3 and 4, 205

B

Baeyer, Adolf von, calixarenes, 2
Benzyl phenol derivatives
crystal structures, 94–96
general synthetic approach, 90
product yield and characterization, 93–94
synthesis of benzyl-α-methyl phenol derivatives, 90
synthesis of benzyl phenol derivatives, 90
synthesis of tert-octyl benzyl phenol (BPh2), 93
See also Self-assembly of benzyl phenol in binding Cs^+

Bimetallic lanthanide complexes
  absorption spectra of [Eu$_2$(b-LH$_2$)(DMF)$_5$] and [Eu(NO$_3$)$_3$(DMSO)$_4$], 185f
  achieving good sensitization for Tb(III) by antenna effect, 182–183
  antenna effect, 180
  antenna effect in sensitized Ln(III) complexes, 181f
  bathochromic shift for replacement of p-$tert$-butyl by nitro groups, 187
  p-$tert$-butylcalix[5]arene, 190
  p-$tert$-butylcalix[8]arene, b-LH$_8$, 184–187
  complexes with calix[8]arenes, 184–189
  complexes with p-$tert$-butylcalix[5]arene, b-L'H$_5$, 190
  conformational exchange mechanism in DMF solutions of [Lu$_2$(L)(DMF)$_5$], 188f
  control of structural and photophysical properties, 179–180
  design of lanthanide containing luminescent probes, 180–182
  effects causing luminescence upon excitation, 189
  energy migration processes in sensitized Ln(III) complexes, 182f
  energy transfer between excited ligand molecule and long-lived Ln(III) states, 181–182
  f–f transitions reliance on energy transfer, 180
  interaction of trivalent lanthanide ions (Ln(III)) with hard bases, 179
  ligand conformation in [Eu$_2$(n-LH$_2$)(DMF)$_5$], 188f
  luminescent lanthanide complexes with calixarenes, 182–183
  magnetic interactions in bimetallic Gd complex, 186–187
  molecular structure of [Eu$_2$(b-L'H$_2$)$_2$(DMSO)$_4$] dimer, 191f
  p-nitro-calix[8]arene, n-LH$_8$, 187–188
  partial energy diagram of trivalent Eu and Tb, 181f
  proposed mechanism for complex formation in DMF, 186
  proposed reaction intermediates in 1:1 and 2:1 transformation, 185f
  scheme and notation of investigated calixarenes, 184f
  scope of research, 183
  p-sulfonato-calix[8]arene, s-LH$_8$, 188–189
  titration of s-LH$_8$ by Eu(NO$_3$)$_3$·4DMSO, 189f
  transition for bimetallic and monometallic complexes, 185f
  tuning photophysical properties, 191
Binding cesium. *See* Self-assembly of benzyl phenol in binding Cs$^+$
Buckminsterfullerene
  Archimedean solid, 336
  truncated icosahedron, 336
  X-ray crystal structure, 336f
  See also Fullerenes
t-Butyl-calix[4]arene hydroxamate. *See* Acyclic analogs of *tert*-butyl-calix[4]arene hydroxamate extractants
t-Butyl-calix[6]arene. *See* Alkaline radioactive wastes

C

Cadmium ions
  extraction, 130–131
  See also Heavy metal cation

separations
Calix[4]arene-bis-crown-6 behavior under irradiation
  atomic charge set for molecular dynamics (MD) studies, 60f
  behavior of sodium complexes, 68–69
  calixcrowns as radiolytically stable molecules, 68–69
  cesium structures after 500 ps of MD in gas phase, 66f
  comparison of 1,3-*alt*-calix[4]arene-bis-crown-6 (BC6) and nitro derivatives for complexation of cesium and selectivity of extraction, 63–64
  complexation of BC6 with cation M^+, 59f
  complex BC6–NO_2 with $NaNO_3$ in crown in gas phase and at 40 ps of MD in water with nearest water molecules around Na^+, 66f
  distribution coefficients of Cs^+ and Na^+ with nitro compounds, 64t
  electrospray/mass spectroscopy (ES/MS) for investigation, 57
  energy averages during MD simulations in gas phase at 300K on BC6–NO_2, 67t
  ES/MS of BC6 in *o*-nitrophenyl octyl ether (NPOE) with internal standard, 61f
  ES/MS conditions, 58–59
  experimental materials, 58
  extraction of cesium and sodium by irradiated organic phases, 65
  identification of main irradiation products of BC6 by ES/MS, 62t
  influence of counter-ion on relative stability of complexes, 67–68
  influence of irradiated NPOE on Cs^+ and Na^+ extraction, 65t
  influence of nitro groups on gas phase stability of cation complex, 63
  irradiation process, 58
  molecular dynamics simulations, 65–69
  molecular dynamics studies, 59–60
  qualitative identification of BC6 radiolysis products, 60, 63
  reference calixarene 1,3-*alt*-calix[4]arene-bis-crown-6 (BC6) and nitro compound modeled, 57f
  representation of stability in gas phase of various complexes, 64f
  solid phase separation of radiolysis products, 58
  solvation and complex stability of cesium and sodium complexes in presence of water, 68
  structural averages during MD simulations in gas phase at 300K for BC6–NO_2, 67t
Calix[4]arene-crown-6
  synthesis of derivatives, 15
  synthetic scheme, 16
  See also Cesium extraction
Calix[4]arene-crown-6 ethers. *See* Cesium extraction from acidic nuclear waste
Calix[4]arene di(*N-X*-sulfonyl carboxamides)
  competitive extraction of Pb(II) and competing metal ions, 116f, 119f
  conformationally locked, 120, 121f, 122
  effect of para substituent on Pb(II) extraction, 117, 118f
  extraction of mercury(II) and cadmium(II), 130–131
  extraction of palladium(II) and platinum(II), 130
  extraction of Pb(II), 114, 116
  extraction of soft metal cations, 118
  fluorogenic reagent for Hg(II) recognition, 118, 120, 132
  pH profiles for Pb(II) extraction, 115f

solvent extraction of silver(I), 129–130
synthesis, 114, 115f
See also Metal ion separations
Calix[4]arene lanthanide perchlorate complexes. *See* Lanthanide calix[4]arene complexes
Calix[4]arene mono(*N-X*-sulfonyl carboxamides), metal ion separations, 122–123
Calix[4]arene tetrathioamides, efficient Ag(I) picrate extractants, 125, 127
Calix[4]arenes
composition, 15
conformations, 15f
heterocalixarene analogues, 239
modules for construction of deep cavitands, 271f
proof of vase-like structure, 238
receptors for cationic, anionic, and neutral substrates, 283
See also Carbamoylmethylphosphine oxide-substituted calixarenes; Metal ion separations
Calix[4]arenes, flexible, structural modification of wide rim carbamoylmethylphosphine oxide(CMPO)-calixarenes, 141–142
Calix[4]arenes, rigidified, structural modification of wide rim carbamoylmethylphosphine oxide(CMPO)-calixarenes, 143–144
Calix[4]pyrroles
heterocalixarene analogues, 239
See also Silica gels, calix[4]pyrrole-functionalized
Calix[5]arenes
structural modification of wide rim carbamoylmethylphosphine oxide(CMPO)-calixarenes, 141

See also Bimetallic lanthanide complexes
Calix[8]arenes. *See* Bimetallic lanthanide complexes
Calixarene metalloreceptors
attributes of Pd atom in new metalloreceptor, 292
competition reactions showing molecular recognition of 4-phenylpyridine (4-Phpy), 293t
design strategy, 283–284
experimental procedures, 284–287
^{1}H NMR spectrum indicating binding of 4-Phpy inside calix[4]arene cavity, 289f
molecular mechanics study showing most stable with pyridine (py) outside cavity, 291
molecular recognition of phenyl-substituted pyridines, 291–292
orientations of coordinated 2-Phpy, 3-Phpy, and 4-Phpy, 293f
preparation of [Pd(L^1)(pyridine)][BF$_4$], 286
preparation of [Pd(L^2)(4-Phpy)][BF$_4$], 286–287
preparation of macrobicyclic calix[4]arene ligand HL^1, 284
preparation of macrobicyclic calixarene ligand HL^2, 285
preparation of metalloreceptor [Pd(L^1)(MeCN)][BF$_4$], 284–285
preparation of metalloreceptor [Pd(L^2)(MeCN)][BF$_4$], 285–286
preparation of model ligand HL^3, 287
preparation of model metalloreceptor [Pd(L^3)(MeCN)][BF$_4$], 287
structures of [Pd(L^1)(py)][BF4] and [Pd(L^2)(4-Phpy)][BF$_4$], 289
structures of macrobicyclic calix[4]arene ligands HL^1 and HL^2, 288

synthesis and characterization of [Pd(L^1)(py)][BF$_4$] and [Pd(L^2)(4-Phpy)][BF$_4$], 288–291
synthesis and characterization of macrobicyclic calix[4]arene ligands, 287
synthesis and characterization of metalloreceptors [Pd(L^1)(MeCN)][BF$_4$] and [Pd(L^2)(MeCN)][BF$_4$], 288
variable temperature ^{1}H NMR spectrum of [Pd(L^1)(py)]$^+$ in CD$_3$NO$_2$, 290f

Calixarene partitioning
acid alizarin violet N (AAV) enhancing partitioning, 228
alternative separation with calixarene extractants, 225–226, 232
aqueous biphasic systems, 227–229
calculation of distribution ratios, 226
crystal structure of p-sulfonatocalix[4]arene, 225f
distribution of sulfonated and unsubstituted calixarenes in [Rmim][PF$_6$], 230, 232f
distribution ratios for metal ions between [bmim][PF$_6$] as function of pH, 232f
distribution ratios for p-sulfonatocalix[4]- and calix[6]arene, 229f
distribution values for Fe(III) and Eu(III), 228f
experimental, 226
green chemistry and alternative separations technologies, 224
partitioning and structures of sulfonated dyes in PEG-2000 ABS, 227f
room temperature ionic liquids, 229–231
separations science, 223
speciation diagram for p-sulfonatocalix[4]arene as function of pH, 228, 229f
structure of complex in Cr(AAV)(OH$_2$)$_3$•4H$_2$O, 228f
structure of low melting 1-decyl-3-methylimidazolium hexafluorophosphate ([Rmim][PF$_6$]), 230f

Calixarene phosphine oxides
complexation studies, 161
dependence on length of spacers, 163
extraction of thorium and europium nitrates, 161f
extraction studies, 160–161
structures, 160f
See also Phosphorylated calixarenes

Calixarenes
applications in separations, 112
basic family, 3
basic scaffold for ligating groups, 135
chromogenic, 6
complexation of anions, 5
discovery, 2–3
ion-binding capability, 3–4
ion-complexing with pendant functionalities, 5
molecule-binding capability, 3
platform for ligand introduction, 209
radiolysis of, 53–54
solid state complex, 5
tailoring for extraction, 56–57
techniques establishing structures, 3
traditional cavity-forming modules, 270
uranophile, 5
See also Carbamoylmethylphosphine oxide-substituted calixarenes; Phosphorylated calixarenes

Calixcrowns
  earliest study, 4
  radiolytically stable molecules, 68–69
  See also Calix[4]arene-bis-crown-6 behavior under irradiation
Calixpherands, cyclic moiety on calixarene, 4
Capsules
  cavitand complexing fullerene in toluene, 280f
  cavitands self-assembling into cylindrical capsule, 277f
  cyclic tetraimide dimerizing through hydrogen bonding, 277
  dimerization of cavitand via hydrogen bonding and self-inclusion, 280f
  dimerization of deep cavity tetraamides via intermolecular hydrogen bonding, 278, 280f
  exhibiting complexation of elongated guest-molecules, 278, 279f
  self-assembly, 277–280
  self-inclusion, 278, 280
  unsymmetrically filled capsule, 278
  See also Deep cavities; Hydrogen-bonded cavities
Carbamoylmethylphosphine oxide-substituted calixarenes
  calix[5]arenes, 141
  comparing distribution coefficients for various lanthanides and actinides, 139f
  complexes with diamagnetic Th^{4+} and paramagnetic Yb^{3+} cations, 139–140
  composition and structure of complexes with wide rim, 139–141
  (diphenylphosphoryl)acetic acid and active ester by X-ray diffraction, 137f
  distribution coefficient as function of HNO_3 concentration for extraction with narrow rim, 147f
  extraction of ^{241}Am from aqueous nitric acid, 144t
  extraction of different cations by different carbamoylmethylphosphine oxide (CMPO)-calixarenes, 142f
  extraction of narrow rim, 146–147
  flexible calix[4]arenes, 141–142
  ^1H NMR spectrum of complex with anhydrous $Th(ClO_4)_4$, 140f
  ionophoric properties of wide rim, 137–138
  linear oligomers, 142–143
  narrow rim, 146–147
  partial CMPO-derivatives, 144–145
  rigidified calix[4]arenes, 143–144
  schematic representation of explanations for C_{2v}-symmetry of lanthanide complexes with, 140f
  selectivities of wide rim, 138–139
  single crystal X-ray analysis of 1,3-di-CMPO-calixarene, 145f
  structural modification of wide rim CMPO-calixarenes, 141–146
  synthesis of narrow rim, 146
  synthesis of wide rim CMPO-calixarenes, 136–137
  transport studies of wide rim CMPOs, 138t
  variation of phosphorus function, 145–146
  wide rim, 136–141
  See also Phosphorylated calixarenes
N-Carbobenzyloxy protected amino acids, separation on HPLC-modified column, 245–246
Carcerands, molecule complexation with calixarenes, 5

Cavitands
  calix[4]arene and resorcinarene as modules for construction of deep cavitands, 271f
  complexing fullerene in toluene, 280f
  deeper, 281
  dimerization via hydrogen bonding and self-inclusion, 280f
  host-guest properties, 270–271
  ideal as host molecules, 270
  molecule complexation with calixarenes, 5
  self-folding, 273–275
Cavities
  flexible extender, 329–330
  rigid extenders, 327–329
  See also Deep cavities; Hydrogen-bonded cavities
Cesium, impact of gamma irradiation dose on distribution coefficient, 51f
Cesium complexes. See Calix[4]arene-bis-crown-6 behavior under irradiation
Cesium extraction
  affinity of dialkoxycalix[4]arene-crown-6 derivatives for cesium, 16
  calix[4]arenes and conformations, 15f
  cation fluxes from experiments with simulated waste, 21
  cesium distribution coefficients, 18
  cesium fluxes through calix-crown containing supported liquid membranes (SLMs), 21t
  competitive extraction and transport of cesium in presence of excess sodium, 17t
  composition of high activity liquid waste, 20t
  criteria for optimizing nitrophenyl hexyl ether (NPHE) percent, 22–23
  diisopropoxy-calix[4]arene-crown-6 and dinitrophenyloctyloxy-calix[4]arene-crown-6 as carriers, 21
  distribution coefficients of cesium from HNO_3 solutions, 21t
  distribution coefficients of cesium from $NaNO_3$ solutions, 18t
  extraction from high salinity media, 18–19
  extraction isotherms for dioctyloxycalix C6, 23f
  extraction of cesium and sodium–selectivity Cs^+/Na^+, 17t
  extraction of cesium by calixarene-crown-6, [2,5] dibenzo-crown-7, and [3,4] dibenzo-crown-7 from acidic media, 19f
  extraction of cesium from acidic media by mixtures, 20f
  extraction of cesium from nitric acid solutions, 18
  extraction procedures, 13
  flat sheet SLM device for transport experiments, 14f
  general scheme for reprocessing of spent fuel, 22f
  half life and mobility in nuclear waste, 13
  hydraulic properties of calix-crown containing organic phase, 22t
  hydraulic properties of different organic phases, 23t
  materials and methods, 13–14
  mixing dicarbollide (BrCO) and calixarene-crown-6, 19
  model of mass transfer for ^{137}Cs transport, 14
  scheme to 1,3-dialkoxycalix[4]arene crowns, 16
  selective cesium extraction flow sheet, 24f
  solvent extraction of cesium and cesium selectivity over sodium, 16–17

synthesis of calix[4]arene-crown-6 derivatives, 15
test results for actual radioactive waste, 19–21
test to process, 21–24
third phase formation, 23t
transport procedures, 13–14
See also Alkaline radioactive wastes; Assisted liquid-liquid ion extraction; Self-assembly of benzyl phenol in binding Cs$^+$
Cesium extraction from acidic nuclear waste
addition of trioctylamine (TOA) improving stripping efficiency, 41
batch-equilibrium experiments using ^{137}Cs tracer, 29–30
benefits of centrifugal contacting equipment, 27
cesium distribution behavior for solvent Cs-3BOO/150L, 40f
cesium extraction and stripping behavior for various calixarene-crowns, 38t
cesium selectivity studies, 41
chemical stability of solvent, 39
composition of sodium bearing waste (SBW) simulant, 32t
contacting and analytical procedures, 31
distribution behavior of selected elements, 31
elemental analyses by inductively coupled argon plasma (ICAP) spectroscopy, 30–31
experimental materials, 29
extraction and stripping results from SBW simulant, 37, 39
initial contacting experiments with SBW simulant, 31–35
instrumental detection limit for elements, 31
investigating mono- and bis-crown-6 derivatives of 1,3-alternate calix[4]arenes, 27
lipophilic calixarene-crown ethers, 28f
modifier 1-(1,1,2,2-tetrafluoroethoxy)-3-(4-*tert*-octylphenoxy)-2-propanol (Cs-3), 28f
nitration by electrophilic aromatic substitution reactions, 35, 37
nitric acid stability experiments, 30
percent cesium in solvent, raffinate, and mass balance as function of hours in contact with SBW simulant, 34f
process development, 39, 41
selected metal distribution and Cs selectivity data for extraction from SBW simulant using solvent Cs-3BOO/150L with and without TOA, 42t
solvent extraction in separations-technology development, 27
stability of calixarene-crown ethers to nitric acid, 35, 37
stability of selected calixarene crown ethers to nitric acid as function of exposure time, 36f
Charge matching. See Metal ion separations
Cobalt(II) complexes, use of thiacalixarenes in synthesis of koilands, 307, 309
Complexation constants
Benesi–Hildebrand treatment of spectrophotometric measurements, 315, 322
calix[5]arenes and fullerenes C$_{60}$ and C$_{70}$, 316t
See also *p*-Allylcalix[5]arenes
Computational chemistry, calixarenes, 6

Copper koiland, use of thiacalixarenes in synthesis, 307, 309
Cuboctahedron, Archimedean solid, 335
Cyclodextrins, traditional cavity-forming modules, 270

D

Deep cavities
complexes with adamantane, 275, 276f
deeper cavitands, 280f, 281
flexibility of resorcinarene changing with substitution, 274
^{1}H NMR spectra of cavitand during complexation with substituted adamantanes, 276f
intramolecular hydrogen bonding in vicinal diamides causing self-folding, 275f
novel synthetic strategy for deep cavitands, 273, 274f
preparation of deeper cavitands, 274f
self-folding cavitands, 273–275
slow exchange between complexes and free guest species, 274–275
synthesis, 271–273
synthetic route to deep cavitands and calix[4]arenes, 272f, 273f
unique species of molecular containers, 281–282
See also Capsules; Hydrogen-bonded cavities
Dicarbollide, mixing with calixarene-crown-6, 19, 20f
Distribution coefficients
cesium and sodium ions with nitro compounds, 64t
cesium from HNO$_3$ solutions, 21t
expression for cesium, 18, 48
impact of gamma irradiation dose on cesium distribution coefficient, 51f
impact of gamma irradiation dose on potassium distribution coefficient, 52f
Divalent metal ions. See Metal ion separations

E

Edifices, supramolecular. See Bimetallic lanthanide complexes
Electrospray/mass spectroscopy (ES/MS)
conditions, 58–59
studying behavior of complexes under irradiation, 57
Energy transitions. See Bimetallic lanthanide complexes
Environmentally benign liquid-liquid extraction. See Calixarene partitioning
Europium
extraction, 162
See also Phosphorylated calixarenes
Extraction
impact of increasing radiation exposure, 50–52
See also Acyclic analogs of tert-butyl-calix[4]arene hydroxamate extractants; Cesium extraction; Cesium extraction from acidic nuclear waste; Metal ion separations; Self-assembly of benzyl phenol in binding Cs$^+$

F

Ferritin
rhombic dodecahedron, 337

structure, 338f
Fullerenes
  complexation constants for calix[5]arenes and C_{60} and C_{70}, 316t
  complexation of p-allylcalix[5]arenes with, 315–317
  complexation with bis-calixarenes, 313
  molecule complexation with calixarenes, 5
  X-ray crystal structure of buckminsterfullerene, 336f
  See also p-Allylcalix[5]arenes

G

Gadolinium. *See* Lanthanide calix[4]arene complexes
Gamma irradiation fields
  distribution coefficient for cesium, 48
  experimental, 46–50
  gradient reverse-phase high performance liquid chromatography (HPLC) method for Isopar L, 48t
  HPLC method for composition determination, 46
  HPLC trace for standard at two wavelengths, 49f
  impact of diluent on solvent radiolytic stability, 52–53
  impact of dose on cesium distribution coefficient, 51f
  impact of dose on potassium distribution coefficient, 52f
  impact of increasing radiation exposure on extraction and stripping performance, 50–52
  impact of irradiation on solvent appearance, 48
  impact of stripping on solvent system, 53
  PUREX (plutonium uranium extraction) process, 46
  radiolysis of calixarene system, 53–54
  radiolytic yields from gamma radiolysis studies, 54t
  schematic diagram of experimental protocol, 47f
  simulant composition, 47t
  testing impact of diluent on stability of solvent, 50
  treatment of cesium rich high level waste (HLW) stream, 45–46
Green chemistry
  environmental impact of process and waste streams, 224
  See also Calixarene partitioning
Gutsche, calixarene name, 2

H

Heavy metal cation separations
  calix[4]arene structures with hard oxygen-containing donor groups on lower rim, 128f
  calix[4]arene structures with soft sulfur-containing donor groups on lower rim, 126f
  complexation reaction in extraction system with excess Hg(II) over ionophore, 131
  extraction of mercury(II) and cadmium(II), 130–131
  extraction of palladium(II) and platinum(II), 130
  fluorogenic calix[4]arene *N-X*-sulfonyl carboxamide for selective Hg(II) recognition, 132
  ionophores with pendant thioether groups, 127

lower rim-functionalized calix[4]arenes, 125, 127
pH profiles for Hg(II) extraction from aqueous mercuric nitrate solution, 130f
soft, with calix[4]arene N-X-sulfonyl carboxamides, 129–133
solvent extraction of silver(I), 129–130
structures of calix[4]arene di(N-X-sulfonyl carboxamides), 129f
See also Metal ion separations
Hexahomotriazacalix[3]arene macrocycles. *See* Azacalix[3]arenes
Hollow molecular building blocks. *See* Thiacalixarene derivatives
Hosts, multi-component. *See* Hydrogen-bonded cavities
HPLC-based separations
N-carbobenzyloxy protected amino acids, 245–246
nucleotides, 246–247
oligonucleotides, 248–250
polycarboxylates, 250–251
simple anions, 244–245
synthesis of calix[4]pyrrole-functionalized silica gels, 240, 242–243
See also Silica gels, calix[4]pyrrole-functionalized
Hydrogen-bonded cavities
Archimedean duals and irregular polygons, 336–337
Archimedean duals from corresponding Archimedean solids, 337f
Archimedean solids, 331–332, 335–336
buckminsterfullerene, 336
cavities based upon rigid extenders, 327–329
cavity based upon flexible extender, 329–330

co-crystallization of compound 3 with either 4-picoline or 1,10-phenanthroline, 328f
cross-sectional view of X-ray crystal structure of assembly held together by 60 hydrogen bonds and snub cube, 327f
cuboctahedron, 335
elaborating resorcin[4]arenes supramolecularly, 327–330
icosahedral systems, 334
irregular polygons, 338
Lehn's spherand, 333f
multi-component hosts, 325–326
Murakami's cyclophane-based cube, 334f
octahedral systems, 333
platonic solids, 330–331, 332–334
polyhedron model–snub cube, 330
product of co-crystallization of compound 3 with pyridine, 328f
resorcin[4]arenes for multi-component host design, 326–327
rhombic dodecahedron, 337, 338f
schematic of rhinovirus displaying triangulation, 334f
snub cube, 335, 336f
space-filling view of X-ray crystal structure of 3•4(4-vinylpyridine)•MeNO$_2$ and resorcinol•2(4-vinylpyridine), 329f
spherical molecular assemblies, 330–338
spherical organic hosts from laboratory and nature, 332–338
structure of ferritin, 338f
tetrahedral systems, 333
truncated icosahedron, 336
truncated octahedron, 335
Vogtle's spheriphane, 333f
See also Deep cavities
Hydroxamate, calix[4]arene extractants. *See* Acyclic analogs

of *tert*-butyl-calix[4]arene hydroxamate extractants
Hypermolecules, 296

I

Icosahedral systems, platonic solids, 334
Icosahedron, truncated, Archimedean solid, 336
Inductively coupled argon plasma (ICAP) spectroscopy
  contacting and analytical procedures, 31
  elemental analyses, 30–31
  instrumentation, 30
Interfacial features. *See* Assisted liquid-liquid ion extraction
Ion-binding
  calixarenes, 3–4
  calixarenes with pendant functionalities, 5
Ionophores
  investigating solvent extraction of soft metal cations, 127
  ionophoric properties of carbamoylmethylphosphine oxides, 137–138
  See also Heavy metal cation separations
Iron(III) extraction
  competitive extraction studies, 214–217
  single metal ion extraction studies, 211, 214
  See also Acyclic analogs of *tert*-butyl-calix[4]arene hydroxamate extractants
Irradiation. *See* Calix[4]arene-bis-crown-6 behavior under irradiation; Gamma irradiation fields

Irregular polygons, spherical organic hosts from laboratory and nature, 338

K

Koilands
  design of, 296–297
  design of new, 297, 299
  use of thiacalixarenes in synthesis of, 307, 309
  See also Thiacalixarene derivatives
Koilate. *See* Thiacalixarene derivatives

L

Lanthanide calix[4]arene complexes
  analyzing by nuclear magnetic resonance (NMR), 166
  best fit treatment of experimental data, 169–170
  calculated versus experimental paramagnetic shifts of $Yb^{3+}\bullet 1$ perchlorate complex, 173f
  classical analysis of extraction processes, 165
  correlation time, 167
  correlated spectroscopy (COSY) spectrum of $Yb^{3+}\bullet 1$ perchlorate complex, 172f
  crystallographic structure of $Yb^{3+}\bullet 1$ complex, 172f
  effect of water and nitrate ions on NMR spectra, 176–177
  experimental, 167
  experimental relaxation time of solvent molecule exchange, 167
  formation of oligomeric species with Gd^{3+} calix[4]arene complexes, 170

longitudinal relaxation rate of methyl protons of gadolinium complexes, 168
NMR dispersion curves of mixtures of Gd^{3+} complexes, 169f
NMR spectrum of La^{3+}•2 perchlorate complex, 174f
NMR spectrum of Yb^{3+}•2 perchlorate complex, 174, 175f
paramagnetic shifts dependence on complex structure, 170–171
relaxivity studies, 167–170
resonances in spectrum of La^{3+}•1 complex, 171, 173
schematic model of Yb^{3+}•2 complex configuration, 175, 176f
Solomon–Bloembergen equation, 167
spectral analyses, 170–176
structural analysis of complexes with ligand 2, 173, 174f
structures, 166
substrates interacting with paramagnetic lanthanide β-diketonates, 171
Lanthanides. See Bimetallic lanthanide complexes; Carbamoylmethylphosphine oxide-substituted calixarenes; Lanthanide calix[4]arene complexes; Phosphorylated calixarenes
Lariat ether N-X-sulfonyl carboxamides
monovalent metal ion separations, 114
See also Metal ion separations
Lead ions
effect of para substituent on extraction, 117, 118f
extraction with calix[4]arene di(N-X-sulfonyl carboxamides), 114, 116
pH profiles for extraction, 115f
See also Metal ion separations

Lehn's spherand, tetrahedral system, 333f
Linear oligomers, structural modification of wide rim carbamoylmethylphosphine oxide (CMPO)-calixarenes, 142–143
Liquid-liquid extraction. See Assisted liquid-liquid ion extraction; Calixarene partitioning; Self-assembly of benzyl phenol in binding Cs^+
Luminescence. See Bimetallic lanthanide complexes

M

Macrocycles. See Azacalix[3]arenes
Mass transfer model, transport of ^{137}Cs, 14
Mercury ions
calixarene-based fluorogenic reagent for selective recognition, 118, 120
dependence of Hg loading and relative emission intensity of fluorogenic calixarene after Hg(II) extraction, 133f
extraction, 130–131
fluorogenic calix[4]arene N-X-sulfonyl carboxamide for selective Hg(II) recognition, 132
pH profiles for Hg(II) extraction using calix[4]arene di(N-X-sulfonyl carboxamides), 130f
See also Heavy metal cation separations; Metal ion separations
Metal ion complexation. See Acyclic analogs of tert-butyl-calix[4]arene hydroxamate extractants
Metal ion separations
calix[4]arene di(N-X-sulfonyl carboxamides), 114–122
conformationally locked, 120, 122

effect of para substituent on extraction of Pb(II), 117
extraction of Pb(II), 114, 116
extraction of soft metal cations, 118
fluorogenic reagent for Hg(II) recognition, 118, 120
synthesis, 114, 115f
calix[4]arene mono(N-X-sulfonyl carboxamides), 122–123
charge matching, 112–113
competitive extraction of Pb(II) and competing metal ions, 116f, 119f
extractants with high selectivities for Ca(II) and Na(I), 113f
matching of charges on metal ion and ionized ligand, 113f
Pb(II) extraction selectivity, 116
pH profiles for Pb(II) extraction, 115f
N-X-sulfonyl carboxamide function for carbon-pivot lariat ethers, 113–114
Metalloreceptors
interactions with coordinated substrate, 283
See also Calixarene metalloreceptors
Model of mass transport, transport of ^{137}Cs, 14
Molecular-binding, calixarenes, 3
Molecular dynamics (MD) simulations
atomic charge set, 60f
behavior of sodium complexes, 68–69
calix[4]arene-crown-6, 13
details of studies, 59–60
energy averages in gas phase on for BC6–NO$_2$ (1,3-alt-calix[4]arene-bis-crown-6)–NO$_2$, 67t
influence of nitrate counter-ion in gas phase and in explicit water phase, 65–69

solvation and complex stability (cesium and sodium complexes) in presence of water, 68
structural averages during MD simulations in gas phase for BC6–NO$_2$, 67t
See also Assisted liquid-liquid ion extraction; Calix[4]arene-bis-crown-6 behavior under irradiation
Molecular modeling
benzyl phenol derivatives, 96, 99
calculated structures of free ligand 2-benzyl phenol and cesium-benzyl phenol complex, 97f
rotational potential energy surfaces (PES) for rotation about 2-3-4-5 bond for 2-benzyl phenol, 2-benzyl 3-methyl phenol, and 2-benzyl-α-methyl phenol, 98f
Molecular networks, definition, 296
Molecular recognition
competition reactions showing 4-phenylpyridine, 293t
energy of interactions, 296–297
phenyl-substituted pyridines for metalloreceptors, 291–292
See also Azacalix[3]arenes
Molecular solids
definition, 296
See also Thiacalixarene derivatives
Monovalent metal ions. See Metal ion separations
Multi-component hosts. See Hydrogen-bonded cavities
Murakami's cyclophane-based cube, octahedral system, 334f

N

Niederl, procedures for cyclic tetramers, 3
Nitric acid

cesium extraction from solutions, 18
stability experiments, 30
stability of calixarene-crown ethers, 35, 37
Nitrogen groups. *See* Heavy metal cation separations
Nuclear magnetic resonance (NMR) study of paramagnetic metal ions, 165–166
See also Lanthanide calix[4]arene complexes
Nuclear waste. *See* Cesium extraction
Nuclear wastes, *p-tert*-butyl-calix[8]arene for Cs^+ removal, 6
Nucleotides, separation on HPLC-modified column, 246–247

O

Octaalkylporphyrinogens. *See* Silica gels, calix[4]pyrrole-functionalized
Octahedral systems, platonic solids, 333
Octahedron, truncated, Archimedean solid, 335
Oligomers, linear, structural modification of wide rim carbamoylmethylphosphine oxide(CMPO)-calixarenes, 142–143
Oligonucleotides, separation on HPLC-modified column, 248–250
Oxyions, extraction. *See* Phase transfer extractants for oxyions

P

Palladium ions extraction, 130

See also Heavy metal cation separations
Partitioning. *See* Calixarene partitioning
Patent literature, calixarene applications, 6
Phase transfer extractants for oxyions
approach for designing, 256
compatibilizing amides with alkane phase, 264
effectiveness of amides as extractants, 259, 260f, 261f
effectiveness of amines as extractants, 259, 262f, 263f
extraction by amides from aqueous acid into chloroform, 260f
extraction by amides from aqueous acid into isooctane, 266f
extraction by amides from aqueous acid into toluene, 265f
extraction by amides from water into chloroform, 261f
extraction by amines from aqueous acid into chloroform, 262f
extraction by amines from water into chloroform, 263f
extraction of oxyanions and oxycations, 257, 259, 264
extractions with toluene and isooctane, 264, 265f, 266f
incorporating structural features into host, 256–257
kinetically rapid phase transfer agents, 256
oxy species for study, 256
routes to prepare amines involving borane reduction step, 258
selective extraction from aqueous solution into organic phase, 255–256
solution stoichiometry of oxyanions in acidic and neutral solutions for interpreting data, 259

synthetic strategies, 257
targeting specific solvent systems and individual complexants, 259, 264
Phosphorus groups
variation in carbamoylmethylphosphine oxide(CMPO)-calixarenes, 145–146
See also Heavy metal cation separations
Phosphorylated calixarenes
best extractants for thorium and europium, 162*f*
calix[n]arene phosphine oxides, 160–161
carbamoylmethylphosphine oxide (CMPO) and trioctylphosphine oxide (TOPO) extractants, 150
comparison of percentage extraction of some lanthanide nitrates with CMPO, *p*-CMPO calix[4]arene, and acyclic counterpart, 154*f*
complexation studies of calix[n]arene phosphine oxides, 161
complexation studies of lower rim CMPO-calixarenes, 159–160
complexation studies of upper rim CMPO-calixarenes, 154–157
dependence of spacer length for lower rim phosphine oxide calixarenes, 163
extent of extraction, 152
extractant dependence on nature of phosphorylated functional groups and position on calixarene, 162
extractant efficiency dependence on spacer length, 162
extraction of europium nitrate by lower rim CMPO-calixarenes, 159*f*

extraction of thorium and europium nitrates by calix[n]arene phosphine oxides, 161*f*
extraction percentage of thorium and europium nitrates by two "homo" *p*-CMPO calix[4]arenes and mixed derivatives, 153*f*
extraction studies of calix[n]arene phosphine oxides, 160–161
extraction studies of lower rim CMPO-calixarenes, 157–158
extraction studies of upper rim CMPO-calixarenes, 152–154
factors affecting binding abilities of ligands, 151
flexible calixarenes: "homo" *p*-CMPO calix[4]arene and mixed calix[4]arenes, 156–157
importance of calixarenic structure, 152
influence of calixarene flexibility, 153
intra-lanthanide series selectivities, 153
lower rim CMPO-calixarenes, 157–160
overall stability constants of europium and thorium complexes of "homo" *p*-CMPO calix[4]arenes and related compounds, 155*t*
overall stability constants of europium complexes of "homo" and "mixed" *p*-CMPO calix[4]arenes, 157*t*
participation in complexation with metal ions, 150
percentage extraction of europium and thorium nitrates by "homo" upper rim CMPO calix[4]arenes and acyclic compounds, 152*t*
percentage extraction of thorium and lanthanide nitrates by lower

rim CMPO calixarenes as function of number of CH$_2$ spacers, 158f
rigid calixarene: p-CMPO calix[4]arene, 155–156
stability constants of complexes of lanthanide and thorium complexes with lower rim CMPO calix[4]arenes, 159t
stability constants of lanthanide complexes with CMPO and upper rim CMPO calix[4]arene as function of atomic number Z of cations, 156f
stoichiometry of extracted species, 154
structure of lower rim CMPO calix[4]arenes, 158f
structures of calix[n]arene phosphine oxides, 160f
structures of upper rim CMPO-calixarenes and acyclic counterparts, 151f
upper rim CMPO-calixarenes, 151–157
Photophysical properties. *See* Bimetallic lanthanide complexes
Platinum ions
extraction, 130
See also Heavy metal cation separations
Platonic solids
icosahedral systems, 334
Lehn's spherand, 333f
Murakami's cyclophane-based cube, 334f
octahedral systems, 333
organic hosts from laboratory and nature, 332–334
schematic of rhinovirus displaying triangulation, 334f
spherical molecular assemblies, 330–331
structure of five, 331f
tetrahedral systems, 333

Vogtle's spheriphane, 333f
Plutonium extraction. *See* Alkaline radioactive wastes
Polycarboxylates
chromatogram of separation of isomeric dicarboxylate anions, 252f
chromatogram of separation of related carboxylate anions, 251
separation on HPLC-modified column, 250–251
Polygons, irregular, spherical organic hosts from laboratory and nature, 338
Polyhedron model, snub cube, 330
Polyvalent metal ions. *See* Metal ion separations
Porphyrinogens
calix[4]pyrroles resembling, 239
See also Silica gels, calix[4]pyrrole-functionalized
Potassium, impact of gamma irradiation dose on distribution coefficient, 52f
Process chemistry. *See* Cesium extraction from acidic nuclear waste
Proton-ionizable group. *See* Metal ion separations
PUREX process
gamma irradiation fields, 46
simulation of uranyl extraction by tributyl phosphate, 72
See also Assisted liquid-liquid ion extraction

R

Radioactive wastes
cesium extraction, 19–21
from test to process, 21–24
reprocessing spent fuel, 21f
See also Alkaline radioactive wastes

Radiolysis
  calixarene system, 53–54
  qualitative identification of 1,3-*alt*-calix[4]arene-bis-crown-6 (BC6) products, 60, 63
  solid phase separation of products, 58
  See also Calix[4]arene-bis-crown-6 behavior under irradiation; Gamma irradiation fields
Resorcinarene
  modules for construction of deep cavitands, 271*f*
  traditional cavity-forming modules, 270
Resorcin[4]arene cavitands, extraction of Eu(III), 209
Resorcin[4]arenes
  elaborating supramolecularly, 327–330
  multi-component host design, 326–327
  See also Hydrogen-bonded cavities
Room temperature ionic liquids (RTIL)
  alternative to volatile organic compounds (VOCs), 224
  use in liquid/liquid separations, 224
  See also Calixarene partitioning

S

Salinity media, cesium extraction, 18–19
Sapphyrins, expanded porphyrins for anion binding, 241–242
Selectivity
  cesium extraction flow sheet, 24*f*
  cesium over sodium using calix[4]arene crowns, 16–17
  cesium studies, 41
  size of crown moiety, 4
  See also Cesium extraction

Self-assembly of benzyl phenol in binding Cs^+
  attempted double wrap around effect, 104
  calculated structures of free ligand 2-benzyl phenol and cesium-benzyl phenol complex, 97*f*
  cesium extraction from sodium salt media, 87
  cesium extraction with BPh, BPhM 1, and BPh 2, 101*f*
  cesium extraction with BPh and BPhM 1 as function of equilibrium pH, 100*f*
  comparing substituent effects of ortho benzyl group, 99, 102
  comparison between phenols and ortho-benzyl phenol derivatives, 102*t*
  comparison of dibenzyl phenol derivatives, 104*t*
  comparison of substituents in para position, 102*t*, 103*t*
  crystal structures of complexes ([K(2-benzylphenol)$_3$][2-benzylphenolate]) and ([Cs(2-benzylphenol)$_2$][2-benzylphenolate]), 95*f*
  disruption of ligand arrangement about Cs^+, 104
  effect of addition of methyl group to BPh 3, 103
  effects of substitution on methylene bridge, 103
  experimental materials, 89
  extraction methods, 89
  general synthetic approach to benzyl phenols, 90
  influence of addition of three methyl groups to phenol on Cs^+ extraction, 104
  influence of substituent in para position, 102–103
  influence of substitution in

methylene bridge, 103t
molecular modeling studies, 96, 99
product yield and characterization, 93–94
rotational potential energy surfaces (PES) for rotation about 2-3-4-5 bond for 2-benzyl phenol, 2-benzyl 3-methyl phenol, and 2-benzyl-α-methyl phenol, 98f
selectivity of benzyl phenols and importance of alkyl substituent effects, 105
solvent extraction experiments using benzyl phenol and 4-sec-butyl-2-(α-methylbenzyl)phenol (BAMBP), 99
structural similarity between calixarenes and benzyl phenol derivatives, 87
structures of 2-benzyl phenol, calix[4]arene-crown-6, and didehydroxycalix[4]arene-crown-6, 88f
structures of compounds in study, 91t
synthesis of benzyl α-methyl phenol derivatives, 90
synthesis of benzyl phenol derivatives, 90
synthesis of tert-octyl benzyl phenol (BPh 2), 93
synthetic schemes for 2-benzyl phenol derivatives and 2-benzyl-α-methyl phenol derivatives, 92f
X-ray crystal structures of 2-benzyl phenol (BPh) and complexes with K^+ and Cs^+, 94, 96
Sensitized lanthanide complexes. See Bimetallic lanthanide complexes
Separations
broad interpretation, 3
N-carbobenzyloxy protected amino acids on HPLC-modified column, 245–246
nucleotides on HPLC-modified column, 246–247
oligonucleotides on HPLC-modified column, 248–250
simple anions on HPLC-modified column, 244–245
See also Metal ion separations; Silica gels, calix[4]pyrrole-functionalized
Silica gels, calix[4]pyrrole-functionalized
anion binding affinities of control compounds, 245t
binding anions and neutral substrates, 239–240
chromatogram of HPLC-based separation of isomeric dicarboxylate anions, 252f
chromatogram of separation of related carboxylate anions, 251f
control calix[4]pyrrole monomers, 244
HPLC-based separation of N-carbobenzyloxy protected amino acids on calixpyrrole modified silica gel column, 246f
HPLC-based separation of nucleotides on calixpyrrole column, 247f
HPLC-based separation of oligonucleotides hexamers, 249f
HPLC-based separation of various phenyl substituted anions on calixpyrrole modified silica gel columns, 242f
resembling porphyrinogens, 239
retention times for various anions on HPLC columns, 243t
schematic representations, 241f
separation of N-carbobenzyloxy protected amino acids, 245–246
separation of nucleotides, 246–247

separation of oligodeoxythymidylate fragments containing 12 to 18 nucleotide subunits, 248f
separation of oligonucleotides, 248–250
separation of simple anions on columns, 244–245
separations of oligonucleotides of similar length, 249–250
separations of polycarboxylates, 250–251
studies of sapphyrin-based HPLC supports, 241–242
synthetic scheme of "β-hook" and "meso-hook" calixpyrroles, 240, 242–243
X-ray structures of tetraspirocyclohexylcalix[4]pyrrole•fluoride anion complex and chloride anion complex, 240f
Silver ions
calixarenes with π-coordinating allyl groups on lower rim, 127
solvent extraction with calix[4]arene N-X-sulfonyl carboxamides, 129–130
See also Heavy metal cation separations
Snub cube
Archimedean solid, 335, 336f
polyhedron model, 330
structure, 327f
Sodium bearing waste (SBW)
cesium extraction and stripping results from SBW simulant, 37, 39
composition of SBW simulant, 32t
initial contacting experiments with SBW simulant, 31–35
See also Cesium extraction from acidic nuclear waste

Sodium complexes. See Calix[4]arene-bis-crown-6 behavior under irradiation
Sodium extraction. See Cesium extraction
Soft metal ions. See Heavy metal cation separations
Solomon–Bloembergen equation, relaxation time of solvent molecule exchange, 167
Solvent extraction
cesium, 16–17
See also Cesium extraction; Self-assembly of benzyl phenol in binding Cs$^+$
Solvent stability
impact of diluent, 52–53
See also Gamma irradiation fields
Spherical molecular assemblies
Archimedean solids, 331–332, 335–336
platonic solids, 330–331, 332–334
polyhedron model–snub cube, 330
spherical organic hosts from laboratory and nature, 332–338
See also Hydrogen-bonded cavities
Stability. See Gamma irradiation fields
Stripping
extraction from sodium bearing waste (SBW) simulant, 37, 39
impact of increasing radiation exposure, 50–52
impact on solvent system, 53
See also Cesium extraction from acidic nuclear waste
Strontium extraction. See Alkaline radioactive wastes
Substituent effects. See Self-assembly of benzyl phenol in binding Cs$^+$
N-X-Sulfonyl carboxamide function.

See Metal ion separations
Sulfur groups. See Heavy metal
 cation separations
Supramolecular edifices. See
 Bimetallic lanthanide complexes

T

Technetium (Tc). See Alkaline
 radioactive wastes
Tetrahedral systems, platonic solids,
 333
Tetrahydroxamates. See Acyclic
 analogs of *tert*-butyl-calix[4]arene
 hydroxamate extractants
Tetrasulfinylcalix[4]arenes, synthesis
 and structural analysis, 302–304
Tetrasulfonylcalix[4]arenes,
 synthesis and structural analysis,
 304, 305*f*
Thiacalixarene derivatives
 calix[4]arene derivatives and
  schematic of cone configuration
  and linear koilands by fusion of
  two subunits, 298*f*
 common structural features, 299,
  301
 conformers of compound 12 by ^{1}H
  and ^{13}C NMR spectroscopy,
  306–307
 crystal structure of tetranuclear
  copper complex, 307, 308*f*
 crystal structures of trimeric
  inclusion complex, 302*f*
 design of koilands, 296–297
 design of new koilands, 297, 299
 features of compound 6 in
  hexagonal crystal system, 301–
  302
 functionalization of thiacalixarenes
  at lower rim, 306–307
 lateral views of crystal structures of
  dichloromethane inclusion
  complexes, 300*f*
 portions of structure of
  dichloromethane inclusion
  complex, 301*f*
 procedure for preparing, 299–300
 role of alkali metal cations in
  synthesis and distribution of
  conformers, 306
 schematic of formation of
  centrosymmetric koilate by
  centrosymmetric koilands and
  connectors, 298*f*
 solid state structural analysis of
  Zn(II) and Co(II) complexes with
  tetrathiacalixarene, 309
 solid state structure of compound 7
  by X-ray analysis, 302–303
 solid state structure of compound 8
  by X-ray analysis, 303–304, 305*f*
 structural analysis of
  thiacalix[4]arenes, 299, 301–302
 synthesis and structural analysis of
  tetrasulfinylcalix[4]arenes, 302–
  304
 synthesis and structural analysis of
  tetrasulfonylcalix[4]arenes, 304,
  305*f*
 use of thiacalixarenes in synthesis
  of koilands, 307, 309
 X-ray structure of 3-D network
  between consecutive compounds
  9, 305*f*
 X-ray structure of three conformers
  of compound 12, 308*f*
Thorium
 competitive extraction studies,
  214–217
 extraction, 162
 single metal ion extraction studies,
  211, 214
 Th^{4+} ion complexes with
  carbamoylmethylphosphine
  oxides, 139–140
 See also Acyclic analogs of *tert*-

butyl-calix[4]arene hydroxamate extractants; Phosphorylated calixarenes
Transition metal ions. *See* Metal ion separations
Transport
  wide rim carbamoylmethylphosphine oxides, 138
  See also Carbamoylmethylphosphine oxide-substituted calixarenes
Tributyl phosphate (TBP)
  effect of concentration and synergism with TBP molecules, 76–78
  simulated species, 72
  simulating synergistic and salting out effects, 78
  synergistic effects upon addition, 83
  See also Assisted liquid-liquid ion extraction
Trihydroxamates. *See* Acyclic analogs of *tert*-butyl-calix[4]arene hydroxamate extractants
Trioctylphosphine oxide (TOPO)
  extraction of lanthanides and actinides, 150
  See also Phosphorylated calixarenes
Trivalent lanthanide ions. *See* Bimetallic lanthanide complexes
TRUEX process
  extraction of lanthanides and actinides, 150
  See also Carbamoylmethyl-phosphine oxide-substituted calixarenes
Truncated icosahedron, Archimedean solid, 336
Truncated octahedron, Archimedean solid, 335

U

Uranyl ion (UO_2^{++})
  calixarene based uranophiles, 209
  complexation by calixarenes, 5
  See also Assisted liquid-liquid ion extraction

V

Vogtle's spheriphane, tetrahedral system, 333*f*
Volatile organic compounds (VOCs), separation technology, 224

W

Waste, minimizing volume, 13

X

X-ray crystal structures
  azacalix[3]arenes, 201*t*, 205
  benzyl phenol derivatives and complexes with K^+ and Cs^+, 94–96
  buckminsterfullerene, 336*f*
  cross-sectional view of assembly held together by 60 hydrogen bonds and snub cube, 327*f*
  crystallographic structure of Yb^{3+}•1 perchlorate complex, 172*f*
  (diphenylphosphoryl)acetic acid and active ester by X-ray diffraction, 137*f*
  single crystal X-ray analysis of 1,3-di-CMPO-calixarene, 145*f*
  space-filling view of X-ray crystal structure of 3•4(4-vinylpyridine)•$MeNO_2$ and

resorcinol•2(4-vinylpyridine), 329f
structure of 3-D network between thiacalixarenes, 305f
structure of three conformers thiacalixarene, 308f
tetraspirocyclohexylcalix[4]pyrrole •fluoride anion complex and chloride anion complex, 240f
See also Hydrogen-bonded cavities; Thiacalixarene derivatives

Y

Yb^{3+} ions
complexes with carbamoylmethylphosphine oxides, 139–141
COSY spectrum of Yb^{3+}•1 perchlorate complex, 172f
crystallographic structure of Yb^{3+}•1 perchlorate complex, 172f
NMR spectrum of Yb^{3+}•2 perchlorate complex, 174, 175f
schematic model of Yb^{3+}•2 complex conformation, 175, 176f
spectral analysis of calix[4]arene complexes, 170–176
See also Lanthanide calix[4]arene complexes

Z

Zinc(II) complexes, use of thiacalixarenes in synthesis of koilands, 307, 309
Zinke, Alois, calixarenes, 2